Methods in Molecular Biology

Series Editor
John M. Walker
School of Life and Medical Sciences
University of Hertfordshire
Hatfield, Hertfordshire, UK

For further volumes:
http://www.springer.com/series/7651

For over 35 years, biological scientists have come to rely on the research protocols and methodologies in the critically acclaimed *Methods in Molecular Biology* series. The series was the first to introduce the step-by-step protocols approach that has become the standard in all biomedical protocol publishing. Each protocol is provided in readily-reproducible step-by step fashion, opening with an introductory overview, a list of the materials and reagents needed to complete the experiment, and followed by a detailed procedure that is supported with a helpful notes section offering tips and tricks of the trade as well as troubleshooting advice. These hallmark features were introduced by series editor Dr. John Walker and constitute the key ingredient in each and every volume of the *Methods in Molecular Biology* series. Tested and trusted, comprehensive and reliable, all protocols from the series are indexed in PubMed.

Spermatogonial Stem Cells

Methods and Protocols

Edited by

Jon M. Oatley

School of Molecular Biosciences, Washington State University, Pullman, WA, USA

Brian P. Hermann

Department of Neuroscience, Developmental and Regenerative Biology, University of Texas at San Antonio, San Antonio, TX, USA

Humana Press

Editors
Jon M. Oatley
School of Molecular Biosciences
Washington State University
Pullman, WA, USA

Brian P. Hermann
Department of Neuroscience, Developmental and Regenerative Biology
University of Texas at San Antonio
San Antonio, TX, USA

ISSN 1064-3745 ISSN 1940-6029 (electronic)
Methods in Molecular Biology
ISBN 978-1-0716-3141-6 ISBN 978-1-0716-3139-3 (eBook)
https://doi.org/10.1007/978-1-0716-3139-3

Cover Caption: The image was generated by Jon Oatley's group and shows seminiferous cords from a postnatal day 3 (P3) mouse testis with ID4-EGFP epifluorescence (green) and anti-DDX4 immunostaining (red) counterstained with DAPI (blue).

This Humana imprint is published by the registered company Springer Science+Business Media, LLC, part of Springer Nature.
The registered company address is: 1 New York Plaza, New York, NY 10004, U.S.A.

Preface

In mammals, the actions of spermatogonial stem cells (SSCs), a subset of mitotic spermatogonia, provide the foundation for continual and robust spermatogenesis. However, molecular and functional heterogeneity among SSCs within and between species and across development, biomarkers, and tools to study this key cellular population vary. Furthermore, designing experiments to explore the biology of spermatogonia (and male germ cells in general) using relevant approaches as well as interpreting results from these experiments requires a high-level understanding of each method's advantages and limitations. In this volume, a compendium of chapters guides seasoned and new investigators in the field of mammalian germ cell biology on key experimental approaches and perspectives for critical analysis of spermatogonial subsets including the rare SSC pool. Written in the format of the highly successful Methods in Molecular Biology series, chapters include introductions to the topic, lists of necessary materials and reagents, step-by-step reproducible and rigorous protocols, and tips on troubleshooting and considerations of pitfalls.

Pullman, WA, USA — *Jon M. Oatley*
San Antonio, TX, USA — *Brian P. Hermann*

Contents

Contributors

ANDRE L. CALDEIRA-BRANT • *Laboratório de Biologia Estrutural e Reprodução, Departamento de Morfologia, Instituto de Ciências Biológicas, Universidade Federal de Minas Gerais, Belo Horizonte, MG, Brazil; Department of Obstetrics, Gynecology and Reproductive Sciences, Magee-Womens Research Institute, University of Pittsburgh School of Medicine, Pittsburgh, PA, USA*

CONNOR CASON • *Priority Research Centre for Reproductive Science, Discipline of Biological Sciences, The University of Newcastle, Callaghan, NSW, Australia; Infertility and Reproduction Program, Hunter Medical Research Institute, New Lambton Heights, NSW, Australia*

KEREN CHENG • *Department of Neuroscience, Developmental and Regenerative Biology, University of Texas at San Antonio, San Antonio, TX, USA; International Institutes of Medicine, The Fourth Affiliated Hospital, Zhejiang University School of Medicine, Yiwu, China*

HELIO CHIARINI-GARCIA • *Laboratório de Biologia Estrutural e Reprodução, Departamento de Morfologia, Instituto de Ciências Biológicas, Universidade Federal de Minas Gerais, Belo Horizonte, MG, Brazil*

MICHELA CICCARELLI • *Center for Reproductive Biology, Washington State University, Pullman, WA, USA; Department of Veterinary Clinical Sciences, College of Veterinary Medicine, Washington State University, Pullman, WA, USA*

VICTORIA D. DIAZ • *Department of Neuroscience, Developmental and Regenerative Biology, The University of Texas at San Antonio, San Antonio, TX, USA*

CHRISTOPHER B. GEYER • *Department of Anatomy and Cell Biology, Brody School of Medicine, East Carolina University, Greenville, NC, USA; East Carolina Diabetes and Obesity Institute at East Carolina University, Greenville, NC, USA*

BRIAN P. HERMANN • *Department of Neuroscience, Developmental and Regenerative Biology, The University of Texas at San Antonio, San Antonio, TX, USA*

ROBIN M. HOBBS • *Centre for Reproductive Health, Hudson Institute of Medical Research, Clayton, VIC, Australia; Department of Molecular and Translational Sciences, Monash University, Clayton, VIC, Australia*

YOUNG SUN HWANG • *Department of Biomedical Sciences, University of Pennsylvania School of Veterinary Medicine, Philadelphia, PA, USA*

OLEKSANDR KIRSANOV • *Department of Anatomy and Cell Biology, Brody School of Medicine, East Carolina University, Greenville, NC, USA*

NATHAN C. LAW • *Center for Reproductive Biology, Department of Animal Sciences, College of Agricultural, Human, and Natural Resource Sciences, Washington State University, Pullman, WA, USA*

JULIEN M. D. LEGRAND • *Centre for Reproductive Health, Hudson Institute of Medical Research, Clayton, VIC, Australia; Department of Molecular and Translational Sciences, Monash University, Clayton, VIC, Australia*

TESSA LORD • *Priority Research Centre for Reproductive Science, Discipline of Biological Sciences, The University of Newcastle, Callaghan, NSW, Australia; Infertility and Reproduction Program, Hunter Medical Research Institute, New Lambton Heights, NSW, Australia*

Yanyun Luo • *Department of Surgery, Weill Cornell Medicine, New York, NY, USA*
John R. McCarrey • *Department of Neuroscience, Developmental and Regenerative Biology, University of Texas at San Antonio, San Antonio, TX, USA*
Sarah K. Munyoki • *Department of Obstetrics, Gynecology, and Reproductive Sciences, Integrative Systems Biology Graduate Program, Magee-Women's Research Institute, University of Pittsburgh School of Medicine, Pittsburgh, PA, USA*
Makoto Nagano • *Department of Obstetrics and Gynecology, McGill University and The Research Institute of the McGill University Health Centre, Montreal, QC, Canada*
Jon M. Oatley • *Center for Reproductive Biology, Washington State University, Pullman, WA, USA; School of Molecular Biosciences, College of Veterinary Medicine, Washington State University, Pullman, WA, USA*
Kyle E. Orwig • *Department of Obstetrics, Gynecology, and Reproductive Sciences, Integrative Systems Biology Graduate Program, Magee-Women's Research Institute, University of Pittsburgh School of Medicine, Pittsburgh, PA, USA*
Lorena Roa-de la Cruz • *Department of Neuroscience, Developmental and Regenerative Biology, The University of Texas at San Antonio, San Antonio, TX, USA*
Kotaro Sasaki • *Department of Biomedical Sciences, University of Pennsylvania School of Veterinary Medicine, Philadelphia, PA, USA*
Marco Seandel • *Department of Surgery, Weill Cornell Medicine, New York, NY, USA*
Yasunari Seita • *Department of Biomedical Sciences, University of Pennsylvania School of Veterinary Medicine, Philadelphia, PA, USA; Bell Research Center for Reproductive Health and Cancer, Nagoya, Aichi, Japan*
Gunapala Shetty • *Department of Experimental Radiation Oncology, The University of Texas MD Anderson Cancer Center, Houston, TX, USA*
Anukriti Singh • *Department of Neuroscience, Developmental and Regenerative Biology, The University of Texas at San Antonio, San Antonio, TX, USA*
Jiachen Sun • *State Key Laboratory of Reproductive Medicine, Nanjing Medical University, Nanjing, Jiangsu, China*
Vera van Melis • *Department of Neuroscience, Developmental and Regenerative Biology, The University of Texas at San Antonio, San Antonio, TX, USA*
Xin Wu • *State Key Laboratory of Reproductive Medicine, Nanjing Medical University, Nanjing, Jiangsu, China*
Makiko Yamada • *Department of Surgery, Weill Cornell Medicine, New York, NY, USA*
Fan Yang • *State Key Laboratory of Reproductive Medicine, Nanjing Medical University, Nanjing, Jiangsu, China*
Xiangfan Zhang • *Department of Obstetrics and Gynecology, McGill University and The Research Institute of the McGill University Health Centre, Montreal, QC, Canada*

Chapter 1

Introduction: The Why's and How's for Studying Spermatogenesis and Spermatogonial Stem Cells

Brian P. Hermann and Jon M. Oatley

Abstract

Spermatogenesis is maintained throughout adulthood by a pool of adult stem cells termed spermatogonial stem cells (SSCs). Research investigations into spermatogenesis can provide insight into the etiology of certain types of male infertility (e.g., Sertoli cell only syndrome), elucidate means of improving food animal production, reveal new therapeutic avenues to address naturally occurring defects in sperm production, mitigate iatrogenic male infertility (e.g., arising from cancer therapy), and potentially intervene for male contraception. This chapter will serve as a commentary about why studying spermatogenesis is important, including a high-level overview of spermatogonia and SSCs, and make the case for a critical need for use of stringent definitions for SSCs and experimental platforms that allow for clear distinction of the multiple types of spermatogonia that exist in testes of mammals.

Key words Spermatogonia, Spermatogenesis, Spermatogonial stem cells

1 The Significance of Studying Spermatogenesis

Spermatogenesis is the highly ordered male germ cell differentiation process in adult males, which takes place within the seminiferous tubules of the testis [1]. In the adult, this lineage begins with a population of proliferative germline stem cells and involves the subsequent multi-step differentiation of their progeny to produce large numbers of haploid sperm each day during the reproductive lifespan of male mammals. Spermatogenic development can be divided into three sequential phases that involve distinct cell types: (1) clonal amplification of spermatogonia via mitotic divisions, (2) two rounds of meiosis in spermatocytes, and (3) morphological transformation of post-meiotic spermatids through spermiogenesis. The general organization of the spermatogenic lineage in mammals is considered to be highly conserved between species [2–7].

Jon M. Oatley and Brian P. Hermann (eds.), *Spermatogonial Stem Cells: Methods and Protocols*, Methods in Molecular Biology, vol. 2656, https://doi.org/10.1007/978-1-0716-3139-3_1,

2 What Are Spermatogonia and Why Should You Care?

As noted above, spermatogonia are mitotic male germ cells in the adult testis. Rodent spermatogonia arise from precursor male germ cells termed prospermatogonia during the first week of postnatal life. Prospermatogonia are the products of post-migratory primordial germ cells (PGCs) that first undergo a proliferative period after reaching the testis as M-prospermatogonia between ~E13.5 and E15.5 in mice [8], until entering a period of quiescence as T1-prospermatogonia [9, 10]. Subsequently, between postnatal days (P) 0–3, these prospermatogonia become proliferative again as T2-prospermatogonia, while situated centrally within seminiferous cords, and ultimately migrate to the basement membrane where they become spermatogonia by P6 [9, 10]. There are active debates ongoing in the literature about the mechanisms involved in specifically forming the foundational SSCs from amongst the nascent spermatogonia [11–13].

In the adult testis, spermatogonia can be broadly classified into subsets considered to be undifferentiated and differentiating, which are terms that were initially based upon morphological appearance of spermatogonial nuclei using histological approaches [1, 14, 15]. Undifferentiated spermatogonia have nuclei that appear euchromatic, in which the nuclear morphology of differentiating spermatogonia is more heterochromatic [16]. These histological definitions have been largely replaced in the contemporary literature with molecular markers that indicate states along the differentiation trajectory. Specifically, in rodents, differentiating Types A1, A2, A3, A4, Intermediate (In), and B spermatogonia [17, 18] exhibit expression of KIT receptor tyrosine kinase, which is absent from undifferentiated Type A spermatogonia, including A_{single}, A_{paired}, and $A_{aligned}$ clones of 4–16 spermatogonia [19–21]. Indeed, KIT expression is activated at the Aal-A1 transition as a result of retinoic acid (RA) stimulation and, by definition, is the initiating event of spermatogonial differentiation.

Spermatogonial stem cells (SSCs) are a subset of undifferentiated spermatogonia and function as adult stem cells in mammalian testes that maintain spermatogenesis and are essential for male fertility [22–24]. SSCs sustain spermatogenesis by balancing self-renewal and differentiation fate decisions to maintain a pool of stem cells and produce progenitor spermatogonia that are capable of differentiation in response to RA stimulation to ultimately produce sufficient sperm, respectively [22, 24–26]. Indeed, SSC fate imbalances, which favor either SSC differentiation or self-renewal, will necessarily hinder spermatogenesis and cause male infertility through SSC depletion or failure to seed waves of spermatogenic differentiation, respectively [14]. Consequently, understanding the fundamental characteristics of SSCs and their regulatory underpinnings has extraordinary significance for infertility and contraception. Individual molecular features that distinguish SSCs and progenitors are largely absent, but use of multiple parameters may bring experimentalists close to separating these two classes of undifferentiated spermatogonia [27]. Ultimately, the only way to define SSCs is to use an assay that includes a functional readout of the biological activities of SSCs, perpetual self-renewal, and differentiation.

3 Importance of Using Stringent and Definitive Definitions of SSCs

As noted above, the undifferentiated spermatogonial pool in the adult testis comprises two highly similar, yet functionally distinct subsets—SSCs and progenitor spermatogonia. Although sharing many of the same molecular features, progenitor spermatogonia have a finite transient-amplifying replicative capacity and, thus, fail to meet the functional criteria of perpetual self-renewal that is a defining SSC characteristic. Consequently, any experimental definitions of SSCs must distinguish between SSCs and progenitor spermatogonia (and any other germ cells, for that matter) on the basis of their differing biological activities. Ralph Brinster's group initially reported what has become the gold-standard experimental approach used for detecting and quantifying SSCs in 1994 [28, 29], SSC transplantation. As the name implies, this method involves the transplant of donor SSCs into the seminiferous tubules of recipient testes to measure regenerative capacity that is the result of production and maintenance of complete donor-derived spermatogenesis in the recipient. Only SSCs can produce spermatogenesis that is sustained beyond more than one complete duration of the spermatogenic lineage (~35 days in mice; [30]) and such events are detectable after 2 months in mice and distinguished from single rounds of spermatogenesis produced from progenitors and more advanced spermatogenic cells. Experimentally definitive identification of donor-derived spermatogenesis in the recipient seminiferous tubules is typically facilitated by the use of transgenic donor cells that bear ubiquitous reporter alleles such as *GtRosa26-LacZ* or *Actb-eGfp*. Alternatively, use of recipient testes that are completely devoid of endogenous spermatogenesis (e.g., W/W^{v} KIT mutants [31] or *Nanos2* knockouts [32]) can be used to detect spermatogenesis from unlabeled donor cells, but this limits the quantitative nature of SSC transplant. Importantly, transplant into irradiated or chemically cleared (busulfan-treated) recipient testes [28] does not permit definitive identification of donor-derived spermatogenesis from unlabeled donor cells due to potential for recovery of endogenous spermatogenesis. This seemingly subtle distinction is critical to proper application of transplantation to experimentally detect SSCs and is the difference between valid and invalid data. The purpose of this volume is to present readers with bona fide approaches to study SSCs and their progeny.

4 Organization of This Volume

This volume contains a collection of methodology chapters that are organized into sections to help the reader apply these methods in the laboratory. Chapters 2, 3, 4, 5, and 6 cover methods of phenotyping spermatogenic cells, including using histological approaches (Chapter 2), fluorescence in situ hybridization (Chapter 3), bulk and single-cell transcriptomics (Chapter 4), and epigenomics (Chapters 5 and 6). Chapters 7, 8, and 9 address in vitro approaches to studying the male germline, which include primary cultures of spermatogonia (Chapter 7), in vitro gametogenesis from pluripotent stem cells (Chapter 8), and RNAi-mediated gene knockdown (Chapter 9). The focus of Chapters 10, 11, 12, 13, 14, 15, and 16 is on in vivo approaches to studying spermatogonia, including isolation of primary undifferentiated spermatogonia (Chapter 10), spermatogonial transplantation (Chapter 11), viral transduction of spermatogonia (Chapter 12), pharmacological manipulation of spermatogonial differentiation (Chapter 13), spermatogonial regeneration after cytotoxic insult (Chapter 14), conditional knockouts (Chapter 15), and lineage tracing (Chapter 16). Finally, Chapters 17 and 18 are non-methodological perspective reviews on the state-of-the art methods for studying spermatogonia and SSCs in livestock (Chapter 17) and primates (Chapter 18). Together, these chapters serve as the definitive workbook of approaches for experimentation on spermatogonia and SSCs.

Acknowledgments

This work was supported by NIH grants R01 HD101223 (JMO), R01 HD90007 (BPH), and U01 DA054179 (BPH).

References

1. Russell LD, Ettlin RA, Sinha Hikim AP, Clegg ED (1990) Mammalian spermatogenesis. In: Russell LD, Ettlin RA, Sinha Hikim AP, Clegg ED (eds) Histological and histopathological evaluation of the testis. Cache River Press, Clearwater, pp 1–38
2. Green CD, Ma Q, Manske GL, Shami AN, Zheng X, Marini S, Moritz L, Sultan C, Gurczynski SJ, Moore BB, Tallquist MD, Li JZ, Hammoud SS (2018) A comprehensive roadmap of murine spermatogenesis defined by single-cell RNA-Seq. Dev Cell 46 (5): 651–667. doi:S1534-5807(18)30636-1 [pii]; https://doi.org/10.1016/j.devcel.2018.07.025
3. Guo J, Grow EJ, Mlcochova H, Maher GJ, Lindskog C, Nie X, Guo Y, Takei Y, Yun J, Cai L, Kim R, Carrell DT, Goriely A, Hotaling JM, Cairns BR (2018) The adult human testis transcriptional cell atlas. Cell Res 28 (12): 1141–1157. doi:10.1038/s41422-018-0099-2 [pii];https://doi.org/10.1038/s41422-018-0099-2
4. Hermann BP, Cheng K, Singh A, Roa-De La Cruz L, Mutoji KN, Chen IC, Gildersleeve H, Lehle JD, Mayo M, Westernstroer B, Law NC, Oatley MJ, Velte EK, Niedenberger BA, Fritze D, Silber S, Geyer CB, Oatley JM, McCarrey JR (2018) The Mammalian spermatogenesis single-cell transcriptome, from

spermatogonial stem cells to spermatids. Cell Rep 25(6):1650–1667. doi:S2211-1247(18)31602-4 [pii];https://doi.org/10.1016/j.celrep.2018.10.026

5. Lau X, Munusamy P, Ng MJ, Sangrithi M (2020) Single-cell RNA sequencing of the Cynomolgus macaque testis reveals conserved transcriptional profiles during mammalian spermatogenesis. Dev Cell 54(4):548–566. doi:S1534-5807(20)30593-1 [pii];https://doi.org/10.1016/j.devcel.2020.07.018
6. Shami AN, Zheng X, Munyoki SK, Ma Q, Manske GL, Green CD, Sukhwani M, Orwig KE, Li JZ, Hammoud SS (2020) Single-cell RNA sequencing of human, macaque, and mouse testes uncovers conserved and divergent features of mammalian spermatogenesis. Dev Cell 54(4):529–547. doi:S1534-5807(20)30399-3 [pii];https://doi.org/10.1016/j.devcel.2020.05.010
7. Singh A, Hermann BP (2023) Conserved Transcriptome Features Define Prepubertal Primate Spermatogonial Stem Cells as Adark Spermatogonia and Identify Unique Regulators. Int J Mol Sci 24(5):4755. https://doi.org/10.3390/ijms24054755
8. McLaren A (2003) Primordial germ cells in the mouse. Dev Biol 262(1):1–15
9. Hilscher B, Hilscher W, Bulthoff-Ohnolz B, Kramer U, Birke A, Pelzer H, Gauss G (1974) Kinetics of gametogenesis. I Comparative histological and autoradiographic studies of oocytes and transitional prospermatogonia during oogenesis and prespermatogenesis. Cell Tissue Res 154(4):443–470
10. McCarrey JR (2013) Toward a more precise and informative nomenclature describing fetal and neonatal male germ cells in rodents. Biol Reprod 89 (2):47. doi:biolreprod.113.110502 [pii];https://doi.org/10.1095/biolreprod.113.110502
11. Yoshida S, Sukeno M, Nakagawa T, Ohbo K, Nagamatsu G, Suda T, Nabeshima Y (2006) The first round of mouse spermatogenesis is a distinctive program that lacks the self-renewing spermatogonia stage. Development 133(8):1495–1505
12. Kluin PM, de Rooij DG (1981) A comparison between the morphology and cell kinetics of gonocytes and adult type undifferentiated spermatogonia in the mouse. Int J Androl 4(4):475–493
13. Law NC, Oatley MJ, Oatley JM (2019) Developmental kinetics and transcriptome dynamics of stem cell specification in the spermatogenic lineage. Nat Commun 10(1):2787. doi:10.1038/s41467-019-10596-0 [pii];https://doi.org/10.1038/s41467-019-10596-0
14. de Rooij DG, Grootegoed JA (1998) Spermatogonial stem cells. Curr Opin Cell Biol 10(6):694–701. doi:S0955-0674(98)80109-9 [pii]
15. Chiarini-Garcia H, Hornick JR, Griswold MD, Russell LD (2001) Distribution of type A spermatogonia in the mouse is not random. Biol Reprod 65(4):1179–1185
16. de Rooij DG, Russell LD (2000) All you wanted to know about spermatogonia but were afraid to ask. J Androl 21(6):776–798
17. Roosen-Runge EC, Giesel LO Jr (1950) Quantitative studies on spermatogenesis in the albino rat. Am J Anat 87(1):1–30. https://doi.org/10.1002/aja.1000870102
18. Clermont Y, Leblond CP, Messier B (1959) Duration of the cycle of the seminal epithelium of the rat. Arch Anat Microsc Morphol Exp 48 (Suppl):37–55
19. Busada JT, Chappell VA, Niedenberger BA, Kaye EP, Keiper BD, Hogarth CA, Geyer CB (2015) Retinoic acid regulates kit translation during spermatogonial differentiation in the mouse. Dev Biol 397(1):140–149. doi:S0012-1606(14)00561-2 [pii];https://doi.org/10.1016/j.ydbio.2014.10.020
20. Filipponi D, Hobbs RM, Ottolenghi S, Rossi P, Jannini EA, Pandolfi PP, Dolci S (2007) Repression of kit expression by Plzf in germ cells. Mol Cell Biol 27(19):6770–6781
21. Hermann BP, Mutoji KN, Velte EK, Ko D, Oatley JM, McCarrey JR (2015) Transcriptional and translational heterogeneity among neonatal mouse spermatogonia. Biol Reprod 92(2):54. https://doi.org/10.1095/biolreprod.114.125757
22. Oatley JM, Brinster RL (2008) Regulation of spermatogonial stem cell self-renewal in mammals. Annu Rev Cell Dev Biol 24:263–286
23. Griswold MD, Oatley JM (2013) Concise review: defining characteristics of mammalian spermatogenic stem cells. Stem Cells 31(1):8–11. https://doi.org/10.1002/stem.1253
24. Yang QE, Oatley JM (2014) Spermatogonial stem cell functions in physiological and pathological conditions. Curr Top Dev Biol 107:235–267. doi:B978-0-12-416022-4.00009-3 [pii];https://doi.org/10.1016/B978-0-12-416022-4.00009-3
25. Jaenisch R, Young R (2008) Stem cells, the molecular circuitry of pluripotency and nuclear reprogramming. Cell 132(4):567–582. doi:S0092-8674(08)00115-3 [pii];https://doi.org/10.1016/j.cell.2008.01.015

26. Oatley JM, Brinster RL (2012) The germline stem cell niche unit in mammalian testes. Physiol Rev 92(2):577–595. doi:92/2/577 [pii];https://doi.org/10.1152/physrev.00025.2011
27. Mutoji KN, Hermann BP (2017) Defining the phenotype and function of mammalian spermatogonial stem cells. In: Oatley JM, Griswold MD (eds) The biology of mammalian spermatogonia. Springer, New York, pp 67–90
28. Brinster RL, Avarbock MR (1994) Germline transmission of donor haplotype following spermatogonial transplantation. Proc Natl Acad Sci U S A 91(24):11303–11307
29. Brinster RL, Zimmermann JW (1994) Spermatogenesis following male germ-cell transplantation. Proc Natl Acad Sci U S A 91(24): 11298–11302
30. Hermann BP, Sukhwani M, Hansel MC, Orwig KE (2010) Spermatogonial stem cells in higher primates: are there differences to those in rodents? Reproduction 139(3): 479–493
31. Brinster CJ, Ryu BY, Avarbock MR, Karagenc L, Brinster RL, Orwig KE (2003) Restoration of fertility by germ cell transplantation requires effective recipient preparation. Biol Reprod 69(2):412–420
32. Ciccarelli M, Giassetti MI, Miao D, Oatley MJ, Robbins C, Lopez-Biladeau B, Waqas MS, Tibary A, Whitelaw B, Lillico S, Park CH, Park KE, Telugu B, Fan Z, Liu Y, Regouski M, Polejaeva IA, Oatley JM (2020) Donor-derived spermatogenesis following stem cell transplantation in sterile NANOS2 knockout males. Proc Natl Acad Sci U S A 117(39):24195–24204. doi:2010102117 [pii];https://doi.org/10.1073/pnas.2010102117

Chapter 2

Optimization of Testicular Fixation-Embedding Techniques for Improved Evaluation of Mammalian Spermatogonial Morphology and Function

Helio Chiarini-Garcia and Andre L. Caldeira-Brant

Abstract

Numerous methods have been successfully used to evaluate mammalian spermatogonial biology However, the conventional light microscopy assays present a challenge in precisely identifying spermatogonial phenotypes, which can result in discrepancies between molecular and morphological findings. Such precise association could lead to a more robust interpretation of spermatogonial activity in steady-state spermatogenesis, which may facilitate the translation from basic research to clinical applications. In this chapter, we present two histological processing methods that enable a comprehensive analysis of spermatogonial morphology and function, involving fixation of mammalian testicular tissue in glutaraldehyde and embedding in plastic resin. These techniques have proven to be effective in light microscopy studies.

Key words Aldehyde, Fixation, Plastic resin, Araldite, Glycol methacrylate, Embedding, Light microscopy, Spermatogonia, Testes

1 Introduction

In the twentieth century, the first studies on mammalian spermatogenesis used paraffin-embedded sections or whole-mount preparation to describe spermatogonial biology under light microscopy [1–7]. Such methods were precise enough to morphologically characterize testicular structure, the different types of germ and somatic cells, and their cyclicality during the stages of the seminiferous epithelium cycle (SEC). These approaches and findings enabled subsequent studies to deepen the knowledge about the spermatogenic process, better describing spermatogonial activity under healthy and pathological conditions [8–11].

In mammals, cells of the spermatogonial lineage comprises undifferentiated and differentiating types. The most undifferentiated type A spermatogonia (A_{single}/non-primates; A_{dark}/primates) are classically considered as a subpopulation of reserve

Jon M. Oatley and Brian P. Hermann (eds.), *Spermatogonial Stem Cells: Methods and Protocols*, Methods in Molecular Biology, vol. 2656, https://doi.org/10.1007/978-1-0716-3139-3_2,

spermatogonial stem cells (SSC) due to their self-renewal and low-proliferative capacities. Some other undifferentiated type A spermatogonia (A_{paired} and $A_{aligned}$/non-primate; A_{pale}/primate) are considered as progenitor SSCs, as they have higher mitotic activity and different degrees of commitment to differentiation [12]. After differentiating, the number of subtypes varies depending on the species (A_1, A_2, A_3, A_4, In, B/rodents; B_1-B_4/primate) [12]. Although successful, the use of classical paraffin-embedded methods for the morphological investigation of certain spermatogonial parameters is limited due to histological processing artifacts such as tissue shrinkage and poor resolution [13]. Consequently, one of the major issues of using this technique is the difficulty in accurately distinguishing the different phenotypes/subtypes of spermatogonia due to small cytological variations between them [14, 15].

Currently, with the advancement of genomic and proteomic assays, several genes and proteins have been identified for different spermatogonial subpopulations, shedding light on their developmental dynamics and niche [16–21]. Considering the potential of these new molecular techniques to generate quick and valuable results and the restrictions regarding the use of classical histological processing methods, morphological evaluation of spermatogonia has been gradually left aside throughout the years.

In this sense, it is of interest to demonstrate that sophisticated morphological approaches can still provide unprecedented information on spermatogonial biology. One example is the correlation of spermatogonial morphology with its functionality, which has been called morphofunctional evaluation. Such correlation is possible because spermatogenesis is a very well-organized process that develops in a cyclic and synchronized way along the length of the seminiferous tubules [22, 23]. In this process, spermatogonia, spermatocytes, and spermatids are organized in specific associations in the seminiferous epithelium, comprising/making up the stages of the SEC. These stages are based on the development of the acrosomal system [22]. In each of these stages, which vary between species (e.g., 6 for humans and 12 for mice), the germ cell types carry out specific functions in a time/stage-dependent manner. Within the stages of the SEC, spermatogonial subtypes can be defined by correlating cell morphology with stage specific kinetics (population number, mitosis, and apoptosis) and positioning related to the interstitial compartment (niche). Thus, histological methods can demonstrate spermatogonial behavior in steady-state spermatogenesis, suggesting functionalities even without using molecular approaches.

In the early twenty-first century, studies using histological techniques that apply aldehyde fixative and plastic resin embedding managed to precisely describe minor nuclear details that distinguish spermatogonial generations [14, 15]. These studies also managed

Table 1
Comparison between glycol methacrylate (GMA) and high-resolution light microscopy (HRLM) methods regarding the histological processing and the cytological details obtained

Parameter	GMA	HRLM
Histological Processing		
Fixation and embedding (days)	2	4
Testicular sample size required (mm^2)	5–10	0.5–2
Section thickness (μm)	3–5	1
Staining dies	Multiple	Toluidine Blue-Borate
Immunohistochemistry application	Few	No
Recognition of Cytological Details		
Euchromatin	Partial	Complete
Heterochromatin	Partial	Complete
Nucleoli	Complete	Complete
Spermatogonial subtypes	Partial	Complete

to characterize spermatogonial kinetics and niche [24–30]. Recently, these approaches have enabled researchers to redefine the classical morphological distinction of primate spermatogonia (marmoset and human), describing their population kinetics across the SEC, and demonstrating SSC positioning in particular niche areas [31, 32].

Herein we present our experience in preparing two exceptional methods for light microscopy studies that have brought novel morphofunctional data of mammalian spermatogonia. We labeled them as (a) the glycol methacrylate (GMA) method, after aldehyde fixation and hydrophilic plastic resin embedding based on GMA, and (b) the high-resolution light microscopy (HRLM) method, after aldehyde/osmium fixation and araldite epoxy resin embedding. While both methods provide adequate preservation of histological parameters, it is essential to consider their distinct characteristics before selecting the most appropriate approach for a study (Table 1).

2 Method #1: Aldehyde and Glycol Methacrylate (GMA)

Paraformaldehyde (PFA) and/or glutaraldehyde fixation and GMA embedding is a method that allows satisfactory morphological and, sometimes, immunohistochemical studies. It uses thick sections obtained from large tissue fragments [33]. Unlike paraffin embedding, GMA processing has shown several advantages: (a) easy

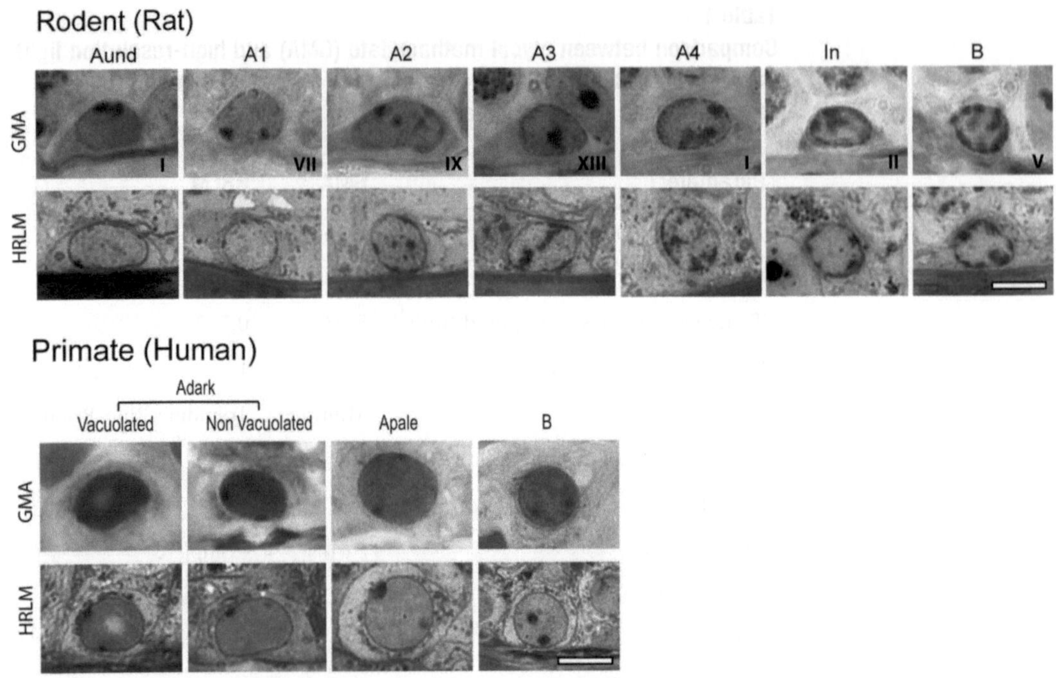

Fig. 1 Photomicrographs comparing the nuclear morphology of different spermatogonial subtypes of rat and human after glycol methacrylate (GMA) embedding and under high-resolution light microscopy (HRLM). The Roman numerals in the GMA rat photomicrographs relate to the stages of the seminiferous epithelium cycle (acrosome system [47]) where they were found. Rat: type A undifferentiated (A_{und}), type A differentiating (A_1, A_2, A_3, A_4), type intermediate (In), and type B (B) spermatogonia. Human: type A dark (A_{dark}), type A pale (A_{pale}), and type B (B) spermatogonia. Bars: 8 μm.

handling, (b) quick processing, (c) water miscibility, (d) infiltration and polymerization at room temperature, (e) semi-thin workable section (0.5 μm), (f) reduced histological artifacts, and (g) better resolution under light microscopy [33–36]. In this sense, the GMA approach has been largely used to describe the testicular and spermatogonial biology of different mammals under diverse conditions [37–48]. However, due to the inability to differentiate minor variations in chromatin features, this method cannot precisely identify spermatogonia subtypes in species with a high number of variants, such as rodents. This cytological discrimination is based mainly on the heterogeneous appearance of the heterochromatin dispersed throughout the nucleus, contrasted with mottled or clear euchromatin spots [14, 15, 22] (Fig. 1). Considering that some rodent spermatogonial subtypes cannot be identified with confidence after GMA embedding, associating their morphological aspects with the stage of SEC in which they are located increases the chance of their identification since each spermatogonium has stage specificity [10, 11]. Thus, classifying cross-sectioned rodent spermatogonia based solely on morphology is not recommended after GMA embedding. In contrast, as the number of spermatogonial generations is lower in primates and as they present fewer nuclear cytological variations between subtypes, their morphological classification using the GMA method is possible regardless of the stage of SEC [13, 31, 32] (Fig. 1).

2.1 GMA Materials

1. Glutaraldehyde and PFA.
2. Phosphate or cacodylate buffer.
3. 2.5% (EM grade) or 5% (biological grade) glutaraldehyde in phosphate/cacodylate buffer 0.05 M pH 7.4 (*see* **Note 1**).
4. Modified Karnovsky's fixative [49]—2% PFA, 2.5% glutaraldehyde (EM grade) in phosphate/cacodylate buffer 0.05 M pH 7.4 (*see* **Note 2**).
5. Ethanol (from 70% to 100% in distilled water).
6. Glycol methacrylate kit.
7. Toluidine blue-borate: 1 g toluidine-blue O and 1 g of anhydrous sodium borate ($Na_2B_4O_7$) dissolved in 100 mL distilled water.

2.2 GMA Methods

2.2.1 GMA: Fixation and Storage

Fixation by Perfusion For spermatogonial evaluation, the best fixative method is the intravascular heart perfusion, as previously described [22, 50]. When properly perfused, testis samples can be obtained as fragments of up to 10 × 10 mm of surface, although they should be cut into thin slabs (1–2 mm thick) and post-fixed with the same fixative by immersion for 12–24 h at 4 °C. This way, the testicular parenchyma will be well preserved, once it has been fixed prior to its manipulation, therefore enabling an accurate spermatogonial evaluation.

Fixation by Immersion When testicular samples cannot be fixed by perfusion, they must be cut into large and thick fragments before being immersed in the fixative for 12–24 h (*see* **Note 3**). In this case, for best results, some precautions must be taken, such as using a mixture of glutaraldehyde with PFA as fixative (*see* **Note 2**). Testes fixed by immersion normally show retractions in the interstitial space and seminiferous tubules dispersed, which impair the study of spermatogonial positioning related to interstitial components.

Storage Considering both methods, testicular fragments can be stored in buffer solution after fixation, at 4 °C for a long-term period (*see* **Note 4**).

2.2.2 GMA: Embedding

For GMA embedding, fixed testicular fragments should be dehydrated with ethanol at crescent concentrations (*see* **Note 5**) before infiltrations, as described below:

Embedding Procedure

1. Dehydrate fragments with baths of ethanol 70%, 85%, and 95% for 30 min each.
2. Further dehydrate fragments with two baths of absolute ethanol for 30 min each.
3. Keep fragments in infiltration resin for at least 12 h.
4. Immerse fragments in pure resin for at least 12 h.
5. Embed samples (*see* **Note 6**).

2.2.3 GMA: Sectioning

Preferentially, use a glass knife in the microtome, cut GMA blocks from 0.25 to 15-μm thick (*see* **Note 7**). For adequate sectioning, excessive resin should be previously trimmed from the block. After cutting and with the help of tweezers, place the sections in a recipient with distilled water to float and allow them to distend, which will remove any wrinkles. Just after distention, the section should be picked up with a histological slide and transferred to hot plate (70–80 °C) until the humidified section dries.

2.2.4 GMA: Staining

To provide adequate contrast and obtain outstanding resolution under light microscopy, the most used dye is the toluidine blue-borate, being elected as one of the best for GMA-embedded tissues. Staining intensity, which is a key factor for an adequate recognition of spermatogonial subtypes, depends on the fixative used and the thickness of the histological section (*see* **Note 8**). Experimental tests may be performed to optimize staining intensity for each specific approach since several dyes and thicknesses can be used, as described by Chiarini-Garcia and colleagues [31].

Staining Procedure

1. Cover the histological section with drops of toluidine blue-borate for 1 min at room temperature.
2. Rinse the slide with running tap water to clean the excess staining for at least 5 min (*see* **Note 9**).
3. With the tissue facing down, press gently the slide over a piece of filter paper to remove excess water (*see* **Note 9**).
4. After drying at room temperature, mount the slide with a coverslip using any mounting medium.

3 Method #2: Aldehyde and Araldite (HRLM)

Glutaraldehyde-fixed and araldite-embedded semi-thin sections are useful for optimizing the identification of spermatogonial subtypes, a process known as HRLM [23, 31, 50]. Unlike method #1, HRLM is a more laborious processing method that takes at least 4 days and requires very thin testicular fragments (Table 1).

Nonetheless, it gives a better resolution of cytological details, therefore allowing a more precise recognition of the different spermatogonial subtypes than GMA does [12] (Fig. 1). However, to accurately identify spermatogonial subtypes under HRLM, the researcher must be previously trained due to the specificity of identifying the minor cytological variations between phenotypes. In addition, these cytological details can only be perceived under large magnifications, when the light microscope is correctly adjusted according to Kohler's methods, as well as the examination of different focal planes within the cross-sections, as previously described by Chiarini-Garcia and Russell [14]. These important cytological nuances may not be readily apparent when identifying/analyzing spermatogonia in "static" photomicrographs. HRLM has allowed unprecedented descriptions of spermatogonial biology in terms of morphology, kinetics, and niche, initially described in rodents [14, 15, 25, 28, 51] and subsequently in other species [23, 26, 27, 29, 31, 32]. Additionally, through HRLM, the precise identification of spermatogonia does not necessarily depend on the correlation between spermatogonial morphology and their position in the stages of SEC, since each spermatogonium can be solely identified by its nuclear characteristics. In a study using Tex14 KO mice, in which intercellular bridges between germ cells were absent and the distribution of these cells in stages was not noticeable, spermatogonial kinetic evaluation was only possible due to the morphological identification of each spermatogonial subtype [52].

3.1 HRLM Materials

1. Glutaraldehyde.
2. Cacodylate or phosphate buffer.
3. 2.5% (EM grade) or 5% (Biological grade) glutaraldehyde in phosphate/cacodylate buffer 0.05 M pH 7.4 (*see* **Note 1**).
4. Osmium tetroxide.
5. Potassium ferrocyanide.
6. Osmium-reduced post-fixation: 1% osmium tetroxide and 1.5% potassium ferrocyanide in distilled water (*see* **Note 10**).
7. Ethanol (from 70% to 100% in distilled water).
8. Propylene oxide (PO) or acetone.
9. Araldite 502 kit (EMS) (*see* **Note 11**).
10. BDMA (benzyldimethylamine) resin.
11. Araldite resin (100 g): 54 g Araldite 502; 46 g dodenyl succinic anhydride (DDSA; Hardener); 2.5–3.0 g BDMA.
12. Toluidine blue-borate: 1 g toluidine blue O and 1 g anhydrous sodium borate ($Na_2B_4O_7$) dissolved in 100 mL distilled water.

3.2 HRLM Methods

3.2.1 HRLM: Fixation and Storage

For adequate testicular and spermatogonia morphological preservation, we strongly recommended choosing fixation by intravascular perfusion [22, 50] rather than by immersion. If immersion is the only viable option, several precautions throughout the procedure must be taken to preserve the histological integrity of the testicular parenchyma (*see* **Note 12**). Before embedding, small samples of ~2 × 2 mm surface and ~1 mm thickness should be obtained from the fixed testicle (Table 1), since the post-fixation step contains osmium tetroxide, which penetrates only tiny tissues.

3.2.2 HRLM: Embedding

After glutaraldehyde and osmium fixation, testicular fragments are progressively dehydrated in ethanol, infiltrated with araldite resin plus propylene oxide (PO) or acetone, and finally embedded in pure araldite, as described below.

Embedding Procedure

1. Dehydrate fragments with baths of ethanol 50%, 70%, and 85% (10 min each), two baths of ethanol 95%, and two baths of absolute ethanol.
2. Perform two baths of propylene oxide, for 30 min each.
3. Perform araldite infiltration in the tissue gradually. Start with 1: 1 propylene oxide: araldite for at least 60 min and subsequently use 1:3 (overnight or more) and 1:4 (overnight or more) mixtures. Finally, keep the tissue in pure resin for at least 60 min before embedding. All vials should be closed during the infiltration process to keep the resin from hardening.
4. Embed samples.
5. To harden the araldite, first keep the vials in oven at 45 °C (overnight or 12 h) and subsequently increase the temperature to 60 °C (overnight or more).

3.2.3 HRLM: Sectioning

To avoid the observation of overlapped nuclear cytological details of spermatogonia under light microscopy—which impairs the distinction of subtypes—tiny histological sections (~1 μm thick) must be used. The sections acquired for HRLM analysis are the ones known as semi-thin sections used for area selection prior to the acquisition of ultra-thin sections in electron microscopy.

For sectioning, use the following steps:

1. Under a stereoscopic microscope, trim the resin block into a trapezoid shape to facilitate the cut during microtomy.
2. Expose the testicular fragment from the resin block using a razor blade or a handheld rotary grinding device [53].
3. Using a microtome (or ultramicrotome) equipped with a glass knife attached to a water boat, cut sections at ~1 μm thickness. Float the section on distilled water in the attached boat.

4. With a small loop, pick up and transfer the section into a droplet of distilled water placed on a glass slide.
5. Transfer the slide to a hot plate (70–90 °C) until the water droplet containing the section evaporates.

3.2.4 HRLM: Staining

To obtain satisfactory contrast of the semi-thin sections, follow the steps outlined below (*see* **Note 10**):

1. Put several drops of toluidine blue-borate over the sections on a slide.
2. Heat on a hot plate (about 60 °C) until the edges of the stain begin to dry and turn into an iridescent green/purple color.
3. To keep the section from releasing from the slide, gently rinse it with distilled water.
4. With the tissue facing down, carefully press the slide over a piece of filter paper to remove excess water.
5. After complete drying at room temperature, mount the slide with a coverslip using any conventional medium.

4 Notes

1. Glutaric aldehydes can be produced at different degrees of purity. EM or grade I is specially purified to be used as fixatives for electron microscopy, while biological grade or grade II, which have a lower degree of purity, is meant for light microscopy studies. If you choose fixation through intravascular perfusion, we advise using the biological grade glutaraldehyde since it does not stiffen the vascular system, therefore allowing a more adequate perfusion of the testis. However, due to its lower degree of purity, it must be used at a concentration of 5%, instead of 2.5%. In case of fixing only via immersion, using 2.5% EM glutaraldehyde is preferable.
2. For satisfactory preservation of cell membrane and cytosol proteins, glutaraldehyde, which is a non-coagulant fixative, is the best option because it fixes the cell as a single interlocking structure [53]. Due to its difficulty in deeply penetrating tissues, glutaraldehyde is often used to fix the testis by intravascular perfusion and single cells in culture. Another aldehyde that has been widely used is PFA. However, PFA produces less cross-linking between proteins, thus reducing the tissue meshwork. Therefore, when choosing fixation by immersion, using a mixture of glutaraldehyde with PFA is recommended, as described by Karnovsky [49]. In this mixture, PFA, which has five times more penetrating capacity than glutaraldehyde, pre-fixes the tissue; glutaraldehyde comes in later and finishes

the process. Fixatives should be prepared by employing a phosphate or cacodylate buffer system to avoid the lowering of the pH, which normally occurs during fixation and might cause tissue artifacts.

3. For proper testis fixation by immersion, the albuginea capsule must be taken out. The albuginea is a dense connective tissue that avoids fixative penetration into the testicular parenchyma by diffusion. If completely removing the albuginea is arduous, small holes or cuts must be made on it around the entire testis. Another option is to cut the testis into large and thick slabs (10 × 10 mm surface and ~4 mm thickness) and put them in the fixative. After 12–24 h, only thin slabs from the surface (1–2 mm thickness) of the immersed tissue should be used for analysis—the rest must be discarded. To ensure appropriate immersion-fixation for the preservation of spermatogonial stem cells at a cytological level under high magnification of light microscopy, it is crucial to use a fixative volume that is at least 30 times greater than the tissue volume. While these precautions are essential for identifying spermatogonial subtypes, they are also recommended, although not mandatory, for histological evaluations of the testicular parenchyma as a whole.
4. In terms of long-term storage, fixed tissues should be kept immersed in buffer solution at 4 °C. Cacodylate buffer can be toxic to the cells due to the presence of arsenic, although it prevents the growth of microorganisms. Therefore, we suggest using phosphate buffer by adding one drop of 2.5% glutaraldehyde or 4% PFA to avoid the proliferation of microorganisms.
5. If for any reason it is not possible to dehydrate the tissue with ethanol, water can be eliminated by using crescent concentrations of resin diluted in distillated water, such as 50%, 70%, 80%, 90%, and then in pure resin.
6. Testicular fragments must be smaller than the cutting surface of the embedding block and it should be centrally positioned on it, considering that blocks' edges might be damaged during sectioning. Therefore, after resin polymerization, it is recommended to firmly attach the resin blocks to a support during microtome sectioning. Aluminum has often been used as a support, though in our lab we have been applying wooden pins attached with dentist resin to the block.
7. The most adequate section thickness depends on the research interest. If the analysis requires low-magnification observation under light microscope (2× to 10× objectives), the sample can be sectioned at a thickness range of 4–8 μm, since tissue's staining intensity is proportional to section thickness. In contrast, sections of 1–3 μm are adequate for high-magnification

(40× to 100× objectives) analyses, considering that the overlapping of cytological structures might impair spermatogonial evaluation. For GMA, we have often used sections of 3–5 μm thick.

8. Staining intensity and the morphological recognition of spermatogonial subtypes depend on the fixative used and the thickness of the histological section. Before starting a study, researchers should perform sectioning optimization at different thicknesses, considering the magnification that will be used for analysis, as **Note 7** describes. To ensure a thorough examination of spermatogonia under light microscopy, researchers should conduct optimization tests on section thicknesses before commencing their study. Note 7 provides guidance on selecting the appropriate thickness, taking into account the magnification used for analysis. Thinner sections are necessary for higher magnifications to achieve a detailed morphological evaluation. In our experience, toluidine blue dissolved in an alkaline solution is the best dye for distinguishing spermatogonial subtypes, and sodium borate is effective for this purpose. To enhance contrast and cytological structure prominence in black-and-white photomicrographs, we recommend using a 550-nm green filter.
9. To remove excess toluidine blue-borate stain from slides, it is recommended to quickly immerse them several times in an acid-water solution. This solution should be prepared using either chloride or acetic acid at a proportion of 0.5-2%. Avoid using acid-alcohol solution as it can cause the GMA to wrinkle. Also, after staining, never dry the sections under a hot plate, as this alters tissue staining and decreases contrast, reducing the observation of small cytological details.
10. As osmium tetroxide is toxic to mucosal membranes, it must be prepared under a proper hood. Using osmium and ferrocyanide (reduced-osmium) as fixatives is essential for transmission electron microscopy (TEM), but not for HRLM, although it yields better resolution of cell membranes.
11. Since it can be sectioned into relatively large sections (1–3 mm^2), araldite resin is the preferential epoxy resin for HRLM. Besides, due to its uniform hardening with little shrinkage, it does not crack when cutting sections at 1 μm thickness. We used the Araldite 502 embedding kit from EMS (Hatfield, PA).
12. When performing fixation by immersion, it is important for researchers to handle and section the testicles carefully, as the interstitial connective tissue and seminiferous epithelium can be easily damaged. To avoid this, precautions such as removing the tunica albuginea, using an adequate volume of fixative, and

sectioning samples with new and sharp blades can help preserve the biological integrity of the material and ensure reliable results.

Acknowledgments

These methods were standardized during the development of different projects that were partially supported by the following Brazilian funding agencies: Coordenação e Aperfeiçoamento de Pessoal de Nível Superior (CAPES), Conselho Nacional de Desenvolvimento Científico e Tecnológico (CNPq), Fundação de Amparo à Pesquisa do Estado de Minas Gerais (FAPEMIG).

References

1. Roosen-Runge EC, Barlow FD (1953) Quantitative studies on human spermatogenesis. I Spermatogonia. Am J Anat 93:143–169. https://doi.org/10.1002/aja.1000930202
2. Heller CG, Clermont Y (1964) Kinetics of germinal epithelium in man. Recent Prog Horm Res 20:545–575
3. Clermont Y (1966) Renewal of spermatogonia in man. Am J Anat 118:509–524
4. Clermont Y, Bustos-Obregon E (1968) Re-examinations spermatogonial renewal in the rat by means of seminiferous tubules mounted "in toto". Am J Anat 122:237–247
5. Huckins C (1971) The spermatogonial stem cell population in adult rats. I. Their morphology, proliferation and maturation. Anat Rec 169:533–558
6. Oakberg EF (1971) Spermatogonial stem-cell renewal in the mouse. Anat Rec 169:515–532
7. de Rooij DG (1973) Spermatogonial stem cell renewal in the mouse. Cell Tissue Kinet 6:281–287
8. Meistrich ML, Eng VWS, Loir M (1973) Temperature effects on the kinetics of spermatogenesis in the mouse. Cell Tissue Kinetics 6:379–393
9. Frankenhuis MT, Kramer MF, de Rooij DG (1982) Spermatogenesis in the boar. Vet Q 4:57–61
10. de Rooij DG (1998) Stem cells in the testis. Int J Exp Pathol 79:67–80. https://doi.org/10.1046/j.1365-2613.1998.00057.x
11. de Rooij DG, Russell LD (2000) All you want to know about spermatogonia but were afraid to ask. J Androl 21:776–798
12. Fayomi AP, Orwig KE (2018) Spermatogonial stem cells and spermatogenesis in mice, monkeys and men. Stem Cell Res 29:207–214. https://doi.org/10.1016/j.scr.2018.04.009
13. Chiarini-Garcia H, Lima MYS, Reis AB et al (2017) Influence of three different histological methods on the morphology and morphometrical data in human testis. Histol Histopathol 32:27–34. https://doi.org/10.14670/HH-11-765
14. Chiarini-Garcia H, Russell LD (2001) High-resolution light microscopic characterization of mouse spermatogonia. Biol Reprod 65:1170–1178
15. Chiarini-Garcia H, Raymer AM, Russell LD (2003) Non-random distribution of spermatogonia in rats: evidence of niches in the seminiferous tubules. Reproduction 126:69–680
16. Guo J, Grow E, Micochova H et al (2018) The adult human testis transcriptional cell atlas. Cell Res 28:1141–1157. https://doi.org/10.1038/s41422-018-0099-2
17. Wang M, Liu X, Chang G et al (2018) Single-cell RNA sequencing analysis reveals sequential cell fate transition during human spermatogenesis. Cell Stem Cell 23:599–614. https://doi.org/10.1016/j.stem.2018.08.007
18. Hermann B, Cheng K, Singh A et al (2018) The mammalian spermatogenesis single-cell transcriptome, from spermatogonial stem cells to spermatids. Cell Rep 25:1650–1667.e8. https://doi.org/10.1016/j.celrep.2018.10.026
19. Sohni A, Tan K, Song H et al (2019) The neonatal and adult human testis defined at the single-cell level. Cell Rep 26:1501–1517.e4. https://doi.org/10.1016/j.celrep.2019.01.045

20. Shami AN, Zheng X, Munyoki SK et al (2020) Single-cell RNA sequencing of human, macaque, and mouse testes uncovers conserved and divergent features of mammalian spermatogenesis. Dev Cell 54:529–547. https://doi.org/10.1016/j.devcel.2020.05.010
21. Kitadate Y, Jorg DJ, Tokue M et al (2019) Competition for mitogens regulates Spermatogenic stem cell homeostasis in an open niche. Cell Stem Cell 24(1):79–92e6. https://doi.org/10.1016/j.stem.2018.11.013
22. Russell LD, Ettlin RA, Sinha Hikim AP et al (eds) (1990) Histological and histopathological evaluation of the testis. Cache River Press, Clearwater
23. Nihi F, Gomes MLM, Carvalho FAR et al (2017) Revisiting the human seminiferous epithelium cycle. Hum Reprod 32:1170–1182. https://doi.org/10.1093/humrep/dex064
24. Russell LD, Chiarini-Garcia H, Korsmeyer SJ et al (2002) Bax-dependent spermatogonia apoptosis is required for testicular development and spermatogenesis. Biol Reprod 66:950–958
25. Chiarini-Garcia H, Hornick JR, Griswold MD et al (2001) Distribution of type A spermatogonia in the mouse is not random. Biol Reprod 65:1179–1185
26. Chiarini-Garcia H, Alves-Freitas D, Barbosa IS et al (2009) Evaluation of the seminiferous epithelial cycle, spermatogonial kinetics and niche in donkeys (Equus asinus). Anim Reprod Sci 116:139–154. https://doi.org/10.1016/j.anireprosci.2008.12.019
27. Nascimento HF, Drumond AL, França LR et al (2008) Spermatogonial morphology, kinetics and niches in hamsters exposed to short- and long-photoperiod. Int J Androl 32:486–497. https://doi.org/10.1111/j.1365-2605.2008.00884.x
28. Drumond AL, Meistrich ML, Chiarini-Garcia H (2011) Spermatogonial morphology and kinetics during testis development in mice: a high-resolution light microscopy approach. Reproduction 142:145–155. https://doi.org/10.1530/REP-10-0431
29. Costa GMJ, Avelar GF, Rezende-Neto JV et al (2012) Spermatogonial stem cell markers and niche in equids. PLoS One 7:e44091. https://doi.org/10.1371/journal.pone.0044091
30. Albuquerque AV, Drumond AL, Shao S et al (2014) A new approach for optimal morphological identification and immunolabeling of spermatogonial cells. Microsc Microanal 20:1304–1311. https://doi.org/10.1017/S1431927614000889
31. Caldeira-Brant AL, Eras-Garcia L, Alves-Freitas D et al (2018) Spermatogonial behavior in marmoset: a new generation, their kinetics and niche. Mol Human Reprod 24:299–309. https://doi.org/10.1093/molehr/gay017
32. Caldeira-Brant AL, Martinelli LM, Marques MM et al (2020) A subpopulation of human Adark spermatogonia behaves as the reserve stem cell. Reproduction 159:437–451. https://doi.org/10.1530/REP-19-0254
33. Chiarini-Garcia H, Parreira GG, Almeida FRCL (2011) High-resolution light microscopic characterization of spermatogonia. In: Chiarini-Garcia H, Melo RCN (eds) Light microscopy. Methods in molecular biology, vol 689. Humana Press, Totowa, pp 3–18. https://doi.org/10.1007/978-1-60761-950-5
34. Bennett HS, Wyrick AD, Lee SW et al (1976) Science and art in preparing tissues embedded in plastic for light microscopy, with special reference to glycol methacrylate, glass knives and simple stains. Stain Technol 51:71–97
35. Cole MB Jr, Sykes SM (1974) Glycol methacrylate in microscopy: a routine method for embedding and sectioning animal tissues. Stain Technol 49:387–400
36. Woodruff JM, Greenfield SA (1979) Advantages of glycol methacrylate embedding systems for light microscopy. J Histotechnol 2:64–167
37. van Haaster LH, de Jong FH, Doctor R et al (1992) The effect of hypothyroidism on Sertoli cell proliferation and differentiation and hormone levels during testicular development in the rat. Endrocrinology 131:1574–1576
38. Meistrich ML, Kangasniemi M (1997) Hormone treatment after irradiation stimulates recovery of rat spermatogenesis from surviving spermatogonia. J Androl 18:80–87
39. França LR, Becker-Silva SC, Chiarini-Garcia H (1999) The length of the cycle of seminiferous epithelium in goats (Capra hircus). Tissue Cell 31:274–280
40. Paula TAR, Chiarini-Garcia H, França LR (1999) Seminiferous epithelium cycle and its duration in capybaras (Hydrochoerus hydrochaeris). Tissue Cell 31:327–334
41. Shuttlesworth GA, de Rooij DG, Huhtaniemi I et al (2000) Enhancement of a spermatogonial proliferation and differentiation in irradiated rats by gonadotropin-releasing hormone antagonist administration. Endocrinology 141:37–49
42. Meistrich ML, Wilson G, Kangasniemi M et al (2000) Mechanism of protection of rat spermatogenesis by hormonal pretreatment: stimulation of spermatogonial differentiation after irradiation. J Androl 21:464–469

43. Neves ES, Chiarini-Garcia H, França LR (2002) Comparative testis morphometry and seminiferous epithelium cycle length in donkey and mules. Biol Reprod 67:247–255
44. Leal MC, Becker-Silva SC, Chiarini-Garcia H et al (2004) Sertoli cell efficiency and daily sperm production in goats (Capra hircus). Anim Reprod 1:122–128
45. Costa GMJ, Chiarini-Garcia H, Morato RG et al (2008) Duration of spermatogenesis and daily sperm production in the jaguar (Panthera onca). Theriogenology 70:1136–1146. https://doi.org/10.1016/j.theriogenology.2008.06.035
46. Drumond AL, Weng CC, Wang G et al (2011) Effects of multiple doses of cyclophosphamide on mouse testes: accessing the germ cells lost, and functional damage of stem cells. Reprod Toxicol 32:395–406. https://doi.org/10.1016/j.reprotox.2011.09.010
47. Melo MC, Almeida FRCL, Caldeira-Brant AL et al (2014) Spermatogenesis recovery in protein-restricted rats subjected to a normal protein diet after weaning. Reprod Fertil Dev 26:787–796. https://doi.org/10.1071/RD13032
48. Auler PA, Moreira GHFA, Hogg CO et al (2017) Testicular parameters and spermatogenesis in different birthweight boars. Reprod Fertil Dev 29:1720–1728. https://doi.org/10.1071/RD16164
49. Karnovsky MJA (1965) A formaldehyde-glutaraldehyde fixative of high osmolarity for use in electron microscopy. J Cell Biol 27: 137A–138A
50. Chiarini-Garcia H, Meistrich ML (2008) High-resolution light microscopic characterization of spermatogonia. In: Hou SX, Singh SR (eds) Germline stem cells. Methods in molecular biology, vol 450. Humana Press, Totowa, pp 95–107. https://doi.org/10.1007/978-1-60327-214-8
51. Albuquerque AV, Almeida FRCL, Weng CC et al (2013) Spermatogonial behavior in rats during radiation-induced arrest and recovery after hormone suppression. Reproduction 146:363–376. https://doi.org/10.1530/REP-12-0494
52. Rezende-Melo CA, Caldeira-Brant AL, Drumond-Bock AL et al (2020) Spermatogonial asynchrony in Tex14 mutant mice lacking intercellular bridges. Reproduction 160:205–215. https://doi.org/10.1530/REP-20-0118
53. Bozzola JJ, Russell LD (eds) (1999) Electron microscopy: principles and techniques for biologists, 2nd edn. Jones and Bartlett, Sudbury

Chapter 3

Single-Molecule Fluorescence In Situ Hybridization for Spatial Detection of mRNAs in Sections of Mammalian Testes

Victoria D. Diaz and Brian P. Hermann

Abstract

Single-molecule fluorescence in situ hybridization (smFISH) enables the detection and localization of individual mRNAs in tissue sections with single-molecule resolution while preserving spatial context, and thus, is a useful tool for examining gene expression in biological systems. In particular, the growing reliance on single-cell RNA sequencing (scRNA-seq) to explore cellular heterogeneity has reinvigorated this approach as a validation tool to spatially re-map mRNA expression patterns described in isolated cells to their parent tissue. While use of antibody-based methods, such as indirect immunofluorescence (IIF), remain popular as validation strategies, smFISH often affords superior specificity and maintains congruency with scRNA-seq. Here, we present a detailed protocol that combines multiplexed smFISH using the RNAscope approach with IIF to co-visualize mRNAs and proteins within sections of mouse testes. We provide step-by-step guidelines from testis preparation through visualization that enables mapping of combinations of up to four mRNA/protein targets in frozen sections on the RNAscope platform.

Key words smFISH, scRNA-seq, Immunohistochemistry, RNAscope, Spermatogenesis

1 Introduction

In the post-genome era, investigation of multicellular organisms and biological networks underlying organismal function relies on sensitive, high-throughput approaches that can profile the molecular states of cells and tissues. These "omics" approaches can interrogate the DNA, RNA, and protein levels in bulk samples of cells or tissues, generating large-scale data sets which reflect the comprehensive molecular profiles of the starting samples. For instance, RNA-seq using bulk-cell or -tissue preparations detects ensemble average mRNA levels for all genes, but the results may not accurately reflect the profile of any given cell population in the sample [1]. Single-cell RNA-seq (scRNA-seq) solves this problem by resolving gene expression profiles among individual cells [2–

Jon M. Oatley and Brian P. Hermann (eds.), *Spermatogonial Stem Cells: Methods and Protocols*, Methods in Molecular Biology, vol. 2656, https://doi.org/10.1007/978-1-0716-3139-3_3,

7]. Dissociation-based single-cell studies, though, ignore the tissue localization of expression profiles, because source tissues are necessarily dissociated into a single-cell suspension. For instance, we and others have performed exhaustive single-cell RNA-seq to identify comprehensive gene expression phenotypes that correlate with SSC transplantation activity [8, 9]. Yet, since this strategy is borne out of dissociated testicular tissue and profiles individual cells, we do not know the location of these SSCs in the testis. Furthermore, it is generally required in the field to validate single-cell transcriptomes using a complementary, tissue-based gene expression method to confirm the reliability of the single-cell transcriptomes and at least partially remap expression profiles in 2D tissue space [8]. Currently, proteins can be readily localized *in situ* with single-cell resolution using antibody-based approaches (e.g., indirect immunofluorescence) and detected with chromogens or fluorescence, but this approach is wholly dependent upon availability of validated antibody reagents. Moreover, any post-transcriptional gene regulation that takes place during spermatogenesis breaks the correlation between mRNA and protein levels, and, thus, immunostaining for a given protein in testis sections may not accurately represent the pattern of mRNA expression [10]. Direct localization of mRNAs *in situ* is also possible using hybridization of complementary (antisense) DNA or RNA probes, a process called *in situ* hybridization (ISH). This has the advantage of direct verification/validation and extension of scRNA-seq results for putative SSCs, any spermatogenic cell, or any cell type in any tissue.

Since the initial reports of the ISH methodology in 1969 by Gall and Pardue [11, 12], numerous advancements have maximized detection speed, sensitivity, and multiplicity of concurrent target detection. RNAscope (Advanced Cell Diagnostics) is a new commercial methodology that provides a single-molecule fluorescent ISH (smFISH) platform to address the challenges of sensitivity and specificity in spatial profiling low-abundance RNA [13–15]. It employs panels of double-Z probes and signal amplification strategy to maximize detection of individual RNA molecules [16]. This protocol describes our optimized strategy to use RNAscope technology to detect transcripts (e.g., *Gfra1* and *Kit*) together with proteins (e.g., DDX4) in testicular cryosections. This strategy readily facilitates detection of 1–4 transcripts at a time using the RNAscope platform and established (off-shelf) or custom probe sets (Fig. 1). Simultaneous detection of 12 RNAs is also possible using the RNAscope HiPlex application, which advances the use of smFISH to dissect the spatial relationships of discrete cell populations in complex tissues like the testis, which we demonstrate using a panel of probes that recognize mRNAs encoding 12 housekeeping gene products (Fig. 2a) versus negative controls (Fig. 2b).

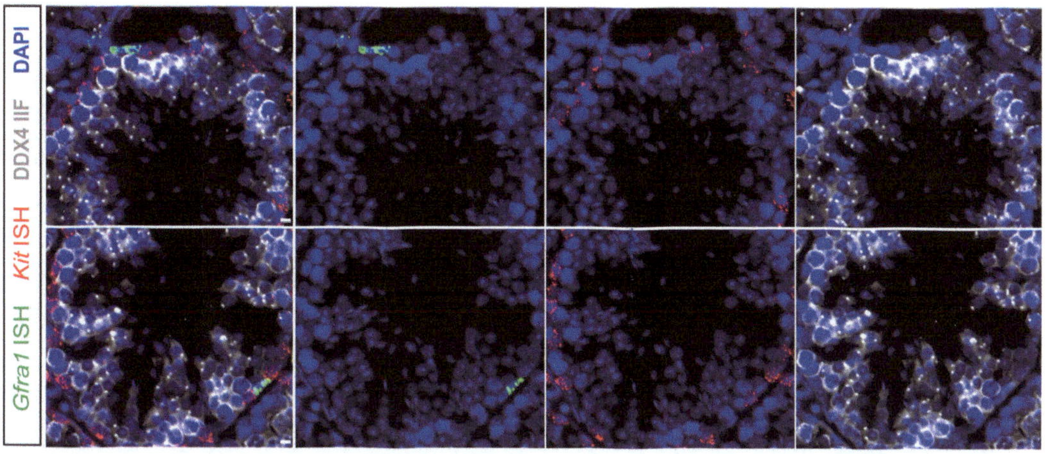

Fig. 1 Example RNAscope results for spermatogonial markers using mouse testes. smFISH-IHC labeling in the adult mouse testis. Using the described RNAscope method, cryosections from fixed adult testicular tissue was labeled for *Gfra1* mRNA [17], *Kit* mRNA (red), immunostained for DDX4 (white), and counterstained for DAPI (blue). Scale = 50 μm

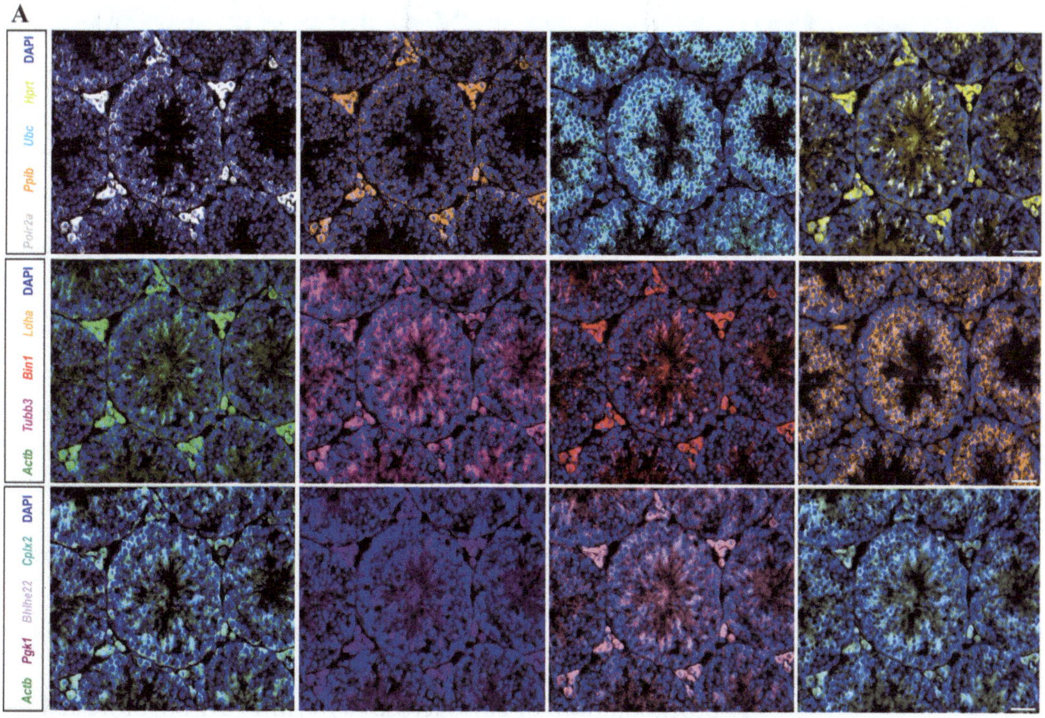

Fig. 2 smFISH-12 Plex labeling using RNAscope HiPlex approach in the adult mouse testis. (**a**) Adult testis tissue sections were hybridized with probes targeting 12 species-specific housekeeping genes and counterstained for DAPI (blue). Scale = 50 μm. (**b**) Testis sections were also hybridized with a negative control probe recognizing bacterial dapB mRNA (an irrelevant gene) and counterstained for DAPI (blue). Scale = 50 μm

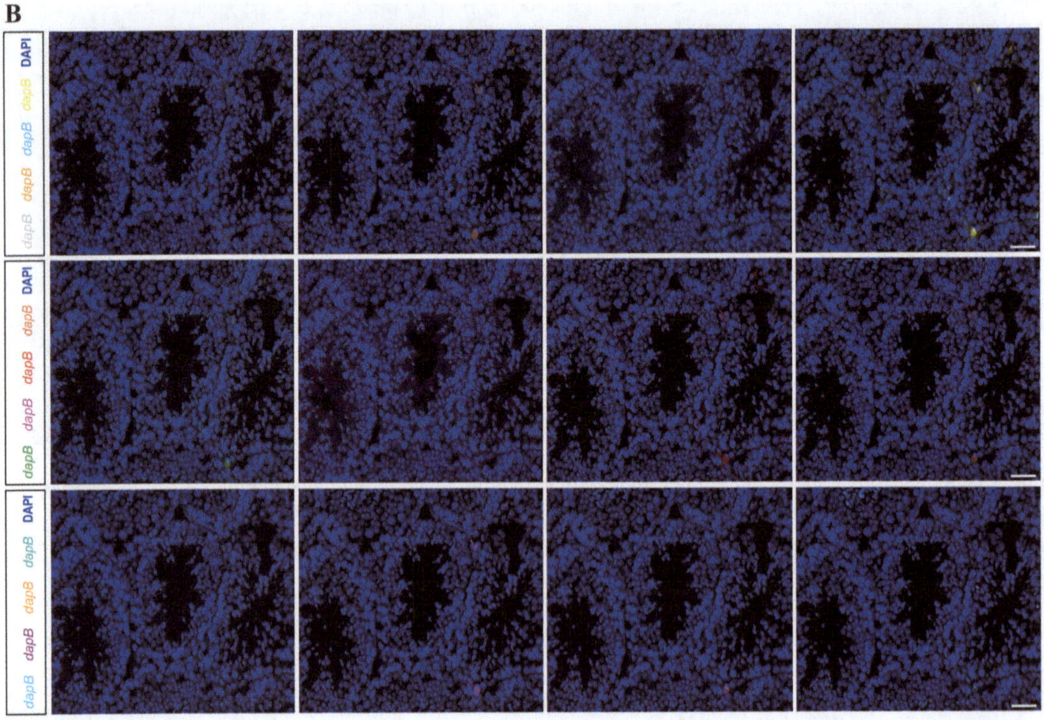

Fig. 2 (continued)

2 Materials

Prepare all solutions using Milli-Q water (prepared by purifying deionized water, to reach a sensitivity of 18 MΩ/cm at 25 °C) unless specified otherwise. Reagents are stored at room temperature, unless specifically noted.

2.1 Testis Preparation and Freezing

1. 10× Dulbecco's phosphate buffered saline, no calcium chloride, no magnesium chloride (DPBS): 26.6 mM KCl, 14.7 mM KH_2PO_4, 1379.3 mM NaCl, 80.5 mM Na_2 $HPO_4{\cdot}7H_2O$, pH 6.7–7.0.
2. 4% (w/v) Paraformaldehyde in 1× DPBS, pH 7.2. Made fresh and stored at 4 °C for 2–3 days (*see* **Note 1**).
3. Prepare 1× DPBS containing DNase/RNase-Free Distilled Water. Store at 4 °C (*see* **Note 2**).
4. Stock solutions of 10% sucrose (w/v) in 1× DPBS. Store at 4 °C (*see* **Note 3**).
5. Stock solutions of 20% sucrose (w/v) in 1× DPBS. Store at 4 °C (*see* **Note 3**).

6. Stock solutions of 30% sucrose (w/v) in 1× DPBS. Store at 4 °C (*see* **Note 3**).
7. Optimal Cutting Temperature Compound (O.C.T.).

2.2 RNAscope Reagents and Buffers

1. RNAscope Multiplex Fluorescent V2 assay kit with AMP 1–3, and TSA Opal dyes for detection of RNA targets [13].
2. Tyramide Signal Amplification (TSA) Opal dyes 520, 570, 620, and 690 (Akoya Biosciences) (*see* **Notes 4** and **31**).
3. RNAscope probes against mRNAs (*see* **Note 5**).
4. RNAscope Target Retrieval buffer: Prepare 200 mL of 1× Target Retrieval buffer by adding 180 mL of water and 20 mL of 10× Target Retrieval buffer into a staining dish (*see* **Note 6**).
5. 1× RNAscope Wash buffer: Prepare 500 mL of wash buffer by transferring 490 mL of water and 10 mL of 50× wash buffer solution into a RNase-free bottle (*see* **Note 7**).
6. Prepare 200 mL of DPBS containing water into a staining dish (*see* **Note 8**).
7. Transfer 200 mL of 100% ethanol (EtOH) into a staining dish.
8. 70% EtOH: Prepare by adding 60 mL of water to 140 mL of 100% EtOH into a staining dish.
9. 50% EtOH: Prepare by adding 100 mL of water to 100 mL of 100% EtOH into a staining dish.
10. 20× Saline Sodium Citrate (SSC): 3 M Sodium Chloride, 0.3 M Sodium Citrate in water, pH 7.0.
11. Prepare 200 mL of 5× SSC buffer containing water into a staining dish (for the optional stopping point in the RNAscope protocol). Store up to 2 months.

2.3 Immunofluorescence (If Necessary)

1. Make 1× DPBS containing 0.1% Tween-20 (1× DPBST).
2. Permeabilization buffer solution: Combine 10% FBS, 5% BSA, and 0.03% TritonX-100 in 1× DPBS. Store 1.5 mL aliquots at −20 °C for long-term storage (*see* **Note 9**).
3. Primary antibody (*see* **Note 10**).
4. Secondary antibody (*see* **Note 11**).

2.4 General Materials and Equipment

1. Dumont forceps.
2. Iridectomy scissors.
3. RNase & DNase decontaminant spray.
4. 1.5 L RNase-free bottle (*see* **Note 12**).
5. 1.5 mL Eppendorf tubes, nuclease free.
6. 15 mL conical tubes, nuclease free.

7. 15 × 15 × 5 mm Cryomolds.
8. Aluminum foil.
9. Cryostat (e.g., Leica CM 1860).
10. Superfrost Plus microscope slides (*see* **Note 13**).
11. 24 × 50 mm coverslips.
12. Tissue-Tek Vertical 24 Slide Rack.
13. Tissue-Tek Staining Dish.
14. Microtome/Cryostat blade.
15. Short razor blade to trim testis blocks.
16. Slide warmer.
17. Microscope slide boxes to store pre-cut sections and labeled slides.
18. Desiccant.
19. Re-sealable bags to store the microscope slide boxes with pre-cut sections.
20. Small plastic trays (e.g., autoclaved lids from empty pipette tips boxes) (*see* **Note 14**).
21. Hydrophobic barrier pen.
22. ProLong Gold antifade mounting medium.
23. Food steamer (*see* **Note 15**).
24. Hybridization incubator (*see* **Note 16**).
25. Humid chamber (*see* **Note 16**).
26. RNAscope hydrogen peroxide solution.

3 Methods

Carry out all procedures at room temperature unless otherwise specified.

3.1 Testis Freezing and Embedding

1. Submerge dissected testes in pre-chilled 4% PFA and leave on a rocker for 24 h at 4 °C. Ensure testes are completely covered and avoid air pockets.
2. Remove fixative and pre-rinse with cold DPBS, then wash twice for 25 min at 4 °C with cold DPBS.
3. Replace DPBS with cold 10% sucrose, keep testes at 4 °C.
4. When testes sink to the bottom of the tube, replace 10% sucrose with cold 20% sucrose. Store at 4 °C.
5. When testes sink to the bottom of the tube, replace 20% sucrose with cold 30% sucrose. Store at 4 °C.

6. Allow the testes to sink to the bottom of the tube overnight in 30% sucrose.
7. Label CryoMold 2 with the orientation of the testes and fill with ice-cold O.C.T. without introducing air bubbles (*see* **Note 17**).
8. Fill CryoMold 1 with ice-cold O.C.T. This cryomold will be used to remove excess sucrose.
9. Using cooled forceps, transfer fixed testes into the CryoMold 1 (*see* **Note 18**).
10. Twirl the testes in the O.C.T., removing as much sucrose solution as possible.
11. Transfer fixed testes into the O.C.T. of CryoMold 2. Cover the entire testis surface with more ice-cold O.C.T.
12. *Immediately* place CryoMold 2 level on dry ice and allow to freeze (*see* **Note 19**).
13. Wrap CryoMold 2 with aluminum foil and store in a labeled resealable plastic bag at −80 °C for long-term storage or wait 2 h before proceeding to cryosectioning (*see* **Note 20**).

3.2 Cryosectioning

1. Before setting the cryostat temperature (−18 to −21 °C), wipe the machine with 70% EtOH.
2. Rinse a new microtome blade with 100% EtOH, pat dry with paper towels, and set aside.
3. Transfer O.C.T.-embedded testis blocks from the −80 °C storage to dry ice (*see* **Note 21**).
4. Once the temperature of the Cryostat chamber has reached −18 to −25 °C, transfer testis blocks, microtome blade, and microscope slides into the Cryostat chamber and equilibrate for 1 h (*see* **Note 22**).
5. Position the testis block on a pre-cooled chuck coated with cold O.C.T. (*see* **Note 23**).
6. Cut 7 μm testis sections and collect on cold Superfrost Plus microscope slides (*see* **Note 24**).
7. Allow the slides to air-dry for 1–2 h inside the cryostat.
8. After drying, immediately store the slides in a slide box with desiccants in a re-sealable bag at −80 °C or proceed immediately to the RNAscope assay (*see* **Note 25**).

3.3 Testis Pretreatment

1. Remove stored slides from the −80 °C freezer or freshly prepared slides from the cryostat and place directly in a staining dish containing DPBS to rinse off the O.C.T. by agitating the slides for 5 min.
2. Bake slides on a slide warmer for 30 min at 60 °C.

3. Post-fix testis sections by immersing slides in a plastic tray filled with cold 4% PFA for 15 min at 4 °C (*see* **Note 1**).
4. Tap slides on a paper towel to remove excess fixative and immerse in 50% EtOH for 5 min.
5. Transfer slides to 70% EtOH for 5 min.
6. Transfer slides to 100% EtOH for 5 min.
7. Repeat **step 6**.
8. Let slides air dry (flat on the benchtop, face-up) for 5 min.
9. Add 5–8 drops of RNAscope hydrogen peroxide solution to each slide and incubate for 10 min.
10. Tap slides on a paper towel to remove excess hydrogen peroxide solution. Submerge slides in a staining dish filled with water and agitate 3–5 times.

3.4 Target Retrieval

1. Place one staining dish containing 200 mL of Target Retrieval buffer and a second staining dish containing 200 mL of distilled water in the heated steamer (95–99 °C) (*see* **Note 26**).
2. Move slides to the boiling water dish for 10 s, then quickly transfer slides to the boiling Target Retrieval buffer dish for 5 min.
3. Immediately transfer slides to a staining dish containing RT water for 15 s.
4. Move slides to a staining dish with fresh 100% EtOH for 3 min. Air-dry the slides.
5. Using a hydrophobic pen, draw a barrier around the sections (*see* **Note 27**). Let it dry for 5 min.
6. Place the slides in the pre-warm hybridization chamber at 40 °C and add 5 drops of Protease III to cover the sections.
7. Incubate the slides for 30 min at 40 °C (*see* **Note 28**).
8. Meanwhile, prepare probe mix and only include probes that will be used. Mix 50 volumes of C1 probe with 1 volume of C2, C3, and C4 probes. Mix thoroughly (*see* **Note 29**).
9. Warm probes for 10 min at 40 °C. Cool the probes to RT, then transfer the probes to a centrifuge and perform a quick spin-down.
10. After the 30 min incubation, wash slides 3–5 times in a dish containing distilled water by moving the slides up and down.

3.5 Hybridization

1. Place the slides back in the hybridization chamber.
2. Add 4 drops (120 μL) of probe mix to cover each section and incubate for 2 h at 40 °C.
3. Wash slides for 2 min in a lid filled with wash buffer. Repeat with fresh buffer.

4. Tap slides on a paper towel to remove excess wash solution.
5. ****optional stopping point****—store the slides overnight in 5× SSC buffer or proceed with the assay (skip to Subheading 3.6).

3.6 Amplification

1. Return the slides to the hybridization oven.
2. Add 4–6 drops of AMP 1 reagent to the slides, and incubate for 30 min at 40 °C (*see* **Note 30**).
3. Wash slides for 2 min in a plastic tray containing wash buffer.
4. Tap slides on a paper towel to remove excess wash solution and return slides to hybridization oven.
5. Once in the oven, add 4–6 drops of AMP 2 reagent to the slides. Incubate for 30 min at 40 °C.
6. Wash slides for 2 min in a plastic tray containing wash buffer.
7. Tap slides on a paper towel to remove excess wash solution.
8. Return the slides to the oven.
9. Add 4–6 drops of AMP 3 reagent and incubate for 15 min at 40 °C.
10. Remove and wash slides for 2 min in a plastic tray containing wash buffer.
11. Tap slides on a paper towel to remove excess wash solution.

3.7 Develop C1 Probe (For Example)

1. Return the slides to the hybridization oven.
2. Cover sections with 4–6 drops HRP-C1. Leave for 15 min at 40 °C. For example, for the C2 probe use HRP-C2, C3 probe use HRP-C3, and C4 probe use HRP-C4.
3. Meanwhile, only prepare the fluorophores that will be used. For example, we will assign Opal 520 to the C1 channel since the lowest expressed gene is targeted in this channel. If Opal 690 is used instead, you may need to use a higher concentration. Prepare a 1:750 stock of Opal 520 (*see* **Note 31**).
4. Wash slides for 2 min in a plastic tray containing wash buffer.
5. Tap slides on a paper towel to remove excess wash solution.
6. Add 200 μL of Opal 520 solution per slide and incubate for 30 min at 40 °C in oven.
7. Wash slides for 2 min in a plastic tray containing wash buffer.
8. Tap slides on a paper towel to remove excess wash solution.
9. Place slides in the oven and add 4–6 drops HRP blocker covering the slides.
10. Incubate for 15 min at 40 °C.
11. Wash slides for 2 min in a plastic tray containing wash buffer.
12. Tap slides on a paper towel to remove excess wash solution.

13. If labeling only the C1-probe, skip to Subheading 3.9 Counterstain or if combined with immunofluorescence (skip to Subheading 3.8). However, if labeling another channel, repeat **steps 1–13** until the desired channels have been developed.

3.8 Immunofluorescence (If Necessary)

1. Add 4 °C cold permeabilization buffer solution to the slides for 2 h at 4 °C inside a humid chamber.
2. Remove the permeabilization buffer solution.
3. Add the primary antibody in permeabilization buffer solution to the slide (*see* **Note 32**).
4. Place slides in a humid chamber and incubate overnight at 4 °C.
5. Wash the slides with DPBST twice for 5 min.
6. Add the secondary antibody in blocking buffer to the slide.
7. Allow it to sit for 45 min at RT in a humid chamber.
8. Tap slides on a paper towel to remove secondary antibody mix (*see* **Note 33**).
9. Wash the slides with DPBST twice for 5 min.
10. Proceed to Subheading 3.9.

3.9 DAPI Counterstain

1. Stain with DAPI by adding 4 drops of DAPI solution to the slide and incubate for 30 s.
2. Quickly tap the slide on a paper towel to remove DAPI and add 1–2 drops of ProLong Gold antifade mounting medium.
3. Coverslip each slide and allow slides to sit in dark for 30 min to overnight.
4. Store slides in a slide box at 4 °C (*see* **Note 34**).

4 Notes

1. Using 4% PFA older than 2 days will decrease staining quality.
2. The DPBS made with nuclease-free water will be used to wash the fixative off the testes and to create different concentrations of sucrose.
3. Vortexing the sucrose solution helps to dissolve the sucrose crystals; follow this with 0.2 μm filtering. Make fresh sucrose solution in DPBS with nuclease-free water for optimal results.
4. To visualize genes of interest, we suggest conjugating probes to the following TSA-linked fluorophore dyes: Opal 520 (FITC range), Opal 570 (Cy3 range), Opal 620 (Texas Red range), and Opal 690 (Cy 5.5 range). In our experience, it is best to

assign lowly expressed genes to the FITC channel and abundantly expressed genes in the Cy5.5 channel (Akoya Biosciences).

5. Target RNA species are detected by Advanced Cell Diagnostics propriety double Z probe design and signal amplification strategy, contributing to its sensitivity and specificity of target-specific signals. For example, we labeled for *Gfra1* and *Kit* transcripts in adult mouse testicular sections (see Fig. 1).
6. Target retrieval buffer is used to expose RNA epitopes masked by the fixation process.
7. Warm 50× wash buffer for 15 min at 40 °C before diluting to 1×.
8. To rinse the O.C.T. off the testis section before starting the assay, dunk the slide several times for 5 min in 1× DPBS.
9. For detection of the target protein antigen, permeabilization of the testes is required for the antibody to gain entry into the cell and recognize the antigen.
10. For experiments involving smFISH and immunostaining, it is important to take into consideration the protease treatment and subsequent wash steps, which may reduce the robustness of antibody labeling. For example, we have found that the antibody dilution required for detecting DDX4 protein (see Fig. 1) was lower (higher antibody concentration) than in our standard immunostaining protocols. Optimization of mRNA-protein co-labeling includes changing the antibody dilution and altering protease incubation times to preserve RNAscope detection reagents and probes. Additionally, for quantitative analysis, DDX4 is a useful marker for image segmentation to delimit individual cells. For more information on quantitative or semi-quantitative assessment of RNAscope staining, please refer to ACD's website (https://acdbio.com/services/quantitative-analysis).
11. Indirect immunofluorescence uses a secondary antibody to detect the primary antibody. The secondary antibody is specific for immunoglobulins derived from the host species for the primary antibody.
12. RNase-free bottles will be used to store wash buffer, 5× SSC buffer, and Milli-Q water.
13. Using Superfrost Plus charged microscope slides is essential to prevent detachment of testis sections during the various processing steps. In our experience, slides older than 3 years are unsuitable.
14. We used lids from empty pipette tip boxes for the post-fixation and wash steps in the RNAscope protocol. Two slides can fit in a single lid. Sterilize lids with 70% EtOH before and after use.

15. For the target retrieval step, we recommend using the Oster Food Steamer (model 5712). After each use, empty the water and wipe down the steamer with 100% EtOH and RNase decontaminant spray to prevent microbial growth.
16. We created a hybridization chamber utilizing a large glass desiccator fitted with a metal plate and an incubator. The metal place serves as a support for the slides inside the hybridization chamber. We place soaking wet filter paper beneath the plate to create a humid environment. It is important to note that there are many potential options for the humid chamber, including commercial options, plastic food storage containers, and pipette tip boxes; all will work, provided the environment maintains high humidity. It is important to wipe down the chamber with 100% EtOH and RNase decontaminant spray before each use to prevent bacterial growth. For immunostaining, we have found that a humid chamber constructed from a pipette tip box is sufficient.
17. Before embedding testes, place O.C.T. on ice for >30 min or store at 4 °C for 2 h.
18. Place sterilized forceps on dry ice for >30 min to minimize temperature changes in the testes that could negatively impact RNA quality within the tissue and reduce the likelihood of successful RNAscope detection.
19. Create a flat surface on the dry ice to prevent freezing testes in an undesired orientation.
20. Storing testis blocks in an air-tight bag and wrapping them with foil beforehand will prevent testes from drying out.
21. Transporting testis blocks on dry ice will prevent temperature fluctuations and ensure RNA quality.
22. Typically, a Cryostat chamber temperature of −18 to −25 °C is sufficiently cold to section mouse testes. However, depending on the model of the cryostat used, set Cryostat chamber temperature according to the manufacturer's recommendations for testicular tissue. Testis blocks, microtome blades, and microscope slides must be equilibrated to Cryostat chamber temperature before sectioning to avoid RNA degradation.
23. Trim testis blocks to a desirable size using a standard razor blade.
24. Before collecting testis cryosections, warm the back (or non-charged side) of the slide with the palm of your hand for a few seconds. Several testis cryosections can be positioned within a central square (0.75″ × 0.75″) on a slide and be sufficiently covered by RNAscope reagents. Be sure to avoid the overlap of O.C.T. from adjacent cryosections with the testis tissue.

25. Store slide boxes inside large resealable bags. From our experience, slides should be used within 1 week of sectioning. Longer storage times may increase autofluorescence and reduce specific labeling.
26. With a clean thermometer, check the temperature. The steamer approximately takes 30 min to reach a temperature of 95–99 °C. Do not allow the reagents to sit longer than 30 min in the warm steamer.
27. Draw a hydrophobic barrier (0.75″ × 0.75″) on a clean slide and use it as a guide for drawing barriers on the experimental slides. During the assay, before you draw the barrier, place the guide slide beneath the experimental slide and draw the barrier.
28. The length of Protease III treatment varies with the developmental age of the animal. In our experience, adult mouse testes are optimally pre-treated with a 30 min Protease incubation at 40 °C, while postnatal day (P)6 mouse testes will be sufficiently pre-treated with only a 15–20 min incubation at RT using 3–4 drops of Protease III. This step should be optimized for each sample type. The kit comes with Protease Plus, Protease III, and Protease IV, each of which vary in digestion strength from weak, moderate, to strong.
29. Gene-specific probes are designed to recognize specific RNA targets and can be visualized by their HRP linkage to one of four fluorescent channels, C1, C2, C3, and C4. These channels are subsequently developed by assigning different TSA-linked fluorophore dyes to the probes and therefore genes of interest can be mixed and matched to Opal 520, Opal 570, Opal 620, and Opal 690. Probes of the same channel should not be combined. Probe mixes can be stored at 4 °C for 6 months.
30. Equilibrate all the amplification reagents and HRP blockers to RT.
31. The Opal dyes have been reconstituted according to the manufacturer's instructions and stored at 2–8 °C. Prepare diluted fluorophore stocks using the provided TSA buffer and store in the dark at 2–8 °C. Diluted stocks may be reused for up to a month. Effective fluorophore dilutions range between 1:750 and 1:1000. If a smaller probe volume is used (e.g., 50 μL), we suggest using a fluorophore concentration of 1:750.
32. For antibody optimization, we recommend carrying out the RNAscope assay until the protease step, and then proceeding with Subheading 3.8. In doing so, one can minimize the use of expensive RNAscope detection reagents and probes. We advise users to include multiple cryosections for optimizing antibody concentrations. Additionally, positive and negative controls should be used in all experiments.

33. We prepared a secondary antibody dilution of 1:200 in blocking buffer.
34. After the assay, slides can be imaged after 8 h and within 2 weeks.

Acknowledgments

This work was supported by NIH grants R01 HD90007 and U01 DA054179.

References

1. Di Carlo D, Tse HTK, Gossett DR (2012) Introduction: Why analyze single cells? Methods Mol Biol 853:1–10. https://doi.org/10.1007/978-1-61779-567-1_1
2. Velte EK, Niedenberger BA, Serra ND, Singh A, Roa-De La Cruz L, Hermann BP et al (2019) Differential RA responsiveness directs formation of functionally distinct spermatogonial populations at the initiation of spermatogenesis in the mouse. 146(12): dev.173088 [pii]. https://doi.org/10.1242/dev.173088
3. Tang F, Barbacioru C, Wang Y, Nordman E, Lee C, Xu N et al (2009) mRNA-Seq whole-transcriptome analysis of a single cell. Nat Methods 6(5):377–382. https://doi.org/10.1038/nmeth.1315
4. Navin N, Kendall J, Troge J, Andrews P, Rodgers L, McIndoo J et al (2011) Tumour evolution inferred by single-cell sequencing. Nature 472(7341):90–94. https://doi.org/10.1038/nature09807
5. Islam S, Kjallquist U, Moliner A, Zajac P, Fan JB, Lonnerberg P et al (2011) Characterization of the single-cell transcriptional landscape by highly multiplex RNA-seq. Genome Res 21(7):1160–1167. https://doi.org/10.1101/gr.110882.110
6. Picelli S, Bjorklund AK, Faridani OR, Sagasser S, Winberg G, Sandberg R (2013) Smart-seq2 for sensitive full-length transcriptome profiling in single cells. Nat Methods 10(11):1096–1098. https://doi.org/10.1038/nmeth.2639
7. Hashimshony T, Wagner F, Sher N, Yanai I (2012) CEL-Seq: single-cell RNA-Seq by multiplexed linear amplification. Cell Rep 2(3): 666–673. https://doi.org/10.1016/j.celrep.2012.08.003
8. Suzuki S, Diaz VD, Hermann BP (2019) What has single-cell RNA-seq taught us about mammalian spermatogenesis? Biol Reprod 101(3): 617–634. https://doi.org/10.1093/biolre/ioz088
9. Hermann BP, Cheng K, Singh A, Roa-De La Cruz L, Mutoji KN, Chen IC et al (2018) The mammalian spermatogenesis single-cell transcriptome, from spermatogonial stem cells to spermatids. Cell Rep 25(6):1650–1667. e8. https://doi.org/10.1016/j.celrep.2018.10.026
10. Hermann BP, Mutoji KN, Velte EK, Ko D, Oatley JM, Geyer CB et al (2015) Transcriptional and translational heterogeneity among neonatal mouse spermatogonia. Biol Reprod 92(2):54. https://doi.org/10.1095/biolreprod.114.125757
11. Gall JG, Pardue ML (1969) Formation and detection of RNA-DNA hybrid molecules in cytological preparations. Proc Natl Acad Sci U S A 63(2):378–383. https://doi.org/10.1073/pnas.63.2.378
12. Pardue ML, Gall JG (1969) Molecular hybridization of radioactive DNA to the DNA of cytological preparations. Proc Natl Acad Sci U S A 64(2):600–604. https://doi.org/10.1073/pnas.64.2.600
13. Newkirk SJ, Lee S, Grandi FC, Gaysinskaya V, Rosser JM, Vanden Berg N et al (2017) Intact piRNA pathway prevents L1 mobilization in male meiosis. Proc Natl Acad Sci U S A 114(28):E5635–E5E44. https://doi.org/10.1073/pnas.1701069114
14. Takei N, Nakamura T, Kawamura S, Takada Y, Satoh Y, Kimura AP et al (2018) High-sensitivity and high-resolution in situ hybridization of coding and long non-coding RNAs in vertebrate ovaries and testes. Biol Proced Online 20:6. https://doi.org/10.1186/s12575-018-0071-z
15. Li K, Xu J, Luo Y, Zou D, Han R, Zhong S et al (2021) Panoramic transcriptome analysis and functional screening of long noncoding RNAs in mouse spermatogenesis. Genome Res 31(1):

13–26. https://doi.org/10.1101/gr.264333.120

16. Wang F, Flanagan J, Su N, Wang LC, Bui S, Nielson A et al (2012) RNAscope: a novel in situ RNA analysis platform for formalin-fixed, paraffin-embedded tissues. J Mol Diagn 14(1): 22–29. https://doi.org/10.1016/j.jmoldx.2011.08.002

17. Green CD, Ma Q, Manske GL, Shami AN, Zheng X, Marini S et al (2018) A comprehensive roadmap of murine spermatogenesis defined by single-cell RNA-Seq. Dev Cell 46(5):651–667. e10. https://doi.org/10.1016/j.devcel.2018.07.025

Chapter 4

Bulk and Single-Cell RNA-Seq Analyses for Studies of Spermatogonia

Anukriti Singh and Brian P. Hermann

Abstract

Robust methods have been developed that leverage next-generation sequencing (NGS) to measure abundance of all mRNAs (RNA-seq) in samples as small as individual cells in order to study the testicular transcriptome in mammals. In this chapter, we present robust options for implementing bioinformatics workflows for the analysis of bulk RNA-seq from aggregate samples of hundreds to millions of cells and single-cell RNA-seq from individual cells. We also provide detailed protocols for using the R packages DESeq2 and Seurat, important parameters for successful implementation, and considerations for drawing conclusions from the results.

Key words Single-cell RNA-seq, Bulk RNA-seq, Bioinformatics, Spermatogenesis

1 Introduction

Mammalian spermatogenesis is the complex cell lineage that begins with spermatogonial stem cells (SSCs) and culminates in the production of spermatozoa. Understanding the fundamental biology underlying the myriad of cellular changes during this developmental process, such as SSC fate regulation, spermatogonial differentiation, male meiosis, and spermiogenesis, relies heavily on inclusive measurements of all expressed genes at the mRNA level. Initially, this was done with gene expression microarrays [1, 2] that are necessarily biased by the repertoire of probes on the arrays, but now, transcriptomes are typically produced with next-generation sequencing measurements of cDNA libraries [3–5]. In either case, measuring mRNA abundance patterns during spermatogenesis is performed using RNA extracted from either whole testes or enriched populations of cells [6–8]. Investigating mRNA changes temporally across the first wave of mouse spermatogenesis permits discovery of transcripts that change as new spermatogenic cell types first emerge during the initial

Jon M. Oatley and Brian P. Hermann (eds.), *Spermatogonial Stem Cells: Methods and Protocols*, Methods in Molecular Biology, vol. 2656, https://doi.org/10.1007/978-1-0716-3139-3_4,

spermatogenic wave [6–9]. Ultimately, such experiments permit identification of mRNAs that exhibit statistically significant changes in abundance (termed "differential gene expression" from here forward) between distinct enriched populations, developmental time-points, or experimental conditions [8, 10, 11]. For instance, bulk RNA-seq has been previously used to compare SSC-enriched/depleted populations isolated from developing mouse testes by fluorescence-activated cell sorting (FACS) selection for TSPAN8 cell-surface labeling or ID4-EGFP epifluorescence intensity [10–12]. Invariably, though, enrichment methods like StaPut gravity sedimentation and FACS fail to purify biologically distinct cell types (hence the term "enrichment") and, instead, produce mixtures of cells with varying purity of the cell types of interest. Thus, results from bulk RNA-seq investigating cell mixtures constitutes ensemble averaging derived from aggregates of multiple spermatogenic and/or somatic cell types that can mask the phenotypes of less prevalent cell types such as SSCs and ignore heterogeneity among cells. To overcome these limitations, transcriptome profiling at the single-cell level is necessary. Recently, a plethora of single-cell RNA-seq studies have begun to comprehensively characterize the full extent of cellular heterogeneity in the testis, revealing rare, transitional, or novel cell types [4, 5, 9, 13, 14]. In the following sections, we will outline the typical steps taken to produce transcriptomes from bulk and single-cell samples and their subsequent use for differential gene expression analyses.

The basic workflow for the generation and analysis of both bulk and single-cell RNA-seq data consists of the following steps as shown in Fig. 1: (1) RNA from the cells of interest is reverse transcribed to complementary DNA (cDNA), which are then sequenced by high throughput (typically short-read) platforms; (2) quality control of the raw sequencing data is performed based on metrics generated by the sequencing platform or calculated from the raw reads; (3) the raw reads are aligned to a reference genome or transcriptome; (4) expression quantification is performed to generate a matrix of gene expression values (columns = samples or cells, rows = genes); (5) gene expression values are normalized to account for technical variation/biases; and (6) differential gene expression analysis is performed between samples.

1.1 cDNA Library Preparation

Generation of cDNA libraries consists of RNA capture, fragmentation, reverse transcription into first strand cDNA, second strand synthesis, and cDNA amplification. Most methods for library preparation begin with capture of poly(A)-tailed RNA to enrich for protein-coding transcripts and exclude abundant RNA types such as rRNA and tRNA [15]. Poly(A)+RNA selection using magnetic or cellulose beads coated with oligo-dT molecules is favored due to its relatively low cost and ease of use. Alternatively, oligo-dT priming-based methods combine poly(A)+ selection and reverse

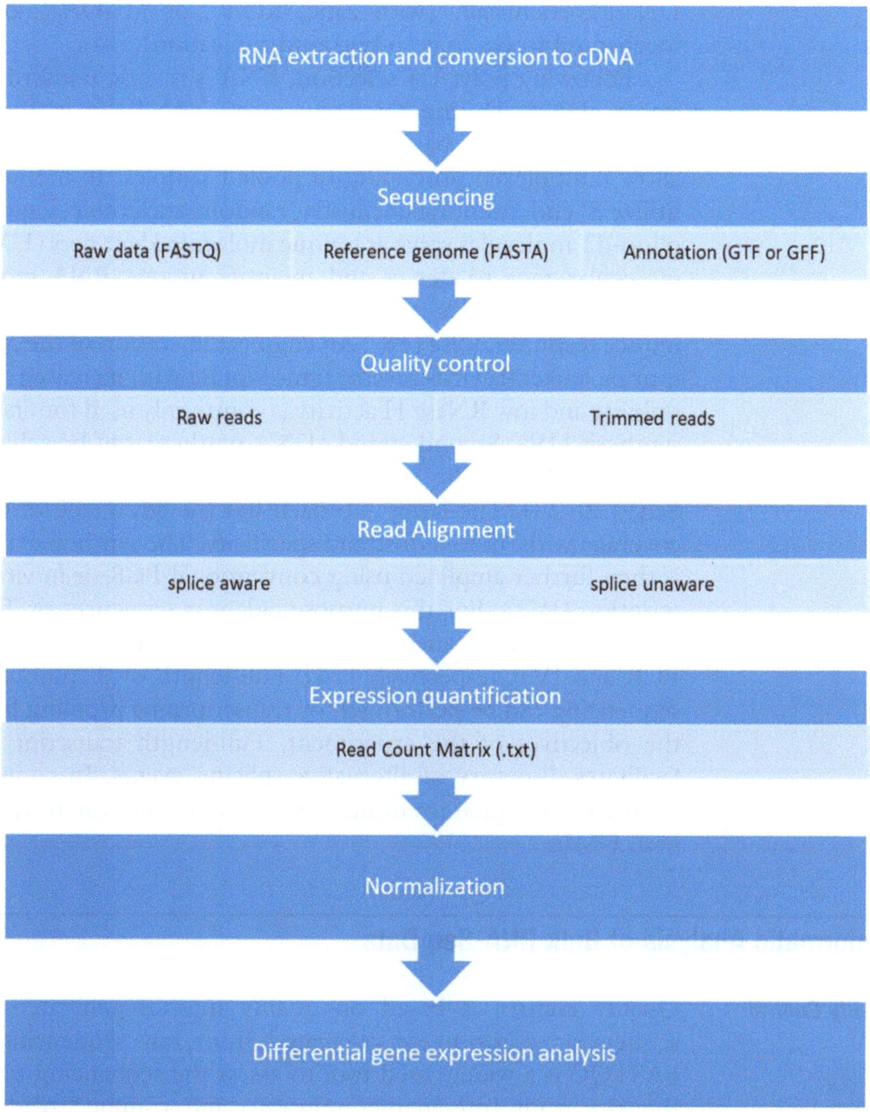

Fig. 1 A generalized bulk and single-cell RNA sequencing workflow is shown

transcription (RT) into one step. However, this method can exhibit 3′ bias, resulting in sequencing reads enriched for the 3′ portion of the transcript. Furthermore, internal poly(A) priming, in which oligo-dT prime at internal A-rich sequences of the transcripts, can generate a high frequency of truncated cDNAs [16, 17]. Following capture, RNA samples are fragmented to a specific size range before RT based on the size limitation of sequencing platform (<600 bp for Illumina sequencers). When the goal is to sequence full-length RNA transcripts, intact (i.e., unfragmented) RNA is first reverse-transcribed into cDNA and then subjected to fragmentation. Fragmentation can be accomplished enzymatically (with tagmentase or

DNase), chemically (with Zinc, KOAc, or MGOAc), or using mechanical forces (e.g., ultrasound sonication).

Following poly(A)+ selection, RNA is reverse transcribed into stable cDNA. During this step, most scRNA-seq methods add single-cell specific barcodes within the oligo-dT molecules to allow multiplexed processing of pooled samples. In methods that utilize 3′ end-counting chemistry, random nucleotide-sequences in oligo-dT molecules serve as unique molecule identifiers (UMIs), or molecular tags to detect and quantify unique RNA molecules. Incorporating UMIs can help correct for amplification bias and reduce technical noise [18]. An engineered version of the Moloney murine leukemia virus reverse transcriptase with increased thermostability and low RNase H activity is commonly used for first strand synthesis [19]. Second strand cDNA synthesis can be achieved by poly(A) tailing or by a template-switching mechanism (SMART-Seq) [20–23]. The latter approach has the advantage of uniform coverage without loss of strand specificity. The synthesized cDNAs is then further amplified using conventional PCR or in vitro transcription (IVT). For this purpose, adaptor sequences or T7 polymerase promoter sequences are included in the oligonucleotides for PCR and IVT, respectively [24]. Full-length or 3′ transcript end sequencing can be performed for transcriptome profiling based on the objectives of the experiment. Full-length transcript analysis facilitates discovery of alternative-splicing events. In comparison, protocols for digital counting of 3′ transcript ends can be combined with UMIs.

2 Bionformatic Analysis of Bulk RNA-Seq Data

2.1 Quality Control

Quality control is based on quality metrics generated by the sequencing platform or calculated from raw sequencing reads. FASTQC is a widely used tool to assess the sequencing quality of the raw reads (pre-adapter removal) and trimmed reads (post-adapter removal) [25]. FASTQC is run independently for every FASTQ file and provides a summary of the per base and per sequence quality scores, per sequence GC content, per sequence adapter content, per sequence read lengths, and overrepresented sequences. After using FASTQC, quality trimming and adaptor sequence removal is an optional step that can be omitted if the preliminary QC does not reveal any quality bias or adapter sequences. Trimmomatic is a popular tool that performs both quality trimming and adaptor sequence removal, but other tools like CutAdapt and NGS QC Toolkit are also available [26–28]. In addition, RSeQC can be used to assess post-alignment quality [29]. RSeQC calculates the distribution of insert size and enables junction annotation. Table 1 provides a list of the commonly used quality control software.

Table 1
Analysis steps and relevant software tools for generating bulk RNA-seq transcriptomes

Analysis step	Software tool	Reference
Quality control	Fastqc	[25]
Trimming	RSeQC	[29]
	Trimmomatic	[26]
	CutAdapt	[27]
	NGS QC Toolkit	[28]
Read alignment	TopHat2	[32]
	STAR	[33]
	HISAT2	[34]
	Bowtie2	[30]
	BWA	[31]
	Salmon	[36]
	Kallisto	[37]
Expression quantification	Cufflinks	[41]
	RSEM	[42]
	Stringtie	[43]
	featureCounts	[44]
	HTSeq	[82]

2.2 Read Alignment

Following quality control, reads are assigned to transcripts either by mapping to a reference genome or transcriptome. Reference-based alignment tools are broadly categorized as "splice aware" and "splice unaware" aligners. When using a transcriptome as reference, splice unaware tools including Bowtie2 and Burrow–Wheeler Aligner (BWA) can be used [30, 31]. When a genome is used as reference, using splice aware tools such as TopHat2, STAR and HISAT2 are advantageous as they can handle gapped (or spliced) alignments [32–34]. While all three tools are generally considered to produce fast results, they differ in computational speed and memory usage. Among these tools, TopHat2 outperforms STAR in terms of high-alignment rates when using a genome annotation file [35]. For de novo assembly, alignment-free mappers such as Salmon and Kallisto are useful [36, 37]. Although they have been shown to perform as well as the reference-based tools on simulated data, Salmon and Kallisto are still under development and are prone to spurious alignments, especially for lowly expressed genes [38]. Most of the tools discussed above are available on the open source, web-based Galaxy platform [39].

2.3 Transcriptome Reconstruction and Expression Quantification

Transcriptome reconstruction, or the identification of all transcripts in a sample, can be performed using two approaches: reference-based assembly or de novo assembly. Reference-based approaches rely on the availability and accuracy of a reference but are less computationally intensive. In comparison, de novo assembly is applied to organisms that lack a reference genome and overcomes the challenge of mapping uncertainty. However, this strategy requires more computational resources and deeper sequencing, making it typically unsuitable for large and complex mammalian transcriptomes [40]. Reference-guided expression quantification tools such as Cufflinks [41], RSEM [42], and Stringtie [43] use a transcript-based approach. A major drawback of transcript-based quantification is that estimating the expression of individual isoforms is more difficult. The union exon-based approach, used by featureCounts, assigns reads to genes with higher confidence compared with assigning reads to isoforms, which allows for a more accurate estimation of gene expression [44].

2.4 Normalization

Normalization of raw read counts is a critical step that corrects for non-biological (technical) variation introduced between the time that samples were isolated and sequencing data were produced. Such variation may be between and within samples and may arise due to the library preparation, sample sequencing read depth, gene length, read mapping bias, gene sequence composition, and sequence similarity [45]. Table 1 provides a detailed summary of the computational tools used to model technical variations in bulk RNA-seq. One source of technical variation between samples is differences in sequencing depth (often called library size), which represents the total number of NGS reads generated for a given sample. Normalization methods apply various global scaling factors on raw read counts to make library sizes comparable across samples. Global scaling quantile normalization methods include quantile (Q), upper quartile (UQ), median (Med), and TC (per sample total count). Methods which are implemented in the DESeq2 and edgeR packages include relative log expression (RLE) and trimmed mean of M-values (TMM), respectively [46–50]. For full-length transcriptomes, gene length can also impact abundance estimates as longer genes yield a higher read count compared to shorter genes expressed at equal levels. The most widely used approaches for gene length normalization include FPKM (fragments per kilobase per million reads), RPKM (reads per kilobase per million reads), and TPM (transcripts per million reads), which is a slight modification of RPKM [51].

Among these methods, RPKM and TPM are generally not used directly for differential expression analysis. TPM is useful for qualitative comparison such as clustering analysis, but should not be used for comparisons across samples when the total RNA content and distribution are very different [52]. In addition to correcting for known artifacts, numerous methods can identify potential latent factors that capture sources of technical variation. These methods include remove unwanted variation (RUV), surrogate variable analysis (SVA), and principal component analysis (PCA). These are used with statistical approaches (e.g., linear regression model or

ComBat) to generate normalized data. Although the correct usage of these approaches can potentially increase statistical power in differential analysis, distinguishing latent factors from biological factors of interest is crucial [53].

2.5 Differential Gene Expression Analysis

Identification of differentially expressed genes is crucial for interpreting the biological differences between the compared conditions. Computational tools for differential gene expression analysis estimate the magnitude of differential expression between two or more conditions based on normalized read counts (i.e., calculate a fold change value). These tools also estimate the statistical significance of the difference and correct for multiple testing (i.e., calculate a p-value or adjusted p-value for each gene). Thresholds for the fold change and p-value can be chosen to improve the conciseness of top markers for defining groups while mitigating the risk of discarding useful genes. Typically, a cutoff of fold changes greater than 1.5 and p-value less than 0.05 is used. Methods for differential expression analysis can be grouped into parametric and non-parametric. Parametric methods capture all information about the data within the parameters, making it possible to predict the value of unknown data from observing the adopted model. Parametric methods are preferred when repeated biological measures from a defined population follow a normal distribution when sampled several times. However, these models assume a particular data distribution. In contrast, non-parametric methods capture more details about the data distribution by not imposing a rigid model to be fitted. If the modeling assumption is valid, parametric approaches are preferred. When the sample size is small, non-parametric methods are advised.

Computational tools can be further classified into three categories based on the statistical approaches used (Table 2). First, negative binomial model-based methods are implemented in edgeR, DESeq2, Cuffdiff2, baySeq, and EBSeq. Second, non-parametric methods which do not assume any distribution are used in SAMseq and NOIseq. Third, a generalized linear model is used in limma, which tests differential expression using the moderated t-statistics. The performance of 11 DE methods including edgeR, DESeq, baySeq, EBSeq, SAMseq, and limma was evaluated using both simulation studies and real RNA-Seq data [54]. All of these methods demonstrated low power with small sample sizes. DESeq2 and limma tended to be more conservative than edgeR, with better control of false positives. Limma-voom was robust to outliers and computationally efficient, but performed worse when the variances were unequal between the groups. Previous experimental validation of selected DEG from three methods (CuffDiff2, edgeR, and DESeq2) reported high-false discovery rates using the CuffDiff2 method [55]. A review of eight DE analysis methods (baySeq, DESeq, DESeq2, EBSeq,

Table 2
Normalization and differential gene expression analysis tools for bulk RNA-seq

Software	Normalization method	Reach count distribution model	Differential expression test	Reference
edgeR	TMM/Upper quartile/RLE/None (all scaling factors set to one)	Negative binomial	Exact test	[83, 84]
DESeq2	DESeq size factors	Negative binomial	Wald test	[85]
baySeq	Scaling factors (quantile/TMM/total)	Negative binomial	Posterior probability through Bayesian approach	[86]
EBSeq	DESeq median normalization	Negative binomial	Posterior probability through Bayesian approach	[87]
Cuffdiff2	Geometric/quartile/FPKM	Negative binomial/ Beta	*t*-test analogical method	[88]
NOIseq	RPKM/TMM/Upper quartile	Non-parametric	Corresponding log of fold change and absolute expression differences have a higher probability than noise values	[89, 90]
SAMseq	Based on mean read count over null features of dataset	Non-parametric	Wilcoxon rank statistics-based permutation test	[91]
limma	TMM	Voom transformation of counts (linear)	Moderated *t*-test	[92]

edgeR, voom, NOIseq, and SAMseq) evaluated precision, accuracy, and sensitivity using qRT-PCR data as reference [56]. According to this study, voom, NOIseq, and DESeq2 showed more consistent results than the other methods. In addition, SAMseq required larger sample sizes to detect significant DE genes. Overall, DESeq2, edgeR, and limma are the best performing tools [57], but none were completely infallible. One strategy to maximize the robustness of DEG sets is to identify consensus DEGs obtained from multiple approaches, which can improve accuracy and reduce false discovery rates; however, it is worth noting that this approach may also increase the false negative rate.

2.6 Analysis of Bulk RNA-Seq Data Using DEseq2

This section describes a basic workflow for the analysis of bulk RNA-seq matrix data (transcriptomes) using the R package DESeq2, which is included in the Bioconductor collection of R libraries. The code provided below can be directly used in R.

2.6.1 Resources Required

1. *Software:* R version 3.6.2 or newer: https://www.r-project.org/.
2. *Hardware*: Computer with Windows, Unix, Mac OS X, or Linux Operating systems with sufficient RAM and storage space for output files.

2.6.2 Initialization, Quality Control, and Filtering

1. To start using DESeq2, download and install the DESeq2 package within the R environment using the first two commands. Once installed, load the packages using the library command.

Box 1 Install BiocManager and load required packages

```
if (!requireNamespace("BiocManager", quietly = TRUE))
 install.packages("BiocManager")
BiocManager::install("DESeq2", "ggplot2", "apeglm",
"gplots")
library(DESeq2)
library(ggplot2)
library(apeglm)
library(gplots)
```

2. Load the matrix of read counts which we named cts from the file counts.csv and the group assignments (e.g., Tspan8high or Tspan8Low sample) which we named *coldata* for each sample from the file metadata.csv. Examine the count matrix and column data using the *head* function. Fix the order of the samples in the read count matrix to be consistent with the sample annotation file. Verify that the columns of the count matrix and rows of the column data are in the same order.

Box 2 Load read count matrix

```
cts <- read.csv('counts.csv', header=TRUE, sep = ",")
head(cts)
coldata <- read.csv('metadata.csv', row.names=1)
head(coldata)
coldata$condition <- factor(coldata$condition)
cts <- cts[, rownames(coldata)]
all(rownames(coldata) == colnames(cts))
```

3. Construct a DESeqDataSet object in DESeq2 named *dds* using the count matrix *cts*, the sample information *coldata*, and a design formula. The design formula specifies which variables in the sample metadata *coldata* we will use for differential expression testing. The design formula can also include batch information to account for batch effects (*see* **Note 1**).

Box 3 Construct DESeqDataSet object in DESeq2

```
dds <- DESeqDataSetFromMatrix(countData = cts,
 colData = coldata,
 design = ~ condition)
dds
```

4. Perform pre-filtering to include only rows that have at least 10 read counts total.

Box 4 Perform data filtering

```
keep <- rowSums(counts(dds)) >= 10
dds <- dds[keep,]
```

2.6.3 Differential Gene Expression Analysis

1. Set the factor levels to specify the comparison to make during differential expression analysis.

Box 5 Differential expression testing

```
dds$condition <- factor(dds$condition, levels = c
("Tspan8low", "Tspan8high"))
```

2. Perform differential expression analysis using the function DESeq on raw read counts. Create a results table for a comparison of the last level over the first level specified in step 5 using

the function results and inspect it. The first column, *baseMean*, contains the average of normalized count values divided by size factors taken over all samples. The remaining columns are *log2FoldChange* (log2 fold change estimates), *lfcSE* (standard error estimate for log2 fold change), *stat* (Wald statistic), *p-value* (using Wald test), and *p-adjusted* values.

Box 6 Summary of differential gene expression results

```
dds <- DESeq(dds)
res <- results(dds)
res
resultsNames(dds)
```

3. Perform log fold change shrinkage using the function and specify the *apeglm* (adaptive t prior shrinkage estimator) method. Shrunken log fold changes are useful for ranking of genes and visualization (*see* **Note 2**).

Box 7 Log fold change shrinkage

```
resLFC <- lfcShrink(dds, coef="condition_Tspan8-
high_vs_Tspan8low", type="apeglm")
resLFC
```

4. Reorder the results table by the smallest p-values and view a summary of the results. Export results to an Excel file using the function write.csv.

Box 8 Sort DE results by p-value

```
resOrdered <- res[order(res$pvalue),]
summary(res)
write.csv(as.data.frame(resOrdered),
file="results_orderedbypadj.csv")
```

2.6.4 Sample Clustering and Visualization

1. Create an MA-plot, which is a scatter plot of log2 fold changes (on the y-axis) versus mean of normalized counts (on the x-axis) (Fig. 2). To examine read counts for a single gene across the samples, plot normalized counts plus a pseudocounts of 0.5 using the *plotCounts* function.

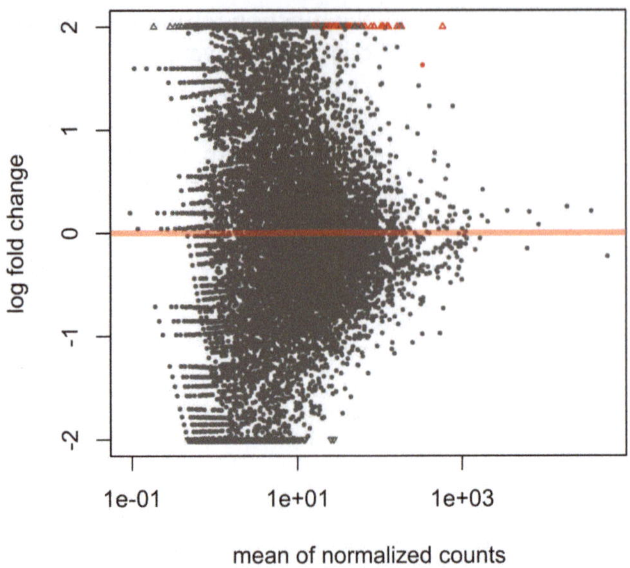

Fig. 2 An example of MA plot, a two-dimensional scatter plot to visualize gene expression changes from two different conditions is shown. The plot shows each gene are points, with log2 fold changes on the y-axis and mean of normalized counts on the x-axis. Genes with similar expression values in both conditions cluster around M-0 line. Points are colored red if the adjusted p-value is less than 0.1. Points which fall outside the plot limits are open triangles pointing either up or down

Box 9 Plot MA-plot and counts

```
plotMA(res, ylim=c(-2,2), labels=FALSE)
plotCounts(dds, gene=which.min(res$padj),
intgroup="condition", labels=FALSE)
```

2. Although differential expression testing is done on raw counts, it is useful to extract transformed values for visualization and clustering. The following commands apply the shifted logarithm transformation using the *normTransform* function, variance stabilizing transformation (VST) using the *vst* function, or regularized logarithm transformation (rlog) using the *rlog* function. Using the assay function, extract a matrix of normalized values. These transformation approaches produce data on the log2 scale which has been normalized with respect to the library size or other normalization factors specified.

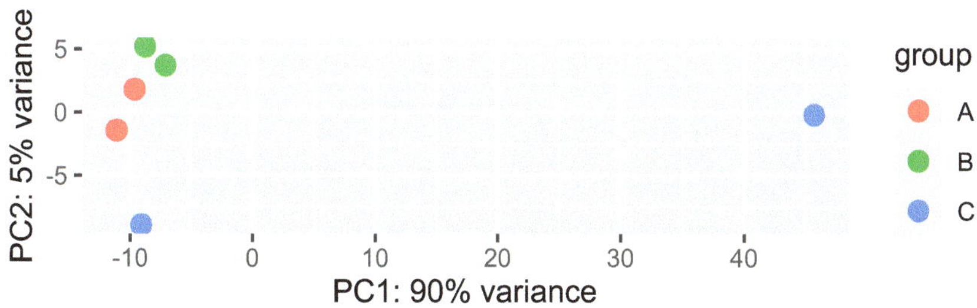

Fig. 3 Example of a Principal component analysis (PCA) plot shows six samples over the first two principal components (PC1 and PC2). Points are colored by group (condition)

Box 10 Data transformation

```
ntd <- normTransform(dds)
vsd <- vst(dds, blind=FALSE)
rld <- rlog(dds, blind=FALSE)
head(assay(vsd), 3)
```

3. Create a principle component analysis (PCA) plot using the transformed data showing the first two principle components, which capture the greatest variance between samples (Fig. 3).

Box 11 PCA Plot

```
plotPCA(vsd, intgroup=c("condition"))
```

4. To generate a heatmap of the transformed count data generated in step 2, the R package *gplots* is used. Select and order the genes to be included in the heatmap using the order function. Plot a heatmap of the shifted logarithm transformed values using the heatmap.2 function (Fig. 4).

Box 12 Plot Heatmap

```
select <- order(rowMeans(counts(dds,normalized=TRUE)),
decreasing=TRUE)[1:100]
heatmap.2(as.matrix(assay(ntd)[select,],),
 scale="row",
 hclust=function(x)
hclust(x,method="average"),
 distfun=function(x) as.dist((1-cor(t(x)))/2),
 trace="none",
 density="none",
 labRow="",
 cexCol=0.7)
```

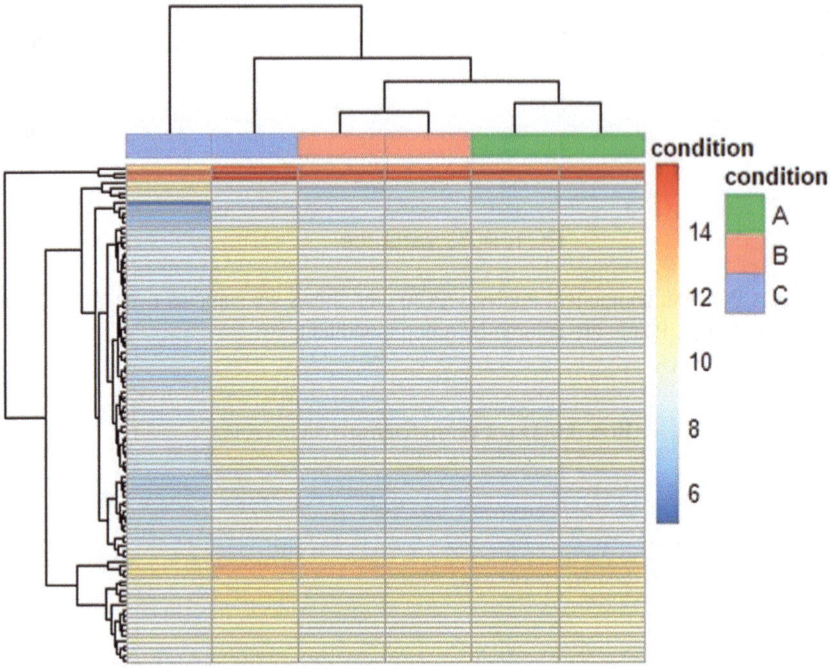

Fig. 4 Heatmap of expression values of 10 differentially expressed genes across individual samples is shown. The columns represent samples, and the rows represent genes. Color indicate expression values (log transformed counts) for each gene

5. The list of differentially expressed genes from step 8 can be used for gene set enrichment analysis using a variety of tools including DAVID, EnrichR and ingenuity pathway analysis (IPA).

3 Bioinformatic Analysis of Single-Cell RNA-Seq Data

Single-cell transcriptome profiling enables an unbiased, high-resolution view of cellular heterogeneity. This allows investigators to identify cell types, investigate rare cell populations, define cellular states, and predict developmental or differentiation trajectories, each of which is hidden by ensemble averaging in bulk RNA sequencing. Two of the most widely used methodologies for assaying individual cells include microfluidic capture SMART-Seq using the Fluidigm C1 platform and droplet-digital RNA-seq using the 10× Genomics platform. The SMART-Seq protocol offers higher sensitivity and generates full-length transcriptomes, which makes it suitable for the detection of splice variants [23]. In addition, the Fluidigm C1 platform allows visual quality control check by microscopic examination. However, this method precludes multiplexing of samples and has a much lower cell throughput (10's to 100's of cell) at a considerable expense (>$50/cell). The increased read depth of the SMART-Seq protocol enables the study of low-abundance transcripts. In comparison, droplet-digital RNA-seq utilizes 3′ end-counting, providing a larger throughput of cells (1000's to 10000's of cells) at a lower sequencing cost per cell (>$1/cell). The ability to profile large amounts of cells, albeit at a reduced RNA capture efficiency, facilitates identification of rare cell populations [58]. Because cell barcodes are introduced randomly, this approach does not allow visual detection of doublets and association of cell properties such as fluorescent signals with transcriptome profiles. Other droplet-based methods include inDrops and Drop-seq, both of which employ 3′ end sequencing, making them cost efficient and suitable for analyzing a larger number of cells. However, these have a lower capture efficiency compared with 10× [59]. In contrast with the methods above, split-pool barcoding-based approaches, including sci-RNA-seq and SPLiT-seq, are highly scalable, with low-cost cDNA library preparation and longer sample storage [60, 61]. A systematic comparison of some scRNA-seq technologies including SMART-seq, SMART-seq2, CEL-seq2, SCRB-seq, and MARS-seq was performed using mouse embryonic stem cells [62]. Overall, SMARTseq2 was found to be the most sensitive and accurate method to analyze full-length transcriptomes and detect alternative splice forms [62]. However, SMARTseq2 is limited by lower throughput and considerable costs. Although less sensitive, SCRB-seq, CEL-seq2, and MARS-seq quantified mRNA levels with less amplification bias due to the use of UMIs [62], while SCRB-seq, MARS-seq, and SMARTseq2 are preferred for analyzing fewer cells [62].

3.1 Quality Control

In droplet-based method, quality filtering must be performed to exclude cell barcodes that are unlikely to represent intact single cells. A commonly used approach is to calculate a dataset-specific threshold for the minimum number of transcripts or UMIs per barcode which is required to consider transcripts arising from a particular cell barcode as actually arising from an individual, live cell. Small quantities of transcripts detected per cell barcode may indicate the presence of a dead cell or ambient RNA molecules, i.e., cell-free transcripts in solution (arising from previously dead/lysed cells). Inversely, a large number of transcripts detected for a barcode may indicate doublets. Alternatively, the EmptyDrops method (implemented in the R package DropletUtils) detects the presence of cells by estimating background levels of RNA present in empty droplets and identifying barcodes with significant deviation from the background. EmptyDrops outperforms methods using UMI count thresholds by recovering cells with low-total RNA content. To identify cell multiplets in droplets, tools such as scrublet and DoubletFinder define a threshold to distinguish the inferred doublets from the assumed singlets. Multiplet rates can be determined empirically and addressed using bioinformatics [63]. Another round of quality control is necessary to distinguish viable single cells from damaged or dying cells based on the number of detected genes, the proportion of RNA derived from the mitochondrial genome and the proportions of unmapped or multi-mapped reads [64]. Cells with high proportions of mitochondrially derived genes, few detected genes, or high proportions of unmapped or multi-mapped reads are often damaged or dying cells. The threshold used for excluding cells with a high-mitochondrial content is specific for each dataset, with cutoffs typically ranging from 5% to 25% mitochondrial gene representation depending on sample type and processing method.

3.2 Normalization

Normalization aims to remove the influence of unwanted technical effects from scRNA-seq data while preserving true heterogeneity. Technical biases such as differences in PCR amplification, mRNA capture, and reverse transcription efficiency must be accounted for prior to downstream analysis. Although bulk and scRNA-seq data share similar features, such as overdispersion of gene expression, they also have unique features, such as the high proportion of zero read counts. Consequently, bulk-based normalization methods may be unsuitable for single cell transcriptomics because they may cause overcorrection of lowly expressed genes. For example, the simplest and most common normalization strategy for bulk RNA-seq data is to calculate a quantity related to the sequencing depth of the samples, known as the "size factor," and divides the expression of all genes by this value [48]. A similar method specifically tailored to single-cell data, called scran, clusters the cells and computes cell-specific size factors more robustly in the presence of zero inflation [65]. This approach yields more accurate estimates of scaling factors with a low runtime. Instead of using cell-specific factors, scnorm and sctransform use gene group-specific factors to perform normalization [66]. The scnorm method uses quantile regression to group genes with similar dependence on sequencing depth and estimates scaling factors for each group. Despite the longer runtime compared to scran, scnorm has better performance, especially for low-throughput high sequencing depth data. For single-cell UMI datasets, sctransform is the preferred statistical approach to perform normalization and variance stabilization transformation [67]. The sctransform method applies gene-specific negative binomial regression to effectively mitigate sequencing depth-dependent differences [66].

Alternatively, spike-in RNAs from the External RNA Control Consortium or housekeeping genes can be used to estimate size factors [68].

3.3 Differential Gene Expression Analysis

The goal of differential expression analysis is to identify genes with statistically significant differences in expression between two or more groups. This analysis can be used to quantify differences in single-cell transcriptomes across different developmental states, treatments, or disease conditions. In contrast to bulk RNA-Seq data, scRNA-seq data have a high proportion of zero or low read counts, which may reflect either cellular expression levels or failure to detect present transcripts (known as technical dropout). Moreover, gene expression distributions show multimodality, reflecting the existence of multiple cell states within the population [11]. Although bulk methods do not account for the aforementioned characteristics, they have been widely used for single cell data [57].

Differential expression analysis tools developed specifically to address technical dropout in scRNA-seq data include MAST, SCDE, DEsingle, and Monocle2. In addition, Seurat provides several tests for differential expression testing. MAST uses a bimodal distribution and proposes a generalized linear model (GLM) to fit the data [69]. It also uses a fraction of genes detected in each cell as a proxy for technical or biological variation. MAST has better performance and computational efficiency than numerous bulk and single-cell methods [57]. Alternatively, SCDE uses a Bayesian approach to calculate differential expression and accounts for dropout and amplification events [70]. DEsingle employs a zero-inflated negative binomial regression model to estimate the proportion of the real and drop-out zeros [71]. Monocle2 proposes a generalized additive model (GAM) to fit the data and a Tobit model is to account for dropouts [72–74]. However, all of the aforementioned tools ignore multimodal distributions of scRNA-seq data.

A recently developed tool called D3E specifically addresses the multimodal distributions of single-cell data, but does not consider dropout events [75]. D3E fits the bursting model of transcriptional regulation [76, 77] to the data and compares the gene expression distribution in one group with respect to another giving estimates of burst size, duty cycle, frequency, and mean of transcription.

3.4 Single Cell Trajectory Inference

Single-cell transcriptome data can be used to model dynamic biological processes such as spermatogenesis by computationally inferring the order of cells along developmental trajectories. Trajectory inference, also called pseudotime analysis, places cells along a trajectory based on similarities in their expression patterns. One of the first trajectory inference tools was Monocle (now updated to Monocle3), which uses the reversed graph embedding machine learning technique to order single cells in an unsupervised manner without a priori knowledge of marker genes [72]. In addition, Monocle can also perform clustering and differential gene expression analysis to reveal novel cell states and their relationships.

However, pseudo-time analysis is limited by the use of steady-state mRNA abundances, which are static snapshots of gene expression. Numerous trajectory inference tools have been developed, and several published reviews have compared a portion of them in detail [78–80]. RNA velocity analysis is an alternative approach which utilizes the relative abundance of nascent (unspliced) and mature (spliced) mRNA to derive a high-dimensional vector that empirically predicts the future states of individual cells [81]. The ratio of spiced to unspliced transcripts at a single time point can also be used to infer the dynamics of expression for individual genes.

3.5 Analysis of Single-Cell RNA-Seq Data Using Seurat

Numerous single-cell RNA-seq platforms are available that employ different protocols to generate a gene expression matrix. The computational tools used for bulk RNA-seq read mapping and expressing quantification can also be employed for scRNA-seq analysis (Table 1). This section covers a basic computational workflow for analysis of scRNA-seq data from the widely used 10× Genomics platform. Cell ranger is used for processing sequencing data (FASTQ files) into read counts. The R package Seurat provides a method for integration of multiple datasets across technologies using a normalization method called SCTransform.

3.5.1 Generation of a Gene Expression Count Matrix for Seurat Analysis

1. Install Cell Ranger from the 10× Genomics website in a Unix/Linux environment.
2. Make a directory to store the raw sequencing data using the *mkdir* command. Download the FASTQ files (.tar files) and extract using the *tar* command. This outputs two sets of FASTQ files with the following naming convention: Sample_S1_L00X_R1_001.fastq.gz.

Box 13 Download FASTQ files

```
mkdir ~/human_cellranger_counts
cd ~/human_cellranger_counts
tar -xvf human_1k_v3_fastqs.tar
```

3. Download a prebuilt reference transcriptome from the 10× Genomics support site using wget command and decompress it using the tar command (*see* **Note 3**).
4. To generate read counts from the fastq files and reference transcriptome, run the *cellranger count* command on each FASTQ file (*see* **Note 4**). This outputs a folder called *//RUN_NAME/outs/filtered_feature_bc_matrix.* The *outs* directory contains matrix.mtx, genes.tsv (or features.tsv), and barcodes.tsv files, which can be used as input for the Seurat package.

5. Install and load the packages Seurat, ggplot2, and patchwork within the R environment. Specify the working directory to save outputs from R using *setwd* command.

Box 14 Install Seurat

```
install.packages("Seurat", "ggplot2", "patchwork")
library(Seurat)
library(ggplot2)
library(patchwork)
library(future)
```

6. Import single-cell expression data into Seurat using the *Read10X* function by specifying the directory containing the matrix.mtx, genes.tsv, and barcodes.tsv files generate in step 1.

Box 15 Load single cell gene expression data in Seurat

```
Sample1_1.data <- Read10X(data.dir = "/work/hermannlab/
sample1_1/filtered_feature_bc_matrix/")
Sample1_2.data <- Read10X(data.dir = "/work/hermannlab/
sample1_2/filtered_feature_bc_matrix/")
Sample1_3.data <- Read10X(data.dir = "/work/hermannlab/
sample1_3/filtered_feature_bc_matrix/")
```

7. Create a Seurat object using *CreateSeuratObject*, which contains the non-normalized expression values (raw counts or TPM). The *min.cell* argument includes expression values for genes that are expressed in at least three cells and the *min. features* argument includes only cells with at least 200 expressed genes. Repeat steps 3 and 4 for all datasets to be analyzed, providing unique names for each Seurat object.

Box 16 Create SeuratObject

```
Sample1_1<- CreateSeuratObject(counts = Sample1_1.data,
project = "HumanscRNAseq", min.cells = 3, min.features =
200)
Sample1_2<- CreateSeuratObject(counts = Sample1_2.data,
project = "HumanscRNAseq", min.cells = 3, min.features =
200)
Sample1_3<- CreateSeuratObject(counts = Sample1_3.data,
project = "HumanscRNAseq", min.cells = 3, min.features =
200)
```

8. To create a multisample dataset, merge all data into one single Seurat object (named "testis" herein) using *merge* by passing a vector of multiple Seurat objects to the y parameter. Specify which dataset a cell comes from by setting the *add.cell.ids* parameter. The original cell ID is automatically stored in the object meta data under *orig.ident*.

Box 17 Merge datasets

```
testis <- merge(Sample1_1, y = c(Sample1_2, Sample1_3),
add.cell.ids = c("Sample1_1", " Sample1_2", " Sample1_3",
project = "HumanscRNAseq")
```

3.5.2 Quality Control and Filtering Cells

1. Determine the percentage of reads mapped to mitochondrial genome and store as metadata in the testis R object. Note that gene nomenclature differs among species. Thus, identification of mouse mitochondrial genes, for example, requires setting *pattern* = *"^mt-"*.

Box 18 Calculate mitochondrial quality control metrics

```
testis[["percent.mt"]] <- PercentageFeatureSet(testis,
pattern = "^MT-")
```

2. Visualize QC metrics such as the number of expressed genes (nFeature_RNA), the number of UMI (nCount_RNA), and the percentage of mitochondrial counts as violin plots using the *VlnPlot* function (Fig. 5). Scatter plots of feature-feature relationships are generated using the *FeatureScatter* function (*see* **Notes 5** and **6**).

Box 19 Visualize QC metrics

```
VlnPlot(testis, features = c("nFeature_RNA", "nCount_R-
NA", "percent.mt"), ncol = 3)
FeatureScatter(testis, feature1 = "nCount_RNA", feature2
= "nFeature_RNA")
FeatureScatter(testis, feature1 = "nCount_RNA", feature2
= "percent.mt")
FeatureScatter(testis, feature1 = "nFeature_RNA", fea-
ture2 = "percent.mt")
```

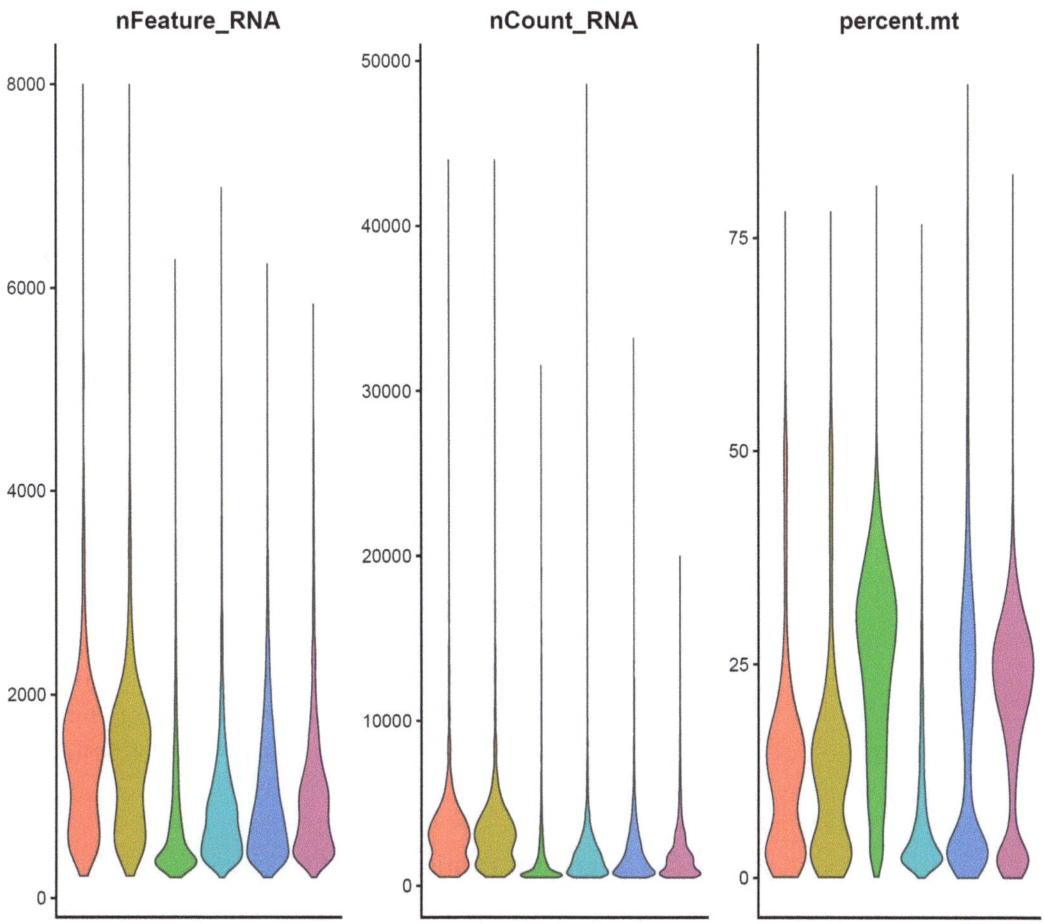

Fig. 5 Violin plots showing quality control metrics including the number of expressed genes (nFeature_RNA), the number of UMI (nCount_RNA) and the percentage of mitochondrial counts (percent.mt) for the datasets. Color indicates the identity of the dataset the cells originate from

3. We can remove outliers, doublets, or potentially apoptotic cells based on read counts, feature counts, or mitochondrial gene content. These values are different for each dataset and should be adjusted appropriately to remove outliers while retaining higher-quality cells. The violin plots and feature scatters generated in the previous step are used to determine appropriate filtering parameters for each dataset. For example, here we use the *subset* function to remove cells that have unique feature counts over 2500 or less than 200 and cells that have greater than 5% mitochondrial counts.

Box 20 Data filtering using QC metrics

```
testis <- subset(testis, subset = nFeature_RNA > 200 &
nFeature_RNA < 2500 & percent.mt < 5)
```

3.5.3 SCTransform Normalization and Integration of Multiple Datasets

1. Setup a list of Seurat objects by splitting the merged object from step 5 into the original datasets using the *SplitObject* function (*see* **Note 7**).

Box 21 Create list of Seurat objects

```
testis.list <- SplitObject(testis, split.by = "orig.
ident")
```

2. Perform SCTransform on each of the datasets individually. This single command performs normalization, variance stabilization, and selection of variable genes. We can regress out uninteresting sources of variation in the expression values in order to improve dimensionality reduction and clustering. Here, we regress out the mitochondrial mapping percentage using the *vars.to.regress* argument (*see* **Notes 8 and 9**).

Box 22 Data transformation using SCTransform

```
for (i in 1:length(testis.list)) {
testis.list[[i]] <- SCTransform(testis.list[[i]],
verbose = FALSE, vars.to.regress=c("percent.mt"))
}
```

3. Select the features to use when integrating multiple datasets ($n = 3000$) using the *SelectIntegrationFeatures* function (*see* **Note 10**). Prepare the object list for integration by calculating the Pearson residuals using the *PrepSCTIntegration* function.

Box 23 Feature selection for Data Integration

```
testis.features <- SelectIntegrationFeatures(object.list
= testis.list, nfeatures = 3000)
options(future.globals.maxSize=9000000000)
testis.list <- PrepSCTIntegration(object.list = testis.
list, anchor.features = testis.features, verbose = FALSE)
```

4. Identify anchors using the *FindIntegrationAnchors* function and integrate the datasets using the *IntegrateData* function,

specifying the SCTransform normalization method in both functions.

Box 24 Perform data integration

```
testis.anchors <- FindIntegrationAnchors(object.list =
testis.list, normalization.method = "SCT",
anchor.features = testis.features, verbose = FALSE)
testis.integrated <- IntegrateData(anchorset = testis.
anchors, normalization.method = "SCT",
verbose = FALSE)
```

3.5.4 Cell Clustering and Visualization

1. Using the integrated object ("testis.integrated") generated in the previous step, perform linear dimensional reduction using PCA and non-linear dimensional reduction using UMAP (Fig. 6) or tSNE embedding. When using *sctransform*, a higher number of principle components can be specified using the

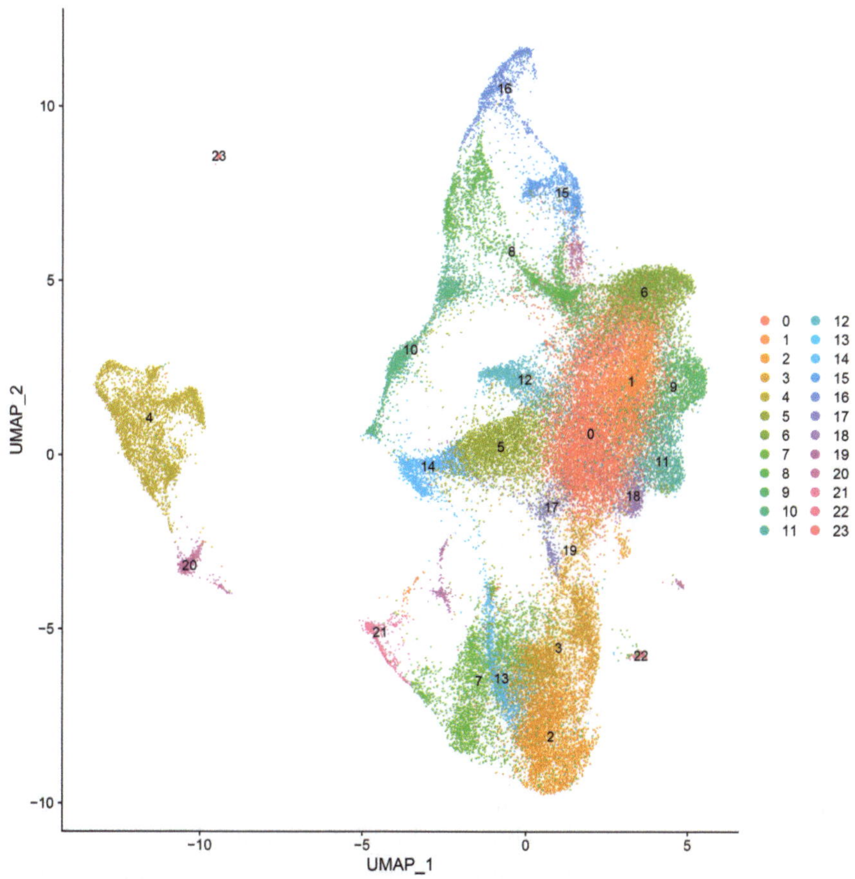

Fig. 6 Unsupervised clustering of all cells in UMAP plot colored by cluster identity

dims argument compared to the standard normalization in Seurat without introducing undue variation.

Box 25 Perform dimensional reduction

```
testis.integrated<- RunPCA(testis.integrated, verbose=-
FALSE)
testis.integrated <- RunUMAP(testis.integrated, dims = 1:
50, label=TRUE)
testis.integrated <- RunTSNE(testis.integrated, dims = 1:
50, label=TRUE)
```

2. Perform shared nearest neighbor (SNN) graph construction using the function *FindNeighbors*, followed by cluster determination using *FindClusters*. The number of clusters can be increased or decreased using the *resolution* parameter. The appropriate number of clusters depends upon sample type, biological context, and goals of the study. The resulting data can be visualized using PCA, UMAP, or tSNE plots generated using the function *DimPlot* and specifying the reduction method in the reduction argument. To generate UMAP plots for each individual dataset, use the *split.by* parameter in the function *DimPlot*,

Box 26 Perform cell clustering

```
testis.integrated <- FindNeighbors(testis.integrated,
dims = 1:50)
testis.integrated <- FindClusters(testis.integrated)
table(Idents(testis.integrated), 5)

DimPlot(testis.integrated, reduction = "PCA", label=-
TRUE)
DimPlot(testis.integrated, reduction = "umap", label=-
TRUE)
DimPlot(testis.integrated, reduction = "tsne", label=-
TRUE)

baseplot<-DimPlot(testis.integrated, reduction = "umap",
pt.size= 0.3, split.by = "orig.ident")
baseplot + NoLegend()
```

3.5.5 Differential Gene Expression Analysis

1. Differential expression analysis is performed on the RNA assay object, which can be changed using the *DefaultAssay* function. Identify markers (differentially expressed genes) for each cluster compared to all remaining clusters using the function

FindAllMarkers (*see* **Note 11**). The resulting table is a ranked list of cluster markers with associated statistics including p-value (p_val), adjusted p-value (p_val_adj), the percentage of cells where the gene is detected in the first group (pct.1), the percentage of cells where the gene is detected in the second group (pct.2), and log fold-change of the average expression between two groups.

Box 27 Perform differential expression testing

```
DefaultAssay(testis.integrated) <- "RNA"
testis.markers<-FindAllMarkers(object=testis.integrated,
only.pos=TRUE, min.pct=0.25)
```

2. Identify the top 10 markers for each cluster and generate an Excel file of the results.

Box 28 Sort DE results by log fold-change

```
top10<- testis.markers %>% group_by(cluster) %>% top_n
(10, avg_logFC)
write.csv(top10, "testis.top10.txt", sep="\t")
```

3. To visualize how gene expression changes across different clusters, various plotting functions are available in Seurat to generated dot plots (Fig. 7), feature plots (Fig. 8), or violin plots (Fig. 9). Input a gene or gene list into the *features* parameter.

Box 29 Visualize differential gene expression using Dot plot, Violin plot and Feature plot

```
DotPlot(object = testis.integrated, features = c("DDX4",
"ACTB", "GATA4", "SOX9", "WT1", "INSL3", "ACTA2", "PE-
CAM1"))
FeaturePlot(testis.integrated, features = c("DDX4"), pt.
size =1)
VlnPlot(testis.integrated, features = c("DDX4"), pt.size
= F)
```

4. Following differential gene expression analysis, assign cell type identity to the clusters based on the expression of known marker genes as well as identify potential novel cell types. First select the clusters using the *WhichCells* function and change the identity of these cells using the *SetIdent* function (*see* **Note 12**).

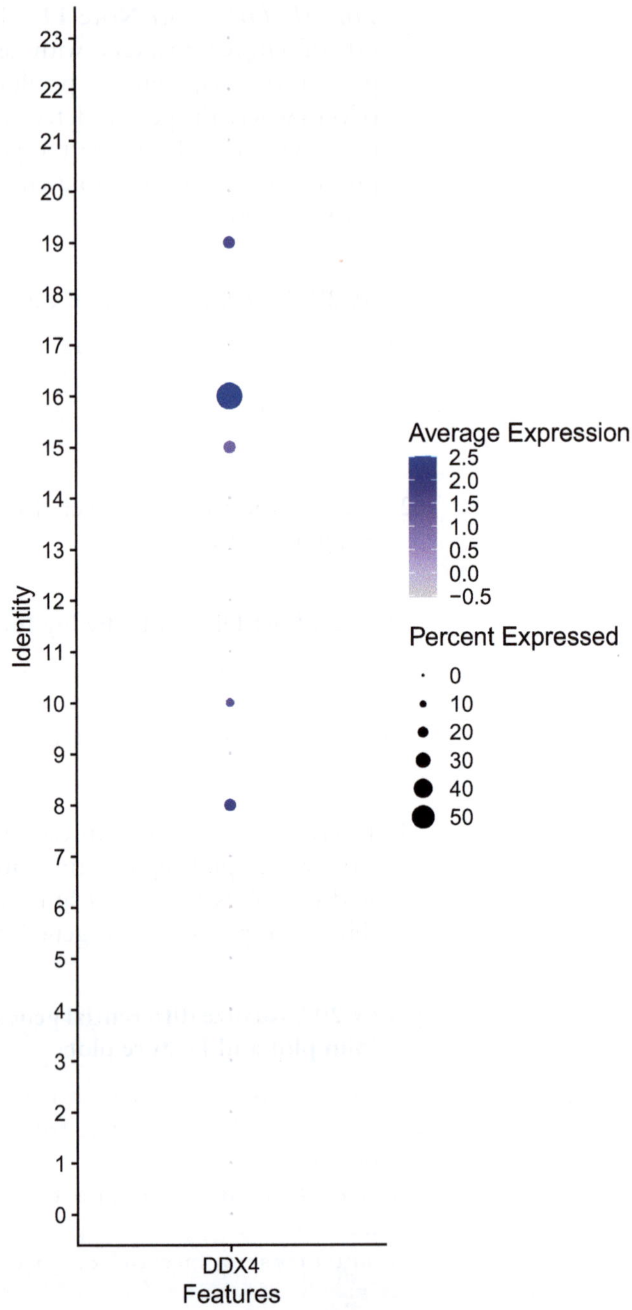

Fig. 7 Dot plot showing the expression levels of a marker genes in each cluster. The size of the dots represents the percentage of cells which express the given gene within a cluster. The color of the dot encodes the average expression level across all cells within a cluster (blue denotes high expression)

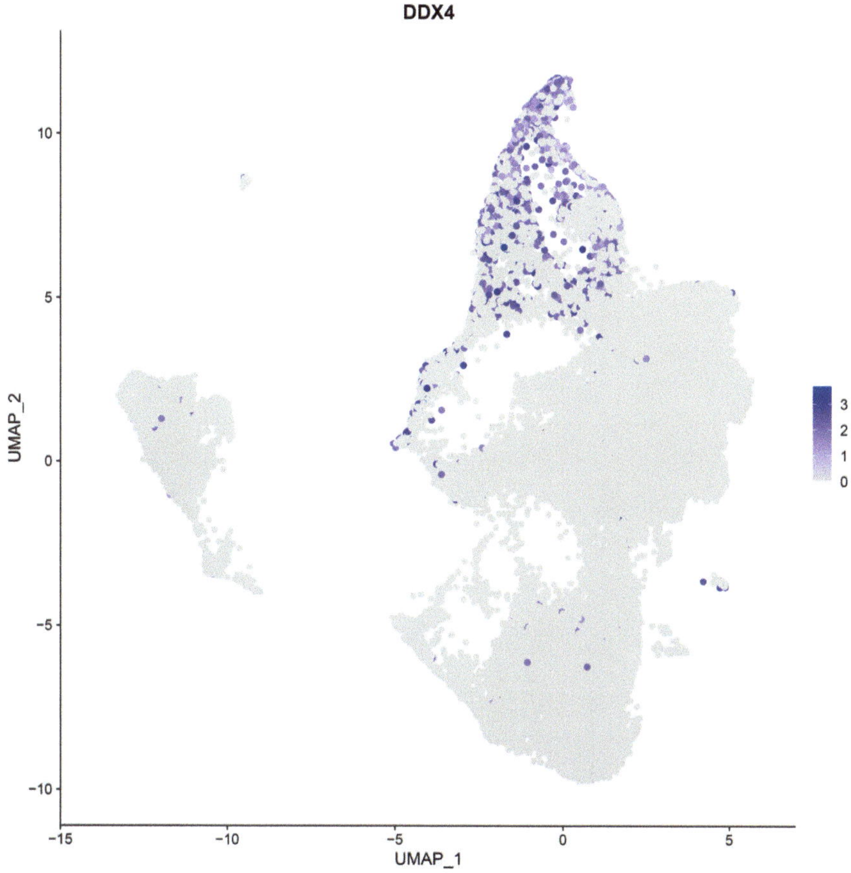

Fig. 8 UMAP plot colored by the expression level of the germ cell marker *DDX4* (blue denotes high expression)

Box 30 Assign cell type identity to clusters

```
cells.use1 <- WhichCells(object = testis.integrated,
idents = c("10", "13", "15"))
testis.integrated <- SetIdent(object = testis.inte-
grated, cells = cells.use1, value = 'Germ cells')
```

5. For focused analyses of a particular cell type, we can select cells based on the identity assigned in the previous step and subset the Seurat object using the *subset* function. The same downstream analysis shown in steps 4–18 can then be rerun on this dataset.

Box 31 Subset data by cell type identity

```
GC<-subset(testis.integrated, idents = "Germ cells"))
```

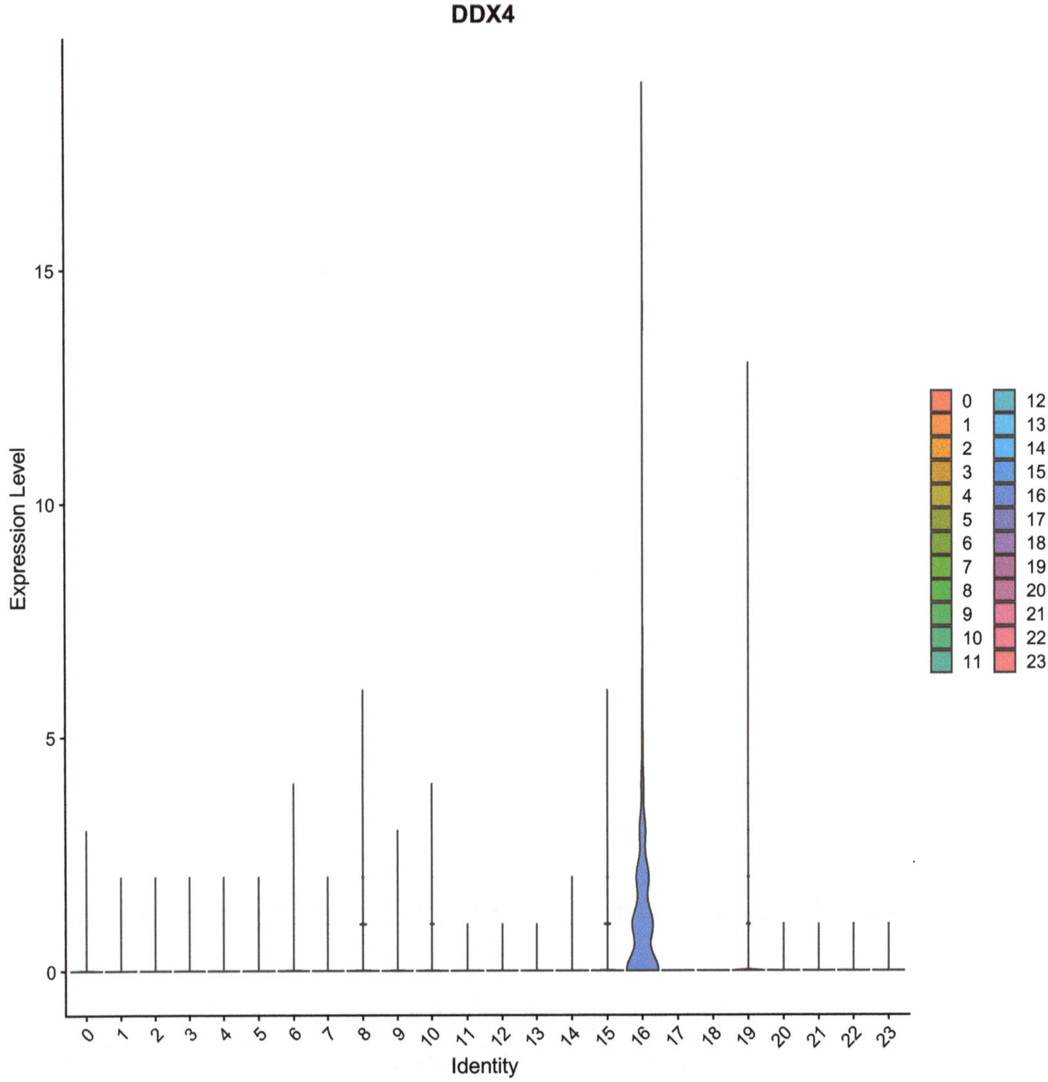

Fig. 9 Violin plot showing the expression levels of selected markers of germ (*DDX4*), Sertoli (*SOX9*, *WT1*), Leydig (*INSL3*), peritubular myoid (*ACTA2*), and endothelial cells (*PECAM1*)

4 Notes

1. In DESeq2, the design formula should be simplified to include only the most important covariates.
2. Using the *lfcShrink* function, if your test is *apeglm*, then you cannot specify a contrast and must specify a coefficient.
3. 10× Genomics provides pre-built custom references for human and mouse genomes. For additional species, a custom reference genome can be built using the *cellranger mkref* command.

4. (Optional) To aggregate or combine two cellranger count runs together, use the *cellranger aggr* command. Specify a .csv file containing the path to hdf5 files.

```
cellranger aggr --id=1k_10K_human_aggr --csv=human_aggr.csv
```

5. In Seurat, visualize QC metrics including nCount_RNA, nFeature_RNA and percent.mt to choose appropriate cutoffs for the data. Alternatively, percent.mt and cell cycle effects can be regressed out during the *SCTransform* step.
6. Following QC, check the dimensions of the expression matrix to calibrate the stringency of the cutoffs before proceeding with downstream analysis.
7. testis@raw.data is a slot in the Seurat object that stores the original gene expression matrix. testis@ident stores the sample ID's of the cells and can be used to store sample metadata.
8. To determine the number of statistically significant PCs, use the *Jackstraw* function. Alternatively, the *PCElbowPlot* function is recommended for larger datasets to reduce computational time. When using *SCTransform*, a greater number of PC's can be used without affecting the results.
9. To evaluate the effects of the cell cycle, run the function *CellCycleScoring* prior to *SCTransform.*
10. By default, *SCTransform* outputs 3000 most variable genes. It may be beneficial to adjust this parameter higher for larger dataset using the variable.features.n argument.
11. The differential expression test can be specified in the *FindMarkers* function using the *test.use* argument. The tests supported are wilcox (default), bimod, roc, t, poisson, negbinom, LR, MAST, and DESeq2.
12. After assigning cell type identity to the clusters, differential expression analysis can be performed on the user-defined groups using the *FindMarkers* function by specifying the groups using the *ident.1* and *ident.2* arguments.

Acknowledgments

The authors graciously acknowledge computational support received from UTSA's HPC cluster SHAMU, operated by the Research Computing Support Group, Office of Information Technology. Some data used in this chapter were generated in the UTSA

Genomics Core, which is supported by NIH grant G12 MD007591 and NSF grants DBI-1337513 and DBI-2018408. This work was supported by NIH grants R01 HD90007 and U01 DA054179.

References

1. Sha J, Zhou Z, Li J, Yin L, Yang H, Hu G, Luo M, Chan HC, Zhou K, Zhu H, Zhu H, Shan Y, Lin M, Wang L, Cheng L, Zhou Y, Wang Y (2002) Identification of testic development and spermatogenesis-related genes in human and mouse testes using cDNA arrays. Mol Hum Reprod 8:511. https://doi.org/10.1093/molehr/8.6.511
2. Feig C, Kirchhoff C, Ivell R, Naether O, Schulze W, Spiess AN (2007) A new paradigm for profiling testicular gene expression during normal and disturbed human spermatogenesis. Mol Hum Reprod 13:33. https://doi.org/10.1093/molehr/gal097
3. Green CD, Ma Q, Manske GL, Shami AN, Zheng X, Marini S, Moritz L, Sultan C, Gurczynski SJ, Moore BB, Tallquist MD, Li JZ, Hammoud SS (2018) A comprehensive roadmap of murine spermatogenesis defined by single-cell RNA-Seq. Dev Cell 46:651–667.e10. https://doi.org/10.1016/j.devcel.2018.07.025
4. Hermann BP, Cheng K, Singh A, Roa-De La Cruz L, Mutoji KN, Chen I-C, Gildersleeve H, Lehle JD, Mayo M, Westernströer B, Law NC, Oatley MJ, Velte EK, Niedenberger BA, Fritze D, Silber S, Geyer CB, Oatley JM, McCarrey JR (2018) The mammalian spermatogenesis single-cell transcriptome, from spermatogonial stem cells to spermatids. Cell Rep 25:1650–1667.e8. https://doi.org/10.1016/j.celrep.2018.10.026
5. Guo J, Grow EJ, Mlcochova H, Maher GJ, Lindskog C, Nie X, Guo Y, Takei Y, Yun J, Cai L, Kim R, Carrell DT, Goriely A, Hotaling JM, Cairns BR (2018) The adult human testis transcriptional cell atlas. Cell Res 28:1141–1157. https://doi.org/10.1038/s41422-018-0099-2
6. Shima JE, McLean DJ, McCarrey JR, Griswold MD (2004) The murine testicular transcriptome: characterizing gene expression in the testis during the progression of spermatogenesis1. Biol Reprod 71:319–330. https://doi.org/10.1095/biolreprod.103.026880
7. Ikami K, Tokue M, Sugimoto R, Noda C, Kobayashi S, Hara K, Yoshida S (2015) Hierarchical differentiation competence in response to retinoic acid ensures stem cell maintenance during mouse spermatogenesis. Development 142:1582–1592. https://doi.org/10.1242/dev.118695
8. Laiho A, Kotaja N, Gyenesei A, Sironen A (2013) Transcriptome profiling of the murine testis during the first wave of spermatogenesis. PLoS One 8:e61558. https://doi.org/10.1371/journal.pone.0061558
9. Chen Y, Zheng Y, Gao Y, Lin Z, Yang S, Wang T, Wang Q, Xie N, Hua R, Liu M, Sha J, Griswold MD, Li J, Tang F, Tong M-HH (2018) Single-cell RNA-seq uncovers dynamic processes and critical regulators in mouse spermatogenesis. Cell Res 28:879–896. https://doi.org/10.1038/s41422-018-0074-y
10. Helsel AR, Yang Q-E, Oatley MJ, Lord T, Sablitzky F, Oatley JM (2017) ID4 levels dictate the stem cell state in mouse spermatogonia. Development 144:624–634. https://doi.org/10.1242/dev.146928
11. Mutoji K, Singh A, Nguyen T, Gildersleeve H, Kaucher AV, Oatley MJ, Oatley JM, Velte EK, Geyer CB, Cheng K, McCarrey JR, Hermann BP (2016) TSPAN8 expression distinguishes spermatogonial stem cells in the prepubertal mouse testis. Biol Reprod 95:117. https://doi.org/10.1095/biolreprod.116.144220
12. Cheng K, Chen IC, Cheng CHE, Mutoji K, Hale BJ, Hermann BP, Geyer CB, Oatley JM, McCarrey JR (2020) Unique epigenetic programming distinguishes regenerative Spermatogonial stem cells in the developing mouse testis. iScience 23:101596. https://doi.org/10.1016/j.isci.2020.101596
13. Sohni A, Tan K, Song HW, Burow D, de Rooij DG, Laurent L, Hsieh TC, Rabah R, Hammoud SS, Vicini E, Wilkinson MF (2019) The neonatal and adult human testis defined at the single-cell level. Cell Rep 26:1501–1517. https://doi.org/10.1016/j.celrep.2019.01.045
14. Suzuki S, McCarrey JR, Hermann BP (2021) An mTORC1-dependent switch orchestrates the transition between mouse spermatogonial stem cells and clones of progenitor spermatogonia. Cell Rep 34:108752. https://doi.org/10.1016/j.celrep.2021.108752

15. Cockrum C, Kaneshiro KR, Rechtsteiner A, Tabuchi TM, Strome S (2020) A primer for generating and using transcriptome data and gene sets. Development 147:dev193854. https://doi.org/10.1242/dev.193854
16. Nam DK, Lee S, Zhou G, Cao X, Wang C, Clark T, Chen J, Rowley JD, Wang SM (2002) Oligo(dT) primer generates a high frequency of truncated cDNAs through internal poly(a) priming during reverse transcription. Proc Natl Acad Sci U S A 99:6152. https://doi.org/10.1073/pnas.092140899
17. Adiconis X, Borges-Rivera D, Satija R, Deluca DS, Busby MA, Berlin AM, Sivachenko A, Thompson DA, Wysoker A, Fennell T, Gnirke A, Pochet N, Regev A, Levin JZ (2013) Comparative analysis of RNA sequencing methods for degraded or low-input samples. Nat Methods 10:623. https://doi.org/10.1038/nmeth.2483
18. Islam S, Zeisel A, Joost S, La Manno G, Zajac P, Kasper M, Lönnerberg P, Linnarsson S (2014) Quantitative single-cell RNA-seq with unique molecular identifiers. Nat Methods 11:163. https://doi.org/10.1038/nmeth.2772
19. Arezi B, Hogrefe H (2009) Novel mutations in Moloney Murine Leukemia Virus reverse transcriptase increase thermostability through tighter binding to template-primer. Nucleic Acids Res 37:473–481. https://doi.org/10.1093/nar/gkn952
20. Tang F, Barbacioru C, Wang Y, Nordman E, Lee C, Xu N, Wang X, Bodeau J, Tuch BB, Siddiqui A, Lao K, Surani MA (2009) mRNA-Seq whole-transcriptome analysis of a single cell. Nat Methods 6:377. https://doi.org/10.1038/nmeth.1315
21. Sasagawa Y, Nikaido I, Hayashi T, Danno H, Uno KD, Imai T, Ueda HR (2013) Quartz-Seq: a highly reproducible and sensitive single-cell RNA sequencing method, reveals nongenetic gene-expression heterogeneity. Genome Biol 14:1–17. https://doi.org/10.1186/gb-2013-14-4-r31
22. Islam S, Kjällquist U, Moliner A, Zajac P, Fan JB, Lönnerberg P, Linnarsson S (2011) Characterization of the single-cell transcriptional landscape by highly multiplex RNA-seq. Genome Res 21:1160–1167. https://doi.org/10.1101/gr.110882.110
23. Ramsköld D, Luo S, Wang YC, Li R, Deng Q, Faridani OR, Daniels GA, Khrebtukova I, Loring JF, Laurent LC, Schroth GP, Sandberg R (2012) Full-length mRNA-Seq from single-cell levels of RNA and individual circulating tumor cells. Nat Biotechnol 30:777. https://doi.org/10.1038/nbt.2282
24. Hashimshony T, Wagner F, Sher N, Yanai I (2012) CEL-Seq: single-cell RNA-Seq by multiplexed linear amplification. Cell Rep 2:666–673. https://doi.org/10.1016/j.celrep.2012.08.003
25. Andrews S (2010) FastQC, Babraham Bioinforma
26. Bolger AM, Lohse M, Usadel B (2014) Trimmomatic: a flexible trimmer for illumina sequence data. Bioinformatics 30:2114–2120. https://doi.org/10.1093/bioinformatics/btu170
27. Martin M (2011) Cutadapt removes adapter sequences from high-throughput sequencing reads. EMBnet J 17:10–12. https://doi.org/10.14806/ej.17.1.200
28. Williams CR, Baccarella A, Parrish JZ, Kim CC (2016) Trimming of sequence reads alters RNA-Seq gene expression estimates. BMC Bioinf 17:103. https://doi.org/10.1186/s12859-016-0956-2
29. Wang L, Wang S, Li W (2012) RSeQC: quality control of RNA-seq experiments. Bioinformatics 28:2184–2185. https://doi.org/10.1093/bioinformatics/bts356
30. Langmead B, Salzberg SL (2012) Fast gapped-read alignment with Bowtie 2. Nat Methods 9: 357. https://doi.org/10.1038/nmeth.1923
31. Li H, Durbin R (2009) Fast and accurate short read alignment with Burrows-Wheeler transform. Bioinformatics 25:1754–1760. https://doi.org/10.1093/bioinformatics/btp324
32. Trapnell C, Pachter L, Salzberg SL (2009) TopHat: discovering splice junctions with RNA-Seq. Bioinformatics 25:1105–1111. https://doi.org/10.1093/bioinformatics/btp120
33. Dobin A, Davis CA, Schlesinger F, Drenkow J, Zaleski C, Jha S, Batut P, Chaisson M, Gingeras TR (2013) STAR: Ultrafast universal RNA-seq aligner. Bioinformatics 29:15–21. https://doi.org/10.1093/bioinformatics/bts635
34. Kim D, Langmead B, Salzberg SL (2015) HISAT: a fast spliced aligner with low memory requirements. Nat Methods 12:357–360. https://doi.org/10.1038/nmeth.3317
35. Engström PG, Steijger T, Sipos B, Grant GR, Kahles A, Rätsch G, Goldman N, Hubbard TJ, Harrow J, Guigó R, Bertone P, Alioto T, Behr J, Bohnert R, Campagna D, Davis CA, Dobin A, Gingeras TR, Jean G, Kosarev P, Li S, Liu J, Mason CE, Molodtsov V, Ning Z, Ponstingl H, Prins JF, Ribeca P, Seledtsov I, Solovyev V, Valle G, Vitulo N, Wang K, Wu TD, Zeller G (2013) Systematic evaluation of spliced alignment programs for RNA-seq data.

Nat Methods 10:1185–1191. https://doi.org/10.1038/nmeth.2722

36. Patro R, Duggal G, Kingsford C (2015) Salmon: accurate, versatile and ultrafast quantification from RNA-seq data using lightweight-alignment. BioRxiv
37. Bray NL, Pimentel H, Melsted P, Pachter L (2016) Near-optimal probabilistic RNA-seq quantification. Nat Biotechnol 34:525. https://doi.org/10.1038/nbt.3519
38. Srivastava A, Malik L, Sarkar H, Zakeri M, Almodaresi F, Soneson C, Love M I, Kingsford C, Patro R (2019) Alignment and mapping methodology influence transcript abundance estimation. BioRxiv. https://doi.org/10.1101/657874
39. Goecks J, Nekrutenko A, Taylor J, Afgan E, Ananda G, Baker D, Blankenberg D, Chakrabarty R, Coraor N, Goecks J, Von Kuster G, Lazarus R, Li K, Taylor J, Vincent K (2010) Galaxy: a comprehensive approach for supporting accessible, reproducible, and transparent computational research in the life sciences. Genome Biol 11:1–13. https://doi.org/10.1186/gb-2010-11-8-r86
40. Benjamin AM, Nichols M, Burke TW, Ginsburg GS, Lucas JE (2014) Comparing reference-based RNA-Seq mapping methods for non-human primate data. BMC Genomics 15:570. https://doi.org/10.1186/1471-2164-15-570
41. Trapnell C, Williams BA, Pertea G, Mortazavi A, Kwan G, Van Baren MJ, Salzberg SL, Wold BJ, Pachter L (2010) Transcript assembly and quantification by RNA-Seq reveals unannotated transcripts and isoform switching during cell differentiation. Nat Biotechnol 28:511–515. https://doi.org/10.1038/nbt.1621
42. Li B, Dewey CN (2011) RSEM: accurate transcript quantification from RNA-Seq data with or without a reference genome. BMC Bioinf 12:323. https://doi.org/10.1186/1471-2105-12-323
43. Pertea M, Pertea GM, Antonescu CM, Chang TC, Mendell JT, Salzberg SL (2015) StringTie enables improved reconstruction of a transcriptome from RNA-seq reads. Nat Biotechnol 33:290–295. https://doi.org/10.1038/nbt.3122
44. Liao Y, Smyth GK, Shi W (2014) FeatureCounts: An efficient general purpose program for assigning sequence reads to genomic features. Bioinformatics 30:923–930. https://doi.org/10.1093/bioinformatics/btt656
45. Bullard JH, Purdom E, Hansen KD, Dudoit S (2010) Evaluation of statistical methods for normalization and differential expression in mRNA-Seq experiments. BMC Bioinf 11:1–13. https://doi.org/10.1186/1471-2105-11-94
46. Robinson MD, Oshlack A (2010) A scaling normalization method for differential expression analysis of RNA-seq data. Genome Biol 11:1–9. https://doi.org/10.1186/gb-2010-11-3-r25
47. Vallejos CA, Risso D, Scialdone A, Dudoit S, Marioni JC (2017) Normalizing single-cell RNA sequencing data: challenges and opportunities. Nat Methods 14:565. https://doi.org/10.1038/nmeth.4292
48. Anders S, Huber W (2010) Differential expression analysis for sequence count data. Genome Biol 11. https://doi.org/10.1186/gb-2010-11-10-r106
49. Smyth GK (2005) Limma: linear models for microarray data. Bioinforma Comput Biol Solut Using R Bioconductor. https://doi.org/10.1007/0-387-29362-0_23
50. Bolstad BM, Irizarry RA, Åstrand M, Speed TP (2003) A comparison of normalization methods for high density oligonucleotide array data based on variance and bias. Bioinformatics 19:185–193. https://doi.org/10.1093/bioinformatics/19.2.185
51. Wagner GP, Kin K, Lynch VJ (2012) Measurement of mRNA abundance using RNA-seq data: RPKM measure is inconsistent among samples. Theory Biosci 131:281–285. https://doi.org/10.1007/s12064-012-0162-3
52. Conesa A, Madrigal P, Tarazona S, Gomez-Cabrero D, Cervera A, McPherson A, Szcześniak MW, Gaffney DJ, Elo LL, Zhang X, Mortazavi A (2016) A survey of best practices for RNA-seq data analysis. Genome Biol 17:1–19. https://doi.org/10.1186/s13059-016-0881-8
53. Abbas-Aghababazadeh F, Li Q, Fridley BL (2018) Comparison of normalization approaches for gene expression studies completed with highthroughput sequencing. PLoS One 13:e0206312. https://doi.org/10.1371/journal.pone.0206312
54. Soneson C, Delorenzi M (2013) A comparison of methods for differential expression analysis of RNA-seq data. BMC Bioinf 14:91. https://doi.org/10.1186/1471-2105-14-91
55. Rajkumar AP, Qvist P, Lazarus R, Lescai F, Ju J, Nyegaard M, Mors O, Børglum AD, Li Q, Christensen JH (2015) Experimental validation of methods for differential gene expression analysis and sample pooling in RNA-seq. BMC

Genomics 16:548. https://doi.org/10.1186/s12864-015-1767-y
56. Costa-Silva J, Domingues D, Lopes FM (2017) RNA-Seq differential expression analysis: an extended review and a software tool. PLoS One 12:e0190152. https://doi.org/10.1371/journal.pone.0190152
57. Van den Berge K, Perraudeau F, Soneson C, Love MI, Risso D, Vert JP, Robinson MD, Dudoit S, Clement L (2018) Observation weights unlock bulk RNA-seq tools for zero inflation and single-cell applications. Genome Biol 19:24. https://doi.org/10.1186/s13059-018-1406-4
58. Zheng GXY, Terry JM, Belgrader P, Ryvkin P, Bent ZW, Wilson R, Ziraldo SB, Wheeler TD, McDermott GP, Zhu J, Gregory MT, Shuga J, Montesclaros L, Underwood JG, Masquelier DA, Nishimura SY, Schnall-Levin M, Wyatt PW, Hindson CM, Bharadwaj R, Wong A, Ness KD, Beppu LW, Deeg HJ, McFarland C, Loeb KR, Valente WJ, Ericson NG, Stevens EA, Radich JP, Mikkelsen TS, Hindson BJ, Bielas JH (2017) Massively parallel digital transcriptional profiling of single cells. Nat Commun 8:14049. https://doi.org/10.1038/ncomms14049
59. Zhang X, Li T, Liu F, Chen Y, Yao J, Li Z, Huang Y, Wang J (2019) Comparative analysis of droplet-based ultra-high-throughput single-cell RNA-Seq systems. Mol Cell 73:130. https://doi.org/10.1016/j.molcel.2018.10.020
60. Cao J, Packer JS, Ramani V, Cusanovich DA, Huynh C, Daza R, Qiu X, Lee C, Furlan SN, Steemers FJ, Adey A, Waterston RH, Trapnell C, Shendure J (2017) Comprehensive single-cell transcriptional profiling of a multicellular organism. Science (80-) 357:661–667. https://doi.org/10.1126/science.aam8940
61. Rosenberg AB, Roco CM, Muscat RA, Kuchina A, Sample P, Yao Z, Graybuck LT, Peeler DJ, Mukherjee S, Chen W, Pun SH, Sellers DL, Tasic B, Seelig G (2018) Single-cell profiling of the developing mouse brain and spinal cord with split-pool barcoding. Science (80-) 360:176–182. https://doi.org/10.1126/science.aam8999
62. Ziegenhain C, Vieth B, Parekh S, Reinius B, Guillaumet-Adkins A, Smets M, Leonhardt H, Heyn H, Hellmann I, Enard W (2017) Comparative analysis of single-cell RNA sequencing methods. Mol Cell 65:631–643.e4. https://doi.org/10.1016/j.molcel.2017.01.023
63. Bloom JD (2018) Estimating the frequency of multiplets in single-cell RNA sequencing from cell-mixing experiments. PeerJ:6–e5578. https://doi.org/10.7717/peerj.5578
64. Ilicic T, Kim JK, Kolodziejczyk AA, Bagger FO, McCarthy DJ, Marioni JC, Teichmann SA (2016) Classification of low quality cells from single-cell RNA-seq data. Genome Biol 17:1–15. https://doi.org/10.1186/s13059-016-0888-1
65. Lun ATL, McCarthy DJ, Marioni JC (2016) A step-by-step workflow for low-level analysis of single-cell RNA-seq data with Bioconductor. F1000Research 5:2122. https://doi.org/10.12688/f1000research.9501.2
66. Hafemeister C, Satija R (2019) Normalization and variance stabilization of single-cell RNA--seq data using regularized negative binomial regression. Genome Biol 20:296. https://doi.org/10.1186/s13059-019-1874-1
67. Zhang Y, Ma Y, Huang Y, Zhang Y, Jiang Q, Zhou M, Su J (2020) Benchmarking algorithms for pathway activity transformation of single-cell RNA-seq data. Comput Struct Biotechnol J 18:2953–2961. https://doi.org/10.1016/j.csbj.2020.10.007
68. Brennecke P, Anders S, Kim JK, Kołodziejczyk AA, Zhang X, Proserpio V, Baying B, Benes V, Teichmann SA, Marioni JC, Heisler MG (2013) Accounting for technical noise in single-cell RNA-seq experiments. Nat Methods 10:1093. https://doi.org/10.1038/nmeth.2645
69. Finak G, McDavid A, Yajima M, Deng J, Gersuk V, Shalek AK, Slichter CK, Miller HW, McElrath MJ, Prlic M, Linsley PS, Gottardo R (2015) MAST: a flexible statistical framework for assessing transcriptional changes and characterizing heterogeneity in single-cell RNA sequencing data. Genome Biol 16:278. https://doi.org/10.1186/s13059-015-0844-5
70. Kharchenko PV, Silberstein L, Scadden DT (2014) Bayesian approach to single-cell differential expression analysis. Nat Methods 11: 740. https://doi.org/10.1038/nmeth.2967
71. Miao Z, Deng K, Wang X, Zhang X (2018) DEsingle for detecting three types of differential expression in single-cell RNA-seq data. Bioinformatics 34:3223–3224. https://doi.org/10.1093/bioinformatics/bty332
72. Trapnell C, Cacchiarelli D, Grimsby J, Pokharel P, Li S, Morse M, Lennon NJ, Livak KJ, Mikkelsen TS, Rinn JL (2014) The dynamics and regulators of cell fate decisions are revealed by pseudotemporal ordering of single cells. Nat Biotechnol 32:381–386. https://doi.org/10.1038/nbt.2859
73. Qiu X, Mao Q, Tang Y, Wang L, Chawla R, Pliner HA, Trapnell C (2017) Reversed graph embedding resolves complex single-cell

trajectories. Nat Methods 14:979. https://doi.org/10.1038/nmeth.4402

74. Qiu X, Hill A, Packer J, Lin D, Ma YA, Trapnell C (2017) Single-cell mRNA quantification and differential analysis with Census. Nat Methods 14:309. https://doi.org/10.1038/nmeth.4150
75. Delmans M, Hemberg M (2016) Discrete distributional differential expression (D3E) - a tool for gene expression analysis of single-cell RNA-seq data. BMC Bioinformatics, 17(1): 110. https://doi.org/10.1186/s12859-016-0944-6
76. Chubb JR, Trcek T, Shenoy SM, Singer RH (2006) Transcriptional Pulsing of a Developmental Gene. Current Biology, 16(10): 1018–1025. https://doi.org/10.1016/j.cub.2006.03.092
77. Raj A, Peskin CS, Tranchina D, Vargas DY, Tyagi S (2006) Stochastic mRNA synthesis in mammalian cells. PLoS Biology, 4(10): 1707–1719. https://doi.org/10.1371/journal.pbio.0040309
78. Saelens W, Cannoodt R, Todorov H, Saeys Y (2019) A comparison of single-cell trajectory inference methods. Nat Biotechnol 37:547. https://doi.org/10.1038/s41587-019-0071-9
79. Chen G, Ning B, Shi T (2019) Single-cell RNA-seq technologies and related computational data analysis. Front Genet:317. https://doi.org/10.3389/fgene.2019.00317
80. Cannoodt R, Saelens W, Saeys Y (2016) Computational methods for trajectory inference from single-cell transcriptomics. Eur J Immunol 46:2496. https://doi.org/10.1002/eji.201646347
81. La Manno G, Soldatov R, Zeisel A, Braun E, Hochgerner H, Petukhov V, Lidschreiber K, Kastriti ME, Lönnerberg P, Furlan A, Fan J, Borm LE, Liu Z, van Bruggen D, Guo J, He X, Barker R, Sundström E, Castelo-Branco G, Cramer P, Adameyko I, Linnarsson S, Kharchenko PV (2018) RNA velocity of single cells. Nature 560:494–498. https://doi.org/10.1038/s41586-018-0414-6
82. Anders S, Pyl PT, Huber W (2015) HTSeq-A Python framework to work with high-throughput sequencing data. Bioinformatics 31:169–169. https://doi.org/10.1093/bioinformatics/btu638
83. Robinson MD, McCarthy DJ, Smyth GK (2009) edgeR: a bioconductor package for differential expression analysis of digital gene expression data. Bioinformatics 26:139–140. https://doi.org/10.1093/bioinformatics/btp616
84. McCarthy DJ, Chen Y, Smyth GK (2012) Differential expression analysis of multifactor RNA-Seq experiments with respect to biological variation. Nucleic Acids Res 40: 4288–4297. https://doi.org/10.1093/nar/gks042
85. Love MI, Huber W, Anders S (2014) Moderated estimation of fold change and dispersion for RNA-seq data with DESeq2. Genome Biol 15:1–21. https://doi.org/10.1186/s13059-014-0550-8
86. Hardcastle TJ, Kelly KA (2010) BaySeq: empirical Bayesian methods for identifying differential expression in sequence count data. BMC Bioinf 11:1–14. https://doi.org/10.1186/1471-2105-11-422
87. Leng N, Dawson JA, Thomson JA, Ruotti V, Rissman AI, Smits BMG, Haag JD, Gould MN, Stewart RM, Kendziorski C (2013) EBSeq: an empirical Bayes hierarchical model for inference in RNA-seq experiments. Bioinformatics 29:1035–1043. https://doi.org/10.1093/bioinformatics/btt087
88. Trapnell C, Hendrickson DG, Sauvageau M, Goff L, Rinn JL, Pachter L (2013) Differential analysis of gene regulation at transcript resolution with RNA-seq. Nat Biotechnol 31:46–53. https://doi.org/10.1038/nbt.2450
89. Tarazona S, García-Alcalde F, Dopazo J, Ferrer A, Conesa A (2011) Differential expression in RNA-seq: a matter of depth. Genome Res 21:2213–2223. https://doi.org/10.1101/gr.124321.111
90. Tarazona S, Furió-Tarí P, Turrà D, Di Pietro A, Nueda MJ, Ferrer A, Conesa A (2015) Data quality aware analysis of differential expression in RNA-seq with NOISeq R/Bioc package. Nucleic Acids Res 43:e140. https://doi.org/10.1093/nar/gkv711
91. Li J, Tibshirani R (2013) Finding consistent patterns: a nonparametric approach for identifying differential expression in RNA-Seq data. Stat Methods Med Res 22:519–536. https://doi.org/10.1177/0962280211428386
92. Law CW, Chen Y, Shi W, Smyth GK (2014) Voom: precision weights unlock linear model analysis tools for RNA-seq read counts. Genome Biol 15:1–17. https://doi.org/10.1186/gb-2014-15-2-r29

Chapter 5

Profiling the Epigenetic Landscape of the Spermatogonial Stem Cell—Part 1: Epigenomics Assays

Keren Cheng and John R. McCarrey

Abstract

Epigenomics encompasses analyses of a variety of different epigenetic parameters which, collectively, make up the epigenetic programming that dictates cell fate and function. Here, protocols are provided for four different epigenomic methods including whole-genome bisulfite sequencing (WGBS) to assess DNA methylation patterns, chromatin immunoprecipitation-sequencing (ChIP-seq) to assess genomic patterns of either specific histone modifications or bound transcription factors, the assay for transposase-accessible chromatin-sequencing (ATAC-seq) to assess genomic patterns of chromatin accessibility, and high-throughput chromosome conformation capture-sequencing (Hi-C-seq) to assess three-dimensional interactions among distant genomic regions, plus computational methodology to integrate data from those four methodologies using Chromatin State Discovery and Characterization (ChromHMM) to obtain the most comprehensive overall assessment of epigenetic programming.

Key words Whole-genome bisulfite sequencing (WGBS), Chromatin immunoprecipitation-sequencing (ChIP-seq), Assay for transposase-accessible chromatin-sequencing (ATAC-seq), High-throughput chromosome conformation capture (Hi-C), Chromatin State Discovery and Characterization (ChromHMM), Multiparametric integrative analysis, Epigenomic profiling

1 Introduction

Spermatogonial stem cells (SSCs) are unique among all germ cell types in either sex because they are the only germ cell type capable of self-renewal and regenerative capacity as demonstrated by transplantation to recipient gonads [1]. Upon mitosis, SSCs can give rise to daughter cells that either sustain SSC fate and function, thereby maintaining the SSC pool, or lose SSC fate and adopt progenitor fate which includes loss of self-renewing and regenerative capacities and gain of commitment to enter the spermatogenic differentiation pathway [2]. These distinct cellular fates are characterized by differential gene expression [3] which is regulated largely by differential epigenetic programming [4].

Jon M. Oatley and Brian P. Hermann (eds.), *Spermatogonial Stem Cells: Methods and Protocols*, Methods in Molecular Biology, vol. 2656, https://doi.org/10.1007/978-1-0716-3139-3_5,

Mitotically or meiotically heritable epigenetic programming is manifested in a variety of forms. These include modification of DNA in the form of 5-methyl cytosine at 5′-CpG-3′ dinucleotides [5] and modifications of histones—particularly histone tails—associated with nucleosomes in the form of acetylation or methylation or any of various other post-transcriptional modifications of specific amino acids in nucleosomal histone tails [6]. These modifications function in conjunction with tissue-specific transcription factor networks, noncoding RNAs and chromatin remodeling factors to establish a two-dimensional (2D) epigenetic landscape that regulates chromatin structure and accessibility. In turn, chromatin accessibility regulates the potential for the transcriptional machinery to reach specific gene promoters or intergenic enhancers required to initiate transcription [7]. The 2D epigenetic landscape is organized by three-dimensional (3D) interactions between gene promoters surrounding the transcriptional start site(s) of each gene and regulatory enhancer sequences that are typically located in cis, but that can be positioned either nearby or very distant from the promoter(s) with which they interact [8]. These cell-type-specific promoter–enhancer interactions form a unique, cell-type specific, 3D "interactome" that regulates the cell-type-specific transcriptome [9].

Because epigenetic programming is manifested in vivo as a multiparametric mechanism, it is ultimately important to view these parameters as an integrated code or program. Indeed, this concept has spawned its own field now known as "multiomics" [10, 11]. Current methodologies typically facilitate assessment of each distinct epigenetic parameter individually, with deduction of the integrated multiparametric profile then being accomplished by combining results from these individual assessments. Below we describe methodologies for assessing each individual epigenetic parameter. In each case, we present methodology for assessment of the relevant epigenetic parameter on a genome-wide basis. In addition, we describe methodology for combining individual datasets representative of genome-wide patterns of different epigenetic parameters to deduce the integrated, multiparametric profile that regulates fate and function within each cell type. Taken together, these assays form the basis of the approach termed "epigenomics" [12, 13].

Methods relevant to genome-wide epigenomic profiling related to gene expression include the following: RNA-sequencing (*RNA-seq*) to assess gene expression patterns, whole-genome bisulfite sequencing (*WGBS*) to assess DNA methylation, chromatin immunoprecipitation-sequencing (*ChIP-seq*) to assess specific histone modifications, the assay for transposase-accessible chromatin-sequencing (*ATAC-seq*) to assess chromatin accessibility/structure, high-throughput chromosome conformation capture-sequencing (*Hi-C-seq*) to simultaneously assess all

Table 1
Cell Numbers needed for epigenonic assays

Bulk Assay	Minimum number of cells	Optimal number of cells
RNA-seq	1000	1,000,000
WGBS	1000	1,000,000
ChIP-seq (for histone mods)	1000–2500	1,000,000
ChIP-seq (for transcription factors)	500,000	10,000,000
ATAC-seq	10,000	75,000
HiC-seq	20,000	10,000,000

3D interactions among distant chromosomal regions, and chromatin state discovery and characterization multiparametric integrative analysis (*ChromHMM*) to combine datasets describing multiple different individual epigenetic parameters into a single, integrated epigenomic profile. In Table 1, we provide the minimum and optimal numbers of cells required for each epigenomic profiling assay we describe in this chapter.

2 Materials

2.1 Whole-Genome Bisulfite Sequencing (WGBS)

2.1.1 Reagents

1. Pico Methyl-Seq Library Prep Kit (Zymo D5456).
2. DNA Clean-up and Concentrator (Zymo D4033).

2.1.2 Equipment

1. Thermocycler.
2. Microfuge.

2.2 Chromatin Immunoprecipitation-Sequencing (ChiP-seq)

2.2.1 Reagents

1. 100 mM dATP solution (New England Biolabs, N0440S).
2. 10× MNase digestion buffer (New England Biolabs, M0247).
3. 25 mM dNTP solution (New England Biolabs, N0447L).
4. 3 M Sodium acetate solution (Sigma-Aldrich, 71196).
5. 50% Polyethylene glycol 6000 (PEG 6000) solution (Sigma-Aldrich, 81304).
6. Agencourt AMPure XP DNA purification kit (Beckman Coulter, A63880).
7. Elution buffer (EB) (Qiagen, 19086).
8. DL-Dithiothreitol (DTT, Sigma-Aldrich, D9779).

9. Ethanol.
10. Ethylenediaminetetraacetic acid (EDTA, Sigma-Aldrich EDS).
11. Glycerol (Sigma-Aldrich, G5516).
12. Klenow fragment (3′ to 5′ exo-, New England Biolabs, M0212L).
13. Linear polyacrylamide (LPA, Sigma-Aldrich, 56575).
14. MaxTract High Density tubes (Qiagen, 129046).
15. Micrococcal nuclease (MNase, New England Biolabs, M0247).
16. Nuclear isolation buffer (Sigma-Aldrich, NUC-101).
17. Phosphate-buffered saline (Life Technologies, 10010-023)
18. Phusion HF Buffer Master Mix (New England Biolabs, F531S).
19. PMSF (Thermo Scientific, Catalog number: 36978).
20. Polymerase II large fragment (Klenow, New England Biolabs, M0210L).
21. Protein A Dynabeads (Life Technologies, 1006D).
22. Protein G Dynabeads (Life Technologies, 1007D).
23. Quick DNA ligation kit (New England Biolabs, M2200L).
24. Sodium chloride (Sigma-Aldrich, 746398).
25. Sodium deoxycholate (Sigma-Aldrich, D6750).
26. T4 DNA ligase reaction buffer (New England Biolabs, B0202S).
27. T4 DNA polymerase (New England Biolabs, M0203L).
28. T4 polynucleotide kinase (T4 PNK, New England Biolabs, M0201L).
29. Trizma base (Sigma-Aldrich, T1503).
30. Triton X-100 (Sigma-Aldrich, T8787).
31. Ultrapure H_2O (Life Technologies, 10977-023).
32. UltraPure phenol: chloroform: isoamyl alcohol (25:24:1, v/v) (Life Technologies, 15593-031).

2.2.2 Equipment

1. Vortex.
2. Rotator.
3. Cold room.
4. Magnetic rack for 200-μL PCR strip tubes.
5. Magnetic rack for 1.5-μL sample tubes.
6. Agilent TapeStation or Agilent Bioanalyzer.

2.2.3 Buffers

1. Complete nuclear isolation buffer:

   1× Nuclear isolation buffer Sigma EZ nuclei isolation lysis buffer (1 mM PMSF, 1× Protease inhibitor cocktail), 0.1% Triton, 0.1% deoxycholate.

2. MNase dilution buffer:

   (a) M Tris–Hcl, pH 8.0, 0.01 M NaCl, 0.001 M EDTA, 50% glycerol.

3. MNase digestion buffer:

   (a) 10× MNase digestion buffer (New England Biolabs, M0247).

4. Complete immunoprecipitation buffer:

   (a) 20 mM Tris–HCl, pH 8.0, 2 mM EDTA, 150 mM NaCl, 0.1% Triton X-100, 1× Protease inhibitor cocktail, 1 mM PMSF.

5. Low salt wash buffer:

   (a) 20 mM Tris–HCl, pH 8.0, 2 mM EDTA, 150 mM NaCl, 1% Triton X-100, 0.1% SDS.

6. High salt wash buffer:

   (a) 20 mM Tris–HCl, pH 8.0, 2 mM EDTA, 500 mM NaCl, 1% Triton X-100, 0.1% SDS.

7. ChIP elution buffer:

   (a) 100 mM NaHCO3, 1% SDS.

2.3 Assay for Transposase-Accessible Chromatin-Sequencing (ATAC-seq)

2.3.1 Reagents

1. 1× PBS.
2. Nuclease-free H_2O.
3. NP-40 10% (Sigma/Roche, 11332473001), store at 4 °C.
4. Tween-20 10% (Sigma/Roche, catalog # 11332465001), store at 4 °C.
5. Digitonin (Promega, catalog # G9441)—Dilute 1:1 with water to make 1% working stock, aliquot, store at −20 °C for up to 6 months, do not freeze/thaw more than 5 times.
6. 1 M Tris–HCl, pH 7.5 (Thermo Fisher, catalog # 15567-027).
7. 5 M NaCl 1 M MgCl2 (Sigma, catalog # M1028).
8. Tagment DNA Enzyme 1 (TDE1) (Illumina, catalog # 15027865).
9. Tagment DNA Buffer (Illumina, catalog # 15027866).
10. MinElute Reaction Cleanup Kit (Qiagen, catalog # 28204).
11. NEBNext High-Fidelity 2× PCR Master Mix (NEB, catalog # M0541S).
12. SYBR Green I (Thermo Fisher, catalog # S7563).
13. Agencourt AMPure XP magnetic beads (Beckman Coulter, catalog # A63880).

14. 80% Ethanol (made fresh).
15. Agilent high-sensitivity DNA Bioanalysis Kit (Agilent, catalog # 5067-4626).
16. Qubit dsDNA HS assay kit (Thermo Fisher, catalog # Q32851).
17. Primers (Table 2).

2.3.2 Equipment

1. Qubit fluorometer (Thermo Fisher).
2. Agilent Bioanalyzer.

2.3.3 Buffers

1. Lysis buffer:
   (a) 50 mM Tris–Hcl, pH 7.5, 10 mM EDTA, 1% SDS, 1× Protease Inhibitor cocktail.
   (b) Store at 4 °C.
2. Resuspension buffer:
   (a) 10 mM Tris–HCl (pH 7.5).
   (b) 5 mM NaCl.
   (c) 3 mM $MgCl_2$.
   (d) Nuclease-free H_2O.
   (e) Store 4 °C.
3. Cold lysis buffer:
   (a) 48.5 μL Resuspension buffer.
   (b) 0.5 μL 10% NP-40
   (c) 0.5 μL 10% Tween-20.
   (d) 0.5 μL 1% Digitonin
   (e) Store 4 °C.
4. Wash buffer:
   (a) 990 μL Resuspension buffer.
   (b) 10 μL 10% Tween-20.
   (c) Store 4 °C.
5. 2× TD buffer:
   (a) 20 mM Tris–Hcl, pH 7.5 m, 10 mM $MgCl_2$, 20% dimethylformamide, store at 4 °C.

2.4 High-Throughput Chromosome Conformation Capture-Sequencing Hi-C-seq

2.4.1 Reagents

1. Halt Protease Inhibitor Cocktail (100×) (Thermo, 78429).
2. Adenosine 5′-triphosphate magnesium salt from bacterial source (ATP) (Sigma, A9187-1G).
3. Biotin-14-dATP (Invitrogen, 19524-016).
4. Biotin-14-dCTP (Invitrogen, 19518-018).
5. Proteinase K (Fungal) (Invitrogen, 25530-031).

Table 2
Illumina/Nextera i5 common adapter and i7 index adapter primers

Primer designation	Primer sequence
Ad1_noMX	AATGATACGGCGACCACCGAGATCTACACTCGTCGGCAGCGTCAGATGTG
Ad2.1_TAAGGCGA	CAAGCAGAAGACGGCATACGAGATTCGCCTTAGTCTCGTGGGCTCGGAGATGT
Ad2.3_AGGCAGAA	CAAGCAGAAGACGGCATACGAGATTTCTGCCTGTCTCGTGGGCTCGGAGATGT
Ad2.4_TCCTGAGC	CAAGCAGAAGACGGCATACGAGATGCTCAGGAGTCTCGTGGGCTCGGAGATGT
Ad2.5_GGACTCCT	CAAGCAGAAGACGGCATACGAGATAGGAGTCCGTCTCGTGGGCTCGGAGATGT
Ad2.6_TAGGCATG	CAAGCAGAAGACGGCATACGAGATCATGCCTAGTCTCGTGGGCTCGGAGATGT
Ad2.7_CTCTCTAC	CAAGCAGAAGACGGCATACGAGATGTAGAGAGGTCTCGTGGGCTCGGAGATGT
Ad2.8_CAGAGAGG	CAAGCAGAAGACGGCATACGAGATCCTCTCTGGTCTCGTGGGCTCGGAGATGT
Ad2.9_GCTACGCT	CAAGCAGAAGACGGCATACGAGATAGCGTAGCGTCTCGTGGGCTCGGAGATGT
Ad2.10_CGAGGCTG	CAAGCAGAAGACGGCATACGAGATCAGCCTCGGTCTCGTGGGCTCGGAGATGT
Ad2.11_AAGAGGCA	CAAGCAGAAGACGGCATACGAGATTGCCTCTTGTCTCGTGGGCTCGGAGATGT
Ad2.12_GTAGAGGA	CAAGCAGAAGACGGCATACGAGATTCCTCTACGTCTCGTGGGCTCGGAGATGT
Ad2.13_GTCGTGAT	CAAGCAGAAGACGGCATACGAGATATCACGACGTCTCGTGGGCTCGGAGATGT
Ad2.14_ACCACTGT	CAAGCAGAAGACGGCATACGAGATACAGTGGTGTCTCGTGGGCTCGGAGATGT
Ad2.15_TGGATCTG	CAAGCAGAAGACGGCATACGAGATCAGATCCAGTCTCGTGGGCTCGGAGATGT
Ad2.16_CCGTTTGT	CAAGCAGAAGACGGCATACGAGATACAAACGGGTCTCGTGGGCTCGGAGATGT
Ad2.17_TGCTGGGT	CAAGCAGAAGACGGCATACGAGATACCCAGCAGTCTCGTGGGCTCGGAGATGT
Ad2.18_GAGGGGTT	CAAGCAGAAGACGGCATACGAGATAACCCCTCGTCTCGTGGGCTCGGAGATGT
Ad2.19_AGGTTGGG	CAAGCAGAAGACGGCATACGAGATCCCAACCTGTCTCGTGGGCTCGGAGATGT

(continued)

Table 2
(continued)

Primer designation	Primer sequence
Ad2.20_GTGTGGTG	CAAGCAGAAGACGGCATACGAGATCACCACACGTCTCGTGGGC TCGGAGATGT
Ad2.21_TGGGTTTC	CAAGCAGAAGACGGCATACGAGATGAAACCCAGTCTCGTGGGC TCGGAGATGT
Ad2.22_TGGTCACA	CAAGCAGAAGACGGCATACGAGATTGTGACCAGTCTCGTGGGC TCGGAGATGT
Ad2.23_TTGACCCT	CAAGCAGAAGACGGCATACGAGATAGGGTCAAGTCTCGTGGGC TCGGAGATGT
Ad2.24_CCACTCCT	CAAGCAGAAGACGGCATACGAGATAGGAGTGGGTCTCGTGGGC TCGGAGATGT

6. DNA polymerase I, large (Klenow) fragment (NEB, M0210S).
7. T4 DNA ligase 1 U/ μL (Invitrogen, 15224090).
8. T4 DNA polymerase (NEB, M0203L).
9. 5× ligation buffer (Invitrogen, 46300-018).
10. T4 polynucleotide kinase (NEB, M0201).
11. 5× T4 DNA ligase buffer (Invitrogen, Y90001).
12. Klenow fragment (3′ → 5′exo-) (NEB, M0212L).
13. Dynabeads® MyOne™ Streptavidin C1 (Invitrogen, 65001).
14. PfuUltra II Fusion DNA polymerase (Stratagene, 600670).
15. Agencourt AMPure® XP (Beckman Coulter, A63881).
16. DNA LoBind tubes, 1.5 mL (Eppendorf, 022431021).
17. Dow Corning® high-vacuum silicone grease colorless (Sigma, Z273554-1EA).
18. Dounce homogenizer, pestle A (Kimble Chase, 885301-0002).
19. Amicon® Ultra-0.5 mL 30 K (Millipore, UFC5030BK).
20. Low molecular weight DNA marker (NEB, N3233S).
21. 1 kb DNA marker (NEB, N3232S).
22. Dynal MPC™-S Magnetic Particle Concentrator (Invitrogen, 12020D).
23. Illumina PE adapter.
24. Illumina PE Primer 1.0, PE Primer 2.0.
25. NEBuffer 2.1 (NEB, B7202S).
26. 37% formaldehyde(FA) (Sigma-Aldrich, 252549).
27. DSG (Thermo Fisher, 20593).
28. EGS (Thermo Fisher, 21565).

2.4.2 Equipment

1. Qubit fluorometer (Thermo Fisher).
2. Agilent bioanalyzer.

2.4.3 Buffers

1. 5× Ligation buffer:
   (a) 250 mM Tris–HCl (pH 7.6).
   (b) 50 mM MgCl2, 5 mM ATP.
   (c) 5 mM DTT and 25% (w/v) polyethylene glycol-8000 in Milli-Q dH_2O.
   (d) Store 50 mL and 1 mL aliquots without DTT/ATP at −80 °C or at 4 °C and add DTT/ATP just before using.
2. 1× Lysis buffer:
   (a) 10 mM Tris–HCl (pH = 8.0).
   (b) 10 mM NaCl.
   (c) 0.2% Igepal CA-630 (NP-40) in Milli-Q dH_2O
   (d) Store at 4 °C.
3. Tween wash buffer (TWB):
   (a) 5 mM Tris–HCl (pH = 8.0).
   (b) 0.5 mM EDTA.
   (c) 1 M NaCl
   (d) 0.05% Tween in Milli-Q dH_2O.
   (e) Store at 4 °C.
4. 2× Binding buffer (BB):
   (a) 10 mM Tris–HCl (pH = 8.0).
   (b) 1 mM EDTA.
   (c) 2 M NaCl in Milli-Q dH_2O
   (d) Store at 4 °C.
5. 5× Annealing buffer (AB):
   (a) 50 mM Tris–HCl (pH = 7.5).
   (b) 50 mM NaCl.
   (c) 0.5 mM EDTA in Milli-Q dH_2O
   (d) Store at 4 °C.
6. TLE (pH 8.0):
   (a) 10 mM Tris–HCl.
   (b) 0.1 mM EDTA in Milli-Q dH_2O
   (c) Store at 4 °C.
7. 2× TD buffer:
   (a) 20 mM Tris–Hcl, pH 7.5 m, 10 mM $MgCl_2$, 20% dimethylformamide, store at 4 °C.

8. Nuclear lysis buffer:
   (a) 50 mM Tris–Hcl, pH 7.5, 10 mM EDTA, 1% SDS, 1 × Protease Inhibitor cocktail. Store at 4 °C.
9. 10× NEB buffer 2 (NEB, B7002S).
10. 10× NEB T4 DNA ligase buffer (NEB, B0202S).

3 Methods

3.1 Gene Expression: RNA-seq

Epigenetic programming dictates cell fate by regulating cell-type-specific gene expression. It is therefore critical that any assessment of epigenetic programming is performed in association with a concurrent or previous assessment of the transcriptome of the relevant cell type(s). It is only by relating data describing epigenetic programming to that describing gene expression that the relevance of cell-type-specific epigenetic programming can be interpreted in a functionally informative manner. Genome-wide patterns of gene expression are best assessed by either bulk (aggregate) [14] or single-cell [15] RNA-sequencing (RNA-seq). Methods of RNA-seq are described in detail in Chapter 4 of this book.

3.2 DNA Methylation: Whole-Genome Bisulfite Sequencing (WGBS)

A variety of methodologies can be used to assess DNA methylation on either a locus-specific [16] or genome-wide [17] basis. The most comprehensive, genome-wide assessment of DNA methylation patterns can be achieved by whole-genome bisulfite sequencing (WGBS) [18] which is described below.

3.2.1 Bisulfite Conversion of Genomic DNA

This protocol is provided with the Pico Methyl-Seq Library Prep Kit from Zymo (Zymo D5456). All buffers and related reagents noted in the protocol below are provided in the Zymo kit.

1. Mix the components of the bisulfite conversion in a 0.2-mL PCR tube as described in Table 3.
2. Incubate the mixture in a thermocycler using the cycling parameters listed in Table 4.
3. Add 600 μL M-Binding Buffer to a Zymo-Spin™ IC Column in a Collection Tube. Add the bisulfite-converted

Table 3
Bisulfite conversion reaction

Components	Volume (μL)
Lightning Conversion Reagent	130
Input DNA (10 pg – 100 ng)	*20*
TOTAL	150

Table 4
Bisulfite conversion PCR

Temperature (°C)	Time
98	8 min
54	1 min
4	<20 h

sample to the column and invert several times to mix. Spin at ≥10,000 × g for 30 s.

4. Discard the flow through from the collection tube and add 100 μL M-wash buffer to the column. Spin at ≥10,000 × g for 30 s.
5. Add 200 μL L-desulfonation buffer to the column and let stand at room temperature (RT) (20–30 °C) for 15–20 min. After the incubation, spin at ≥10,000 × g for 30 s.
6. Discard the flow through. Add 200 μL M-wash buffer to the column and spin at ≥10,000 × g for 30 s. Repeat this wash step.
7. Add 8 μL elution buffer directly to the column matrix and let stand for 1 min. Spin at ≥10,000 × g for 30 s to elute the bisulfite-converted DNA.

3.2.2 Amplification with the PrepAmp Primer

1. Priming Reaction: Mix components of the priming reaction as described in Table 5.
2. In a separate tube on ice, combine components of the PrepAmp mix as described in Table 6.
3. Incubate the PrepAmp priming reaction tubes in a thermocycler using the program shown in Table 7 with the lid temperature set at 25 °C [Important: Make sure that the lid temperature is set at 25 °C. Thermocycler capabilities vary!]:

Note: The PrepAmp Polymerase is not heat stable and will be denatured at high temperatures. Make sure the thermocycler lid is set at 25 °C throughout the reaction.

Note: The thermocycler program can be adjusted to fit thermocycler capabilities as long as the temperature ramps from 8 °C to 37 °C span 8 min.

Note: Primer dimers may appear at 126 bp as a result of a high PrepAmp Primer concentration relative to the amount of input DNA. The PrepAmp Primer (40 μM) can be diluted to 20 μM to alleviate this issue.

3.2.3 Purification with DNA Clean-Up & Concentrator™ (DCC™) (Zymo D4033)

1. In a 1.5-mL tube, add a 7:1 ratio of DNA binding buffer to the product from the PrepAmp polymerase reaction (= 175 μL DNA binding buffer +25 μL product), mix well, and transfer to a Zymo-Spin™ IC column in a Collection Tube. Spin at ≥10,000 g for 30 s.

Table 5
PrepAmp priming reaction components

Components	Volume (μL)
PreAmp Buffer (5×)	2
PreAmp Primer (40 μM)	1
Bisulfite-converted DNA	7
TOTAL	10

Table 6
PrepAmp priming reaction components

Components	Volume (μL)
PreAmp Buffer (5×)	1
PreAmp Pre-mix	3.75
PreAmp Polymerase	*0.30*
(13 U/μL)	
TOTAL	5.05

Table 7
PrepAmp priming reaction components

Step	Temperature (°C)	Time (min)[a]
Conduct 2 cycles of the following two steps:		
1.	98	2
2.	8	1
After cycle 1, pulse-spin tubes and add 5.05 μL PreAmp mix to each tube, then conduct cycle 2. After cycle 2, pulse-spin tubes, then add 0.3 μL PreAmp polymerase and proceed to the remainder of the priming reaction below:		
3.	8	4
4.	16	1 (Ramp rate: 0.1°C/sec)
5.	22	1 (Ramp rate: 0.1°C/s)
6.	28	1 (Ramp rate: 0.1°C/s)
7.	36	1 (Ramp rate: 0.1°C/s)
8.	36.5	1 (Ramp rate: 0.1°C/s)
9.	37	1 (Ramp rate: 0.1°C/s)
After Step 9, return to Step 1 and repeat 1×		

[a]From the Zymo (D5456) Pico Methyl-Seq Library Prep Kit: Refer to Note 4 in Appendix, pg. 8, for alternative ramping rates

2. Add 200 μL DNA wash buffer to the column. Spin at ≥10,000 g for 30 s. Repeat this wash step.
3. Transfer the Zymo-Spin™ IC column to a 1.5-mL microfuge tube. Add 12 μL DNA elution buffer directly to the column matrix and let stand for 1 min at RT. Spin at ≥10,000 g for 30 s to elute.

3.2.4 Library Amplification

1. In a 0.2-mL PCR tube, mix components of the library amplification reaction as shown in Table 8.
2. Incubate the mixture in a thermocycler using the program shown in Table 9 for a total of six (6) amplification cycles:

*Note: This library procedure is designed to be used with a starting input of 100 ng gDNA, but can be scaled up or scaled down as shown in Table 10.

Note: A higher DNA binding buffer to sample ratio (>3:1) may be used when utilizing low DNA input amounts to maximize recovery during purification steps.

Table 8
Library amplification reaction components

Components	Volume (μL)
LibraryAmp Master Mix (2×)	12.5
LibraryAmp Primers (10 μM)	1.0
Purified product from Section 3	*11.5*
(13 U/μL)	
TOTAL	25.0

Table 9
Library amplification reaction conditions

Temperature (°C)	Time
94	30 s
Repeat 5 cycles of the following[a]:	
94	30 s
45	30 s
55	30 s
68	1 min
68	*5 min*
4	Indefinitely

[a]When input = 100 ng gDNA. See below for modifications for different gDNA input levels

Table 10
Alternative cycle numbers

Input gDNA	Number of cycles
10 pg – 1 ng	10
2–50 ng	8
50–99 ng	6

Table 11
Index primer amplification reaction components

Components	Volume (μL)
LibraryAmp Master Mix (2×)	12.5
Index Primer (10 μM)	0.5
Purified product from Section 4	*12.0*
TOTAL	25.0

Table 12
Index primer amplification reaction conditions

Temperature (°C)	Time
94	30 s
Repeat 9 cycles of the following:	
94	30 s
58	30 s
68	*1 min*
68	5 min
4	Indefinitely

3. Purify the PCR product with the DCC™ (refer to Subheading 3.2.3) and elute in 12.5 μL DNA elution buffer.

3.2.5 Amplification with the Index Primer

1. In a 0.2-mL PCR tube: mix components of the index primer reaction as shown in Table 11.
2. Incubate the mixture in a thermocycler using the program shown in Table 12 for a total of ten (10) amplification cycles:
3. Purify the PCR product with DCC™ as described in Subheading 3.2.3 and elute in 12 μL DNA elution buffer.

Table 13
Index sequences

Index	Illumina index	Sequence
A	2	CGATGT
B	4	TGACCA
C	5	ACAGTG
D	6	GCCAAT
E	7	CAGATC
F	12	CTTGTA

3.2.6 Sequencing, Validation, and Quantification of the Library

Note: Libraries prepared with the Pico Methyl-Seq Library Prep Kit are non-directional, and as such, the original-top, original-bottom, and the complementary strands for each will be represented.

Each library should be sequenced to obtain at least 800 million reads for 15–20× coverage of sites. This number is based on the size of the human genome. Sequencing can be scaled up or scaled down according to genome size. Recommended: Treat Pico Methyl-Seq™ libraries as 50-bp single-end reads when mapping back to the reference genome. Sequentially trimming of the first four bases may increase mapping efficiency.

Use the index sequences shown in Table 13, which correspond to Illumina Index sequences for multiplexing and are copyrighted to Illumina, Inc. Oligonucleotide sequences © 2007–2012 Illumina, Inc. All rights reserved.

3.3 Histone Modifications—ChIP-seq (Protocol Modified from Brind'Amour et al. [19])

Genome-wide patterns of histone modifications (or of transcription factor binding) can be assessed by chromatin immunoprecipitation-sequencing (ChIP-seq) [19], which is described here for analysis of histone modification patterns.

3.3.1 Sample Collection by Flow Cytometry

As the ultra-low input native ChIP-seq (ULI-NChIP) method is dilution-based, the conditions of cell isolation and storage are important for the subsequent steps in the procedure. When the expected isolated cell number is sufficient for the formation of a visible cell pellet, we suggest sorting the cells directly in PBS containing protease inhibitor cocktail and then pelleted prior to flash freezing and storage. However, sample loss is much higher when the isolated cell number is insufficient for the formation of a visible pellet, and thus, rarer cell types can be sorted and stored directly in the nuclear isolation buffer that will be used later during micrococcal nuclease (MNase) digestion of chromatin. There is a certain degree of flexibility with the isolation conditions, and cell numbers higher than 2×10^4 cells can also be sorted directly in nuclear isolation buffer, as long as the volume of nuclear isolation buffer is appropriately adjusted [19].

When expecting $>2 \times 10^4$ cells,

1. Label clean 1.5-mL sample tubes.
2. Add 100 μL of PBS containing 1× protease inhibitor cocktail.
3. Sort desired cell population directly into the labeled 1.5-mL sample tubes.
4. Pellet cells by centrifugation (500 × g for 5 min at 4 °C).
5. Remove supernatant.
6. Flash-freeze cell pellets by immersing sample tubes in liquid nitrogen for 0.5 min.
7. Store samples at −80 °C. Cell pellets can be stored from a few weeks to a few years.

When expecting $<2 \times 10^4$ cells,

1. Label a clean 1.5-mL sample tube for each sample to be assayed.
2. Add 10–20 μL of complete nuclear isolation buffer to each tube. Adjust volumes of buffer based on the expected sample size as described in Table 14.
3. Sort desired cell population directly in complete nuclear isolation buffer.
4. Estimate the total volume of the sample plus residual buffer. The contribution of sheath buffer to the final volume should remain below 1/3 of the final volume.
5. [optional] If sheath buffer contribution is above 1/3 of the final volume:
   (a) Pellet cells by centrifugation (500 × g for 5 min at 4 °C).
   (b) Remove supernatant.
   (c) Add 10 μL of complete nuclear isolation buffer.

Table 14
Suggested volumes depending on starting cell number

	Numbers of cells		
	1000–5000	**5000–20,000**	**>20,000**
Nuclear Isolation Buffer[a] (Sigma NUC-101)	10 μL	20 μL	–
PBS	–	–	20–50 μL

[a]Spin down and, if necessary, reduce volume to 10 μL, then flash-freeze, and store at −80 °C

6. Flash-freeze samples in complete nuclear isolation buffer by immersing tubes in liquid nitrogen for 0.5 min.
7. Store samples at −80 °C. Samples can be stored from a few weeks to a few years and can be pooled together in subsequent steps.

3.3.2 Chromatin Preparation

Preparation of MNase fragmented chromatin is one of the most critical steps of native ChIP. Insufficiently digested chromatin will lead to input bias and higher background levels, while the yield of over-digested chromatin will be reduced. Please note that MNase is very sensitive to temperature changes and will lose efficiency with time. Small aliquots of MNase should be prepared and transported to the bench no more than 2–3 times for consistent results. It is recommended to test each new batch of MNase with low numbers of cultured cells prior to use for valuable in vivo samples.

8. Thaw frozen cell pellet or, for lower cell numbers, cells in nuclear isolation buffer (Table 14). At this stage, samples isolated on different days can be pooled together.
9. [optional] If cells are frozen as a pellet, resuspend in appropriate volume of complete nuclear isolation buffer (Table 15).
10. Pipette samples up and down 15–20 times while swirling and place back on ice.
11. Prepare dilute MNAse stock enzyme in MNase dilution buffer at 200 U/μL (0.5 μL stock in 4.5 μL MNase dilution buffer).!

Table 15
Chromatin preparation reaction components

			Numbers of cells	
Component	**>1,000,000 Volume (μL)**	**999,999**	**100,000–9,999**	**1,000–<1,000**
Nuclear isolation Buffer + sheath	50	20	~20	~10
1% Triton/1% DOC solution	–	–	2	1
10× MNase Buffer (NEB)	6	4	10	5
200 mM DTT	0.44	0.37	0.75	0.5
50% PEG6000 (Sigma 81304)	–	–	10	5
1:10 MNase enzyme[a] (NEB M0247)	0.6	0.4	1	0.5
H_2O	2.96	15.23	58.76	29.38
Total	60	40	102.51	51.38

[a]Final MNase concentration = 2 U/μL

Table 16
MNase digestion reaction conditions

	Numbers of cells			
	>1,000,000	999,999	100,000–9999	1000–<1000
Digestion time/temp	7.5 min	5.0 min	7.5 min	7.5 min
	@ 37 °C	@ 37 °C	@ 21 °C	@ 21 °C
Add 100 mM EDTA (μL)	6.6	4.4	11.0	5.5
Add 1% Triton/				
1% DOC solution (μL)	6.6	4.4	8.0	4.0

CAUTION! Add MNase enzyme *only* when you are ready to proceed to digestion.

12. Prepare MNAse digestion buffer as described in Table 13 (always prepare just before digestion).
13. Place samples on rack at 21–25 °C.
14. Add MNase master mix (+MNase enzyme) to each sample. Mix very well (15–20 times) with pipettor.!CAUTION! When digesting multiple samples at the same time, it is important to leave the MNase master mix on ice.
15. Allow reaction to proceed according to the conditions described in Table 16.
16. Stop reaction by adding 10% of the reaction volume of 100 μM EDTA and mix very well with pipette (20–30 times). Place samples immediately on ice.
17. Add 1% Triton/1% deoxycholate solution as indicated in Table 16.
18. Rest samples on ice for 15 min.
19. Vortex samples (medium setting) for approximately 30 s.
20. Add complete immunoprecipitation buffer to the digested chromatin. Digested chromatin should make up <25% of the immunoprecipitation volume. Depending on the amount of cells available and the digestion volume, immunoprecipitation should be performed in a final volume of 100–200 μL.
21. Rotate chromatin at 4 °C for 1 h.
22. Vortex (medium setting) for 30 s.
23. Remove an input control aliquot (Note: for small samples, inputs tend to be larger than for normal ChIP, to reduce error due to low cell numbers). For inputs of <20,000 cells per immunoprecipitation, the input control aliquot should be approximately 10% of the sample.

24. For input control, add 10% volume of 10% SDS, mix well, and add EB to obtain a final volume of 100 μL. Proceed to DNA extraction as described in Subheading 3.3.4.

3.3.3 Chromatin Immunoprecipitation

A challenge of performing ChIP on very small samples is finding the balance between obtaining sufficient material for sequencing and using conditions that are stringent enough to reduce the unavoidable background associated with such samples. A good knowledge of the antigen being pulled down and of the antibody being used should guide the experimental design. Antigens that can be pulled down at a higher frequency allow for use of smaller samples, while it may be difficult to generate high-resolution enrichment profiles of a mark which is present in only a small percentage of a cell population or at only a small number of loci within the genome. This protocol has been successfully used with as few as 1×10^3 cells using the following antibodies; H3K9me3 (Active Motif #39161), H3K27me3 (Diagenode pAb-069-050), H3K36me3 (Abcam #9050) and H3K4me1 (Upstate #07-436). This protocol has been successfully used with as few as 5×10^3 cells using the following antibody: H3K4me3 (Abcam #1012). This protocol remains to be tested for use with smaller input amounts.

1. Pre-wash Protein A/G magnetic beads (Dynabeads, Life Technologies #1006D and 1007D) 3 times in complete immunoprecipitation buffer. Each immunoprecipitation will require 5 μL (if <10,000 input cells) or 10 μL (if 10,000–100,000 input cells) of beads for pre-clearing the chromatin, and 5–10 μL of beads for immunoprecipitation.
2. Prepare antibody beads complexes per Table 17A.

Table 17
Components for native ChIP

	Cell number	
	1000–10,000	**10,000–100,000**
A. Antibody-Bead Complex		
Complete IP Buffer	100 μL	100 μL
Protein A: Protein G Dynabeads 1: 1	5 μL	10 μL
Antibody	0.25 μL	1 μL
B. Chromatin		
Chromatin Lysate	10–50 μL	10–50 μL
Complete IP Buffer	100–200 μL	100–200 μL
Protein A: Protein G Dynabeads 1:1	5 μL	10 μL

3. Pre-clear chromatin per Table 17B: add 5 or 10 μL of pre-washed magnetic beads per immunoprecipitation, depending on input cell number.
4. Rotate for 2–6 h at 4 °C.
5. After incubation, place tubes on a magnetic rack containing antibody-bead complexes per Table 17A.
6. Place tubes containing pre-cleared chromatin per Table 17B on a magnetic rack. Transfer chromatin (supernatant) to antibody-bead complexes.
7. Rotate overnight at 4 °C.
8. Place PCR strip tubes on a magnetic rack.
9. Pipette, aspirate, or pour off unbound chromatin.
10. Resuspend the beads in complete immunoprecipitation buffer and transfer to labeled 1.5-mL sample tubes.
11. Place tubes on a magnetic rack. Remove supernatant.
12. Wash the antibody-bead complexes as follows:
    (a) Resuspend beads in 200 μL low salt wash buffer. Allow beads to settle and then remove the wash buffer and repeat.
    (b) Resuspend beads in 200 μL high salt wash buffer. Allow beads to settle, then remove the high salt wash buffer and repeat.
    (c) Transfer beads to a clean 1.5-mL sample tube.
    (d) Pulse-spin the beads, then use a gel loading pipette tip to remove the last few drops of wash buffer.
13. Resuspend the antibody-bead complexes in 30 μL freshly prepared ChIP elution buffer.
14. Elute DNA for 1–1.5 h in a 65 °C water bath. Vortex tubes regularly.
15. Pulse-spin the sample tubes and place on a magnetic rack.
16. Transfer the eluted chromatin to a clean 1.5-mL sample tube.
17. Wash the beads with 70 μL of elution buffer (EB) and combine the eluted chromatin with other components of the native ChIP reaction as shown in Table 17.
18. Incubate the antibody-bead complex (Tube A in Table 17) >3 h on a rotator at 4 °C.
19. Incubate the Chromatin (Tube B in Table 17) >2 h on a rotator at 4 °C.
20. Remove ChIP buffer from Tube A.
21. Transfer chromatin lysate from tube B to beads in Tube A.
22. Incubate overnight (8–12 h) on a rotator at 4 °C.

3.3.4 DNA Extraction

Various methods can be used for extraction and purification of DNA from chromatin samples. As a general rule, it is preferable to avoid any DNA purification procedure that is column-based, as a significant proportion of smaller samples will not be recovered.

1. Transfer eluted chromatin to a pre-spun phase lock tube.
2. Add 100 μL of phenol:chloroform:isoamyl alcohol 25:24:1 to each sample. Vortex vigorously for 20–30 s.
3. Spin at 13,000 × g for 5 min at 21–25 °C.
4. Transfer the upper phase to a new 1.5 mL tube.!CAUTION! Make sure to take all the liquid and to not touch the new tube with the tip of the pipette (which could transfer some of the solvent into the following reaction).
5. Add 10 μL 3 M sodium acetate and 1 μL LPA (linear polyacrylamide). Mix well. (LPA can be replaced by glycogen as a co-precipitating agent).
6. Add 275 μL cold ultrapure ethanol (stored at −20 °C).
7. Mix very well and allow DNA to precipitate for at least 30 min (can be overnight) at −20 °C.
8. Spin down DNA at 13,000 × g for 30 min at 4 °C.
9. Remove supernatant and wash the pellet with 200 μL of freshly prepared 70% ethanol.
10. Allow pellet to dry.
11. Resuspend in 20 (input <20,000 cells) or 30 (input >20,000 cells) μL of EB.

NOTE: At this point, if the DNA is resuspended in a low volume, the concentration of salts/SDS present will interfere with qPCR or the first step of library construction. We recommend re-purifying the DNA with 1.8 volumes of Agencourt Ampure XP beads.

NOTE: RNase A treatment is recommended prior to qPCR, if unwanted bands appear. For 1,000 cells ChIP, remove 10–15% of the eluate and dilute it for qPCR. Two to three multi-copy amplicons can be assessed with such an aliquot, but there is typically not enough material to assess amplicons of single copy genes.

3.3.5 DNA Purification Using AMpPure XP Beads

1. Make sure the beads are well resuspended. Add 1.8× volume (36 μL if sample was resuspended in 20 μL EB) of bead slurry per sample.
2. Mix very well with a pipette (10–12 times). Once the beads have been added to all samples, pulse-vortex 3–4 times.
3. Incubate at room temperature for 10 min.
4. Place tubes containing the samples on a magnetic rack for 2 min or until the sample is clear of beads.

5. Remove the liquid supernatant from the beads.
6. Wash the beads (leave on the magnetic rack) twice with 100 μL of freshly prepared 70% ethanol.
7. Remove the ethanol, pulse-spin then remove the very last bit of ethanol using a gel loading tip.
8. Allow the beads to air dry.
9. Add 20 (<20,000 cells input) or 30 (>20,000 cells input) μL of EB to the beads, making sure they are well covered.
10. Allow beads to rehydrate at room temperature for approximately 5 min (they are rehydrated when you can resuspend them by flicking the tube). Pipette up and down 15–20 times to fully resuspend.
11. Place the tubes back on the magnetic rack.
12. Transfer the eluted DNA buffer to a new tube, taking care not to transfer any of the magnetic beads.

3.3.6 Library Construction

Raw ChIP material generated by ULI-NChIP-seq will range from a few picograms to a few nanograms of DNA, depending on the input size and antigen. There are various methods and kits that can be used for construction of libraries from low input amounts of material. Described here is an adapter ligation-based protocol followed by a PCR amplification library construction procedure that works with picograms of DNA material. Indexed libraries can be pooled prior to size selection, thus reducing sequencing costs and the number of PCR cycles required to generate sufficient amounts of material for sequencing.

3.3.6.1 End Repair of Fragmented DNA

1. Thaw ChIP samples (in 20 or 30 μL of EB). Keep on ice until the End Repair Master Mix is prepared.
2. In a clean 1.5-mL tube, prepare the End Repair Master Mix as described in Table 18.
3. Mix thoroughly and pulse-spin. Keep master mix on ice while distributing to samples.
4. Add 5 μL (if <10,000 cells per ChIP) or 7.5 μL (if >10,000 cells per ChIP) of End Repair Master Mix to each sample tube. Mix very well (10–15 times) using pipettor.
5. Incubate at room temperature for 30 min.
6. Proceed to DNA purification using Ampure XP beads (3.3.5). In Subheading 3.3.5, **step 9**, elute DNA in 17 μL (if <20,000 cells input) or 25.5 μL (if >20,000 cells input) of EB.

3.3.6.2 A-Tailing

1. Set the heating block or water bath at 37 °C.
2. Thaw and pulse-spin end-repaired (or digested with blunt cutter) samples (in 17 μL (if <20,000 cells input) or 25.5 μL (if >20,000 cells input) EB).

Table 18
Components of end repair reaction

Component	Cell number	
	100–10,000 per ChIP μL of reagent	>10,000 per ChIP μL of reagent
Raw ChIP material in EB	20	30
End Repair Master Mix		
10× Phosphorylation Buffer[a]	2.50	3.75
10 mM dNTP mix (NEB N0447L)	0.50	1.50
T4 DNA polymerase (NEB M0203L)	0.05	0.75
Klenow DNA polymerase	0.10	0.15
Ultrapure H_2O	0.40	0.60
Total	5.00	7.50

[a]10× Phosphorylation buffer is T4 DNA ligase buffer (NEB B0202S)

Table 19
Components of A-tailing master mix

Component	Cell number	
	100–10,000 per ChIP μL of reagent	>10,000 per ChIP μL of reagent
End-repaired/blunt-ended material in EB	17.0	25.5
A-Tailing Master Mix		
10× NEB Buffer 2	2.00	3.00
10 mM dATP (NEB N0440S)	0.50	0.75
Klenow (3′-5′ exo) (NEB M0212L)	0.50	0.75
Total	3.00	4.50

3. In a clean 1.5-mL tube, prepare A-tailing Master Mix as described in Table 19.
4. Vortex and pulse-spin A-tailing Master Mix and add 3.67 (<20,000 cells input) or 4.5 (>20,000 cells input) μL to each sample. Mix thoroughly by pipetting/stirring.
5. Incubate at 37 °C for 30 min.
6. Proceed to DNA purification using Ampure XP beads (3.3.5). Elute DNA in 6.67 (<20,000 cells input) or 10 (>20,000 cells input) μL of EB.

Table 20
Components of adapter ligation reaction

Component	Cell Number	
	100–10,000 per ChIP μL of reagent	>10,000 per ChIP μL of reagent
A-tailed material in EB	6.67	10.00
Adapter-Ligation Master Mix		
2× Quick DNA Ligation Buffer (NEB M2200S)	10.00	15.00
1 μM annealed Illumina adapters	0.67	1.00
Quick DNA Ligase (NEB M2200L)	2.67	4.00
Total	13.34	20.00

3.3.6.3 Adapter Ligation

1. In a clean 1.5-mL tube, prepare the Adaptor Ligation Master Mix as described in Table 20.
2. Add 13.34 (<20,000 cells input) or 20 (>20,000 cells input) μL of Adaptor Ligation Master Mix to each sample, mixing very well.
3. Incubate at 21–25 °C for at least 30 min (can be extended to overnight).
4. Proceed to DNA purification using Ampure XP beads (3.3.5). Add 0.8× volume = 16 (<20,000 cells input) or 24 (>20,000 cells input) μL of Ampure XP beads. Elute DNA in 11 μL of EB.

3.3.6.4 Library Amplification

1. In a clean 1.5-mL tube, prepare the PCR Master Mix as shown in Table 21 (Note: keep on ice until used).
2. Distribute 1 μL of the appropriate indexed primers PE 2.x into a PCR plate or 8-tube strips.
3. Distribute 13.5 μL of PCR master mix into each sample tube.
4. Add 10.5 μL of adaptor-ligated sample to each PCR tube and mix very well (10–20 times) with pipette.
5. Pulse-spin tubes or plate.
6. Run the PCR program shown in Table 22 (*see* Table 23 for appropriate number of amplification cycles to use depending on input amount and chromatin mark being analyzed)
7. Proceed to DNA purification using Ampure XP beads (3.3.5). In Subheading 3.3.5, **step 1**, add 0.8× volume (16 or 24 μL of Ampure XP beads). In Subheading 3.3.5, **step 9**, elute DNA in 15 μL of EB.

Table 21
Components of PCR master mix

	Cell number	
Component	**100–10,000 per ChIP** **μL of reagent**	**>10,000 per ChIP** **μL of reagent**
Adapter-LIgated material in EB	10.50	10.50
10 μM Illumina paired-end indexed primer 2.x (different for each sample)	1.00	1.00
PCR Amplification Master Mix		
2× Phusion HF Master Mix	12.50	12.50
10 μM Illumina paired-end primer 1.0	*1.00*	*1.00*
Total	13.50	13.50

Table 22
Library PCR amplification conditions

Temperature (°C)	**Time**
98	4 min
Repeat the following three steps for 8–15 cycles (see Table 20):	
98	30 s
65	30 s
72	*30 s*
72	5 min
12	Overnight

Table 23
Library PCR amplification—suggested number of amplification cycles

		Cell number
	Histone	
Modification	1000–10,000	10,000–100,000
H3K9me3	10–12	8–10
H3K27me3	10–12	8–10
H3K4me3	14–15	12

8. Evaluate library yield and quality using an Agilent Bioanalyzer (High Sensitivity DNA detection kit) or Agilent TapeStation (D1000 High Sensitivity ScreenTape).

NOTE: Due to the small amount of template in the library construction procedure, it is normal at this stage to observe a large amount of adapter dimers. These will be removed prior to sequencing.

3.4 Chromatin Accessibility: ATAC-seq

Chromatin accessibility, also referred to as chromatin structure, can be assessed by the assay for transposase-accessible chromatin-sequencing (ATAC-seq) [20] as described below.

3.4.1 Cell Lysis

1. Wash 50,000 cells with 50 μL cold 1× PBS in microfuge tube, centrifuge to pellet, discard supernatant.
2. Add 50 μL cold lysis buffer, pipet up, and down 3× gently to resuspend cells.
3. Incubate on ice 3 min.
4. Add 1 mL of wash buffer and invert tube 3 times gently.
5. Centrifuge at 500 × g for 10 min at 4 °C. Discard supernatant (cytoplasm), hold pellet (nuclei) at 4 °C.

3.4.2 Transposition

1. While cells are centrifuging, prepare a transposition reaction mix using the Nextera DNA Library Prep Kit (Illumina, Cat.15032354). For each sample of 50,000 cells, use the volumes shown in Table 24.
2. Add transposition reaction mix to pellet, pipet up, and down 6 times gently to resuspend nuclei.
3. Incubate at 37 °C for 30 min on a thermomixer at 1000 rpm.
4. Isolate DNA using Qiagen MinElute Reaction Cleanup Kit (Qiagen, Cat. 28204). Elute DNA in 10 μL EB.

Table 24
Transposition reaction mix components

Component	Final concentration	Volume (μL)
2× TD Buffer	1×	25.0
1× PBS	–	16.5
10% Tween-20	0.1% v/v	0.5
1% Digitonin	0.01% v/v	0.5
Tn5 Transposase		2.5
Nuclease-free H_2O		*5.0*
Total		50.0

Table 25
PCR amplification reaction components

Component	Volume (μL)
Purified transposed DNA	25.0
Nuclease-free H_2O	10.0
Nextera Ad1.1 (Universal) primer (25 μM)	2.5
Nextera Ad2.[a] (indexing) primer (25 μM)	2.5
NEBNext High-Fidelity 2× PCR Master	25.0
Total	50.0

[a]Ad2.1, Ad2.2, Ad2.3, etc.

Table 26
Library PCR amplification conditions

Temperature (°C)	Time
72	5 min
98	30 s
Repeat the following three steps for 5 cycles:	
98	10 sec
63	30 sec
72	1 min

3.4.3 PCR Amplification (Library Generation)

1. In a clean PCR tube, combine the amplification reaction components shown in Table 25 for each sample.
2. Amplify samples using the PCR program shown in Table 26.
3. Remove tubes from the PCR machine and use 5 μL of each partially amplified library to perform qPCR to determine how many additional PCR cycles are needed for additional amplification. The goal is to stop amplification well prior to saturation to avoid variation among samples due to PCR bias. Combine qPCR amplification reaction components as shown in Table 27.
4. Amplify samples using the PCR program shown in Table 28.
5. Plot R vs Cycle Number. Calculate the number of additional PCR cycles needed for each sample, by determining the number (N) of cycles needed to reach one-third of the maximum R.
6. Continue PCR on the remaining 45 μL of each partially amplified library for the appropriate number (N) of cycles as described in Table 29.

Table 27
qPCR amplification reaction components

Component	Volume (μL)
Partially amplified library	5.00
Nuclease-free H_2O	3.85
Nextera Ad1.1 (Universal) primer (25 μM)	0.50
Nextera Ad2.[a] (indexing) primer (25 μM)	0.50
100× SYBR Green I	0.15
NEBNext High-Fidelity 2× PCR Master	*5.00*
Total	15.00

[a]Ad2.1, Ad2.2, Ad2.3, etc.

Table 28
Library qPCR amplification conditions

Temperature (°C)	Time
98	30 s
Repeat the following three steps for 20 cycles:	
98	10 s
63	30 s
72	1 min

Table 29
Library PCR amplification conditions

Temperature (°C)	Time
98	30 s
Repeat the following three steps for N[a] cycles:	
98	10 s
63	30 s
72	1 min

[a]Number of cycles determined in Subheading 3.4.4, **step 4** above

3.4.4 Library Purification

1. Warm AMPure XP beads to room temperature, and vortex to resuspend.
2. For single left-sided bead purification (to remove primer dimers):

(a) Transfer each PCR sample to a microfuge tube.
(b) Add 1.8× volume (81 μL) AMPure XP beads and pipet up and down 10 times to mix thoroughly.
(c) Incubate at room temperature for 10 min.
(d) Place microfuge tubes in a magnetic rack for 5 min.
(e) Discard supernatant and wash beads with 200 μL 80% EtOH (freshly made), pipet EtOH over beads 10 times, then discard the EtOH.
(f) Leave the tubes on the magnetic rack with caps open for 5 min.
(g) Ensure all EtOH is removed and then resuspend beads in 20 μL nuclease-free H_2O, pipet up and down 10 times to mix thoroughly.
(h) Place the microfuge tubes in a magnetic rack for 1–5 min.
(i) Transfer supernatants to new microfuge tubes.

3.4.5 Assessing library Quality

1. Add 1 μL of each library to 3 μL nuclease-free H_2O (to make a 1:4 dilution).
2. Run 1 μL of each diluted library on an Agilent high-sensitivity DNA bioanalysis chip.
3. Use 1 μL of each diluted library to measure DNA concentration by QuBit.

3.5 Three-Dimensional Interactions: Hi-C-seq

The three-dimensional organization of the genome, also known as the interactome, from which interactions between gene promoters and enhancers can be assessed, can be examined by high-throughput chromosome conformation capture-sequencing (Hi-C-seq) methods [21] as described below.

3.5.1 Crosslinking of Cells

1. Pellet detached adherent or suspension cells and resuspend in freshly made 1% formaldehyde (methanol free) at a volume of 1 mL of formaldehyde solution for every one million cells.
2. Incubate cells at room temperature for 10 min with rotation.
3. Add glycine to a final concentration of 125 mM to quench the formaldehyde and then incubate at room temperature for 5 min with rotation.
4. Pellet cells, wash in PBS, pellet again, and then store in −80 °C or proceed to the next steps.

3.5.2 Cell Lysis and Chromatin Digestion

1. Resuspend up to 15 million crosslinked cells in 500 μL of ice-cold 1× lysis buffer and rotate at 4 for 30 min. For cell amounts greater than 15 million, split the pellet in half for contact generation and then recombine for the sonication.
2. Spin down at 2500 g for 5 min and discard the supernatant.

3. Wash pelleted nuclei once with 500 μL of ice-cold 1× lysis buffer.
4. Remove the supernatant and resuspend pellet in 100 μL of 0.5% SDS.
5. Incubate at 62 °C for 10 min and then add 285 μL of H_2O and 50 μL of 10% Triton X-100 to quench the SDS.
6. Mix well and incubate at 37 °C for 15 min.
7. Add 50 μL of 10× NEB Buffer 2 (NEB #B7202) and 375 U of MboI restriction enzyme (NEB, R0147), and digest chromatin for 2 h at 37 °C with rotation. For lower amounts of starting material, less restriction enzyme should be used: 15 μL should be used for 10–15 million cells, 8 μL for 5 million cells, 1 μL for 250,000 cells.
8. Heat-inactivate MboI at 62 °C for 20 min.

3.5.3 Marking of DNA Ends with Biotin-14-DCTP

1. To fill in the restriction fragment overhangs and mark the DNA ends with biotin, add 19.25 μL of the fill-in master mix described in Table 30.
2. Mix and incubate at 37 °C for 1 h with rotation.
3. Add 948 μL of the ligation master mix described in Table 31.
4. Incubate at room temperature for 4 h with rotation.
5. Pellet nuclei at 2500 rcf for 5 min and remove supernatant.

3.5.4 Sonication

1. Bring pellet up to 880 μL in nuclear lysis buffer.
2. Transfer to a Covaris millitube (e.g., Covaris E220) and shear using the following parameters (fill level 10; duty cycle 5; PIP 140; cycles/burst 200; Time 4 min). Note that sonication time

Table 30
Fill-In reaction master mix components

Component	1× Reaction Volume (μL)
0.4 mM biotin-dATP (Thermo 19524016)	3.75
10 mM dCTP	1.50
10 mM dGTP	1.50
10 mM dTTP	1.50
5 U/μLDNA Polymerase I, Large Klenow Fragment (NEB M0210)	*10.00*
Total	19.25

Table 31
Ligation reaction master mix components

Component	1× Reaction Volume (μL)
10× NEB T4 DNA ligase buffer (NEB, B0202)	150
10% Triton X-100	125
50 mg/mL BSA	3
400 U/μL T4 DNA Ligase (NEB, M0202)	10
H_2O	*660*
Total	948

should be kept constant for different amounts of starting material, but may need to be adjusted for different cell types. An initial sonication time of 4 min is recommended but should be tested for each cell type.

3.5.5 Biotin Pull-Down and Preparation for Illumina Sequencing

1. Prepare for biotin pull-down by washing 5 μL of streptavidin C-1 beads per sample with Tween wash buffer.
2. Resuspend the beads in 10 μL of 2× biotin binding buffer and add to each sample. Incubate at room temperature for 15 min with rotation.
3. Separate on a magnetic rack and discard the supernatant.
4. Wash the beads twice by adding 500 μL of Tween wash buffer and incubating at 55 °C for 2 min with rotation.
5. Wash the beads in 100 μL of 1× (from 2×) TD buffer.
6. Resuspend beads in 25 μL of 2× TD buffer, plus the appropriate amount of Tn5 (e.g., 2.5 μL for 50 ng of total DNA), and H_2O to a total volume of 50 μL. Adjust Tn5 amount linearly for different amounts of DNA.
7. Incubate at 55 °C with interval shaking for 10 min.
8. Place samples on magnetic rack and remove supernatant.
9. Add 50 mM EDTA to samples and incubate at 50 °C for 30 min, then quickly place on magnetic rack, and remove supernatant.
10. Wash samples twice with 50 mM EDTA at 50 °C for 3 min, removing supernatant quickly on magnetic rack each time.
11. Wash samples twice in Tween wash buffer at 55 °C for 2 min, removing supernatant quickly on magnetic rack.
12. Wash samples in 10 mM Tris–HCl.

Table 32
PCR master mix components

Component	1× Reaction Volume (μL)
Phusion HF 2×	25
Nextera Ad1.1 (Universal) primer 12.5 μM	1
Nextera Ad2.[a] (Indexing) primer 12.5 μM	1
H_2O	*23*
Total	50

[a]Ad2.1, Ad2.2, Ad2.3, etc.

Table 33
Library amplification PCR conditions

Temperature (°C)	Time
72	5 min
98	1 min
Repeat the following for N[a] cycles:	
98	15 s
63	30 s
72	1 min

[a]Number of cycles determined as described above

3.5.6 Library Amplification and Post-PCR Size Selection

1. Resuspend beads in 50 μL of the PCR master mix described in Table 32.
2. Run the PCR program described in Table 33. [Note: The cycle number can be estimated using one of two methods: (1) First run 5 cycles of a standard PCR and then remove supernatant from beads. Add 0.25× SYBR green and then run a qPCR up to the beginning of exponential amplification. Or (2) run reactions on a standard PCR and estimate cycle number based on the amount of material from the post-ChIP Qubit (>50 ng should be run for five cycles, ~50 ng should be run for six cycles, 25 ng should be run for seven cycles, and 12.5 ng should be run for eight cycles).
3. Place libraries on a magnetic rack and elute into new tubes. Use DNA Clean & Concentrator Kits (DCC) (Zymo, Cat. D4013) to purify DNA, and elute DNA in 10 μL of water. Purify libraries on a 6% PAGE targeting a fragment size range of 300–700 bp.

[Note – In lieu of PAGE, a two-sided size selection with Ampure XP beads can be performed. After PCR place libraries on a magnetic rack and elute into new tubes. Then, add 25 μL of Ampure XP beads and retain the supernatant to capture fragments less than 700 bp. Transfer supernatant to a new tube and add 15 μL of fresh beads to capture fragments >300 bp.]

4. Perform a final elution (from either PAGE purification or Ampure XP) in 10 μL of water. Quantify libraries with qPCR against Illumina primers and/or by using a Bioanalyzer.

3.6 Integration of Transcriptomic + Epigenomic Profiling Data Using ChromHMM

Integration of genome-wide patterns of multiple different epigenomic programming parameters can be achieved using a variety of software packages, one of the most commonly used is Chromatin State Discovery and Characterization based on use of a multivariate Hidden Markov Model (ChromHMM) as described below.

3.6.1 Installation

1. Install Java 1.5 or later if not already installed.
2. Unzip the file ChromHMM.zip.
3. From the command line, convert the ChromHMM.zip file into the now-unzipped ChromHMM subdirectory containing the ChromHMM.jar file.

Note: ChrommHMM can be installed in conda environment. First, install conda, and in the terminal run the following code.

```
conda install -c bioconda chromhmm
```

3.6.2 Binarization

1. Prepare a tab-delimited text file specifying the design file for the cell-mark-aligned read files according to whether there is a single cell type (option A) or multiple cell types that will be treated independently (option B) or concatenated (option C), or multiple cell types for which the features will be stacked to provide a single-genome annotation (option D).
   (a) option A: Single-cell type,
       Format a single .txt file in the form:

```
cell1 mark1 cell1_mark1_file
cell1 mark2 cell1_mark2_file
```

   (b) option B: Multiple cell types being treated independently,
       Format a single .txt file for each cell type in the form:

```
File 1:
cell1 mark1 cell1_mark1_file
cell1 mark2 cell1_mark2_file
File 2:
```

```
cell2 mark1 cell2_mark1_file
cell2 mark2 cell2_mark2_file
```

(c) option C: Multiple cell types being concatenated,

Format a single .txt file in the form:

```
cell1 mark1 cell1_mark1_file
cell1 mark2 cell1_mark2_file
cell2 mark1 cell2_mark1_file
cell2 mark2 cell2_mark2_file
```

(d) option D: Multiple cell types with stacking features.

Format a single .txt file in the form:

```
genome cell1_mark1 cell1_mark1_file
genome cell1_mark2 cell1_mark2_file
genome cell2_mark1 cell2_mark1_file
genome cell2_mark2 cell2_mark2_file
```

2. If a mark is specified only in a subset of the cell types, then it will be treated as missing in the remaining cell types (encoded with "2") in the binarized files.
3. If there are multiple files for the same cell and mark combination, then each can be specified and their reads will be pooled, or alternatively, the files can be combined outside of ChromHMM.
4. The .bed or .bam files supplied to BinarizeBed or BinarizeBam should normally contain the locations of aligned reads. ChromHMM also allows binarizing data on the basis of already-called peaks, as opposed to aligned reads, in which case the "-peaks" flag needs to be specified when executing the BinarizeBam or BinarizeBed commands. Otherwise, the output will not be meaningful. If you are supplying already-called peaks to ChromHMM, it is not recommended to use broad peak calls. ChromHMM already considers spatial information, such that if you are supplying broad peaks, then ChromHMM could specify states indicative of a combination of marks that do not actually co-occur at the resolution at which this analysis operates.
5. If input control data such as whole-cell extract or IgG ChIP data are available, consider using them either as an additional feature (option A) or by adjusting the binarization threshold locally based on the relative level of the control reads (option B). Option A is recommended if the control data are relatively flat, except in specific positions (often associated with artifacts), or if the control data were not sequenced deeply. Option B is recommended if the control data show peaks in regions other

than artifact-associated regions and are sequenced sufficiently deeply to estimate accurate levels of the local background.

(a) option A: Treat control data as an additional feature,

Format the cell-mark-aligned read file(s) to have the form:

```
cell1 mark1 cell1_mark1_file
cell1 mark2 cell1_mark2_file
cell1 control cell1_control_file
```

(b) option B: Use control data to adjust the binarization threshold locally.

Format the cell-mark-aligned read file(s) to have the form:

```
cell1 mark1 cell1_mark1_file cell1_control_file
cell1 mark2 cell1_mark2_file cell1_control_file
```

The control file is the fourth column. Mark-specific control files can also be specified.

6. Make sure the genome assembly corresponding to the data has a chromosome-length file in the CHROMSIZES directory. Chromosome-length files for the assemblies hg18, hg19, hg38, mm9, mm10, rn5, rn6, danRer7, danRer10, dm3, dm6, ce6, and ce10 are currently provided in the CHROMSIZES directory. If the desired assembly is already included, nothing further needs to be done at this step.

   If the desired assembly is not found in the directory, then it must be prepared and placed in the CHROMSIZES directory. The format is a two-column format in which the first column is the chromosome name and the second is the chromosome length. Such information for a few additional assemblies is available from the UCSC Genome Browser. Linux and Mac users can obtain such files directly by first downloading the fetchChromSizes script from the corresponding directory for their system here: http://hgdownload.cse.ucsc.edu/admin/exe/. At the command prompt, type ./fetchChromSizes assembly > assembly.txt.

7. Execute the command to binarize the data.

   (a) option A Execute BinarizeBed,

   Enter the command:

```
java -mx4000M -jar ChromHMM.jar BinarizeBed chrlengthfile
inputdir cellmarkfiletable outputdir
```

– chrlengthfile contains the chromosome-length file, which can be downloaded from UCSC.

- inputdir is the directory with the aligned reads.
- cellmarkfiletable is the design file described in Step 3.6.2-1 and Step 3.6.2-2. outputdir is the directory in which the binarized data should be written. For each cell type and chromosome.
- "-mx4000M" specifies the amount of memory given to Java and should be adjusted as needed for this and other commands and depends on your own system.

(b) option B Execute BinarizeBam Enter the command:

```
java -mx4000M -jar ChromHMM.jar BinarizeBam chrlengthfile
inputdir cellmarkfiletable outputdir
```

The parameters for BinarizeBam are the same as those listed above for BinarizeBed.

3.6.3 File Preparation for Enrichment Analyses

1. The coordinate files should be in .bed format. This step can also be done after executing the LearnModel command with enrichments then computed by running OverlapEnrichment.
2. Anchor files for external position annotation. This file should be a tab-delimited text file in the subfolder of the corresponding assembly's directory, named ANCHORFILES. The format of this file should be a tab-delimited text file in which the first column is the chromosome, the second column is the coordinate, and, optionally, the third column is the strand ("+" or "−"). This step can also be done after executing LearnModel with enrichments then computed by running NeighborhoodEnrichment, although the neighborhood enrichment analysis would not be automatically included in the report generated by the LearnModel command.

3.6.4 Model Learning

1. To start model learning and have automatic enrichments computed after the model is learned, execute the LearnModel command:

```
java -mx4000M -jar ChromHMM.jar LearnModel inputdir outputdir
numstates assembly
```

In this command, "inputdir" specifies the directory containing the binarized data, which will often be the output directory from previous Steps. "outputdir" is the directory in which the output files from the LearnModel command should go. "numstates" specifies the number of states. "assembly" specifies the genome assembly for the input.

Repeat this step for modeling with different numbers of states. If a computer cluster is available, then model learning can proceed in parallel.

3.6.5 Post-Model-Learning Analysis

1. The individual steps in this section are optional; users can choose any or all steps depending on how they wish to use and analyze the learned models. The full set of options is discussed in the ChromHMM manual.

   To compare models with different numbers of states at the emission parameter level, which can complement comparisons at the enrichment level, use the CompareModels command to generate a heatmap comparison by entering the following command:

```
java -mx4000M -jar ChromHMM.jar CompareModels referencemodel
comparedir outputprefix
```

2. To compute overlap enrichments in addition to those that were already computed after running the LearnModel command, use the OverlapEnrichment command by entering:

```
java -mx4000M -jar ChromHMM.jar OverlapEnrichment inputsegment
inputcoorddir outfileprefix
```

3. To compute neighborhood enrichments other than those that were already computed after applying the LearnModel command, use the NeighborhoodEnrichment command by entering:

```
java -mx4000M -jar ChromHMM.jar NeighborhoodEnrichment input-
segment anchorpositions outfileprefix
```

4. Reorder the states of the model to a different order by entering:

```
java -mx4000M -jar ChromHMM.jar Reorder -o stateorderingfile
inputmodel outputdir
```

5. Generate a segmentation corresponding to the reordered states. To generate a segmentation for data not used in learning an existing model, but in the same format, apply the MakeSegmentation command by entering:

```
java -mx4000M -jar ChromHMM.jar MakeSegmentation modelfile
inputdir outputdir
```

6. To change state colors in the browser files, use the MakeBrowserFiles command by entering:

```
java -mx4000M -jar ChromHMM.jar MakeBrowserFiles -c color-
mappingfile segmentfile segmentationname outputfileprefix
```

Acknowledgments

This work was supported by NIH grants P50 HD98593, U01 DA054179, and R01 HD78679 and by generous gifts from the Robert J. Kleberg, Jr., and Helen C. Kleberg Foundation and the Nancy Smith Hurd Foundation.

References

1. Kubota H, Brinster RL (2018) Spermatogonial stem cells. Biol Reprod 99(1):52–74
2. Law NC, Oatley MJ, Oatley JM (2019) Developmental kinetics and transcriptome dynamics of stem cell specification in the spermatogenic lineage. Nat Commun 10(1):2787
3. Hermann BP, Cheng K, Singh A, Roa-De La Cruz L, Mutoji KN, Chen IC, Gildersleeve H, Lehle JD, Mayo M, Westernströer B, Law NC, Oatley MJ, Velte EK, Niedenberger BA, Fritze D, Silber S, Geyer CB, Oatley JM, McCarrey JR (2018) The mammalian spermatogenesis single-cell transcriptome, from spermatogonial stem cells to spermatids. Cell Rep 25(6):1650–1667.e8
4. Cheng K, Chen IC, Cheng CE, Mutoji K, Hale BJ, Hermann BP, Geyer CB, Oatley JM, McCarrey JR (2020) Unique epigenetic programming distinguishes regenerative spermatogonial stem cells in the developing mouse testis. iScience 23(10):101596
5. Riggs AD (1975) X inactivation, differentiation, and DNA methylation. Cytogenet Cell Genet 14(1):9–25
6. Allis CD, Glover CV, Bowen JK, Gorovsky MA (1980) Histone variants specific to the transcriptionally active, amitotically dividing macronucleus of the unicellular eucaryote, Tetrahymena thermophila. Cell 20(3): 609–617
7. Heinz S, Benner C, Spann N, Bertolino E, Lin YC, Laslo P, Cheng JX, Murre C, Singh H, Glass CK (2010) Simple combinations of lineage-determining transcription factors prime cis-regulatory elements required for macrophage and B cell identities. Mol Cell 38:576–589
8. Spurrell CH, Dickel DE, Visel A (2016) The ties that bind: mapping the dynamic enhancer-promoter interactome. Cell 167:1163–1166
9. Misfud B, Tavares-Cadete F, Young AN, Osborne CS et al (2015) Mapping long-range promoter contacts in human cells with high-resolution capture Hi-C. Nat Genet 47:598–606
10. Guo F, Lin L, Jingyun L, Wu X, Hu B, Zhu P, Wen L, Tang F (2017) Single-cell multi-omics sequencing of mouse early embryos and embryonic stem cells. Cell Res 27:165–183
11. Argelaguet R, Clark SJ, Mohammed H, Stapel LC, Reik W et al (2019) Multi-omics profiling of mouse gastrulation at single-cell resolution. Nature 576:487–491
12. ENCODE Project Consortium (2012) An integrated encyclopedia of DNA elements in the human genome. Nature 489:57–74
13. Roadmap Epigenomics Consortium, Kundaje A, Meuleman W, Ernst J, Bilenky M, Yen A, Heravi-Moussavi A, Kheradpour P, Zhang Z, Wang J et al (2015) Integrative analysis of 111 reference human epigenomes. Nature 518:317–330
14. Wang Z, Gerstein M, Snyder M (2009) RNA-Seq: a revolutionary tool for transcriptomics. Nat Rev Genet 10(1):57–63
15. Eberwine J, Sul JY, Bartfai T, Kim J (2014) The promise of single-cell sequencing. Nat Methods 11(1):25–27
16. Hattori N, Ushijima T (2017) Chapter 8: Analysis of gene-specific DNA methylation. In: Handbook of epigenetics, 2nd edn. Academic Press, Cambridge, MA, pp 113–123
17. Zilberman D, Henikoff S (2007) Genome-wide analysis of DNA methylation patterns. Development 134(22):3959–3965
18. Habibi E, Brinkman AB, Arand J, Kroeze LI, Kerstens HHD et al (2013) Whole-genome bisulfite sequencing of two distinct interconvertible DNA methylomes of mouse embryonic stem cells. Cell Stem Cell 13(3):360–369
19. Brind'Amour J, Liu S, Hudson M, Chen C, Karimi MM, Lorincz MC (2015) An ultra-low-input native ChIP-seq protocol for genome-wide profiling of rare cell populations. Nat Commun 6:6033
20. Buenrostro JD, Giresi PG, Zaba LC, Chang HY, Greenleaf WJ (2013) Transposition of native chromatin for fast and sensitive epigenomic profiling of open chromatin, DNA-binding proteins and nucleosome position. Nat Methods 10(12):1213–1218
21. de Wit E, de Laat W (2012) A decade of 3C technologies: insights into nuclear organization. Genes Dev 26(1):11–24

Chapter 6

Profiling the Epigenetic Landscape of the Spermatogonial Stem Cell: Part 2—Computational Analysis of Epigenomics Data

Keren Cheng and John R. McCarrey

Abstract

The final data-generation step of genome-wide profiling of any epigenetic parameter typically involves DNA deep sequencing which yields large datasets that must then be computationally analyzed both individually and collectively to comprehensively describe the epigenetic programming that dictates cell fate and function. Here, we describe computational pipelines for analysis of bulk mepigenomic profiling data, including whole-genome bisulfite sequencing (WGBS) to detect DNA methylation patterns, chromatin immunoprecipitation-sequencing (ChIP-seq) to detect genomic patterns of either specific histone modifications or bound transcription factors, the assay for transposase-accessible chromatin-sequencing (ATAC-seq) to detect genomic patterns of chromatin accessibility, and high-throughput chromosome conformation capture-sequencing (Hi-C-seq) to detect 3-dimensional interactions among distant genomic regions. In addition, we describe Chromatin State Discovery and Characterization (ChromHMM) methodology to integrate data from these individual analyses, plus that from RNA-seq analysis of gene expression, to obtain the most comprehensive overall assessment of epigenetic programming associated with gene expression.

Key words Whole-genome bisulfite sequencing (wgbs), Chromatin immunoprecipitation-sequencing (chip-seq), Assay for transposase-accessible chromatin-sequencing (atac-seq), High-throughput chromosome conformation capture (hi-c), Chromatin state discovery and characterization (chromhmm), Multiparametric integrative analysis, Epigenomic profiling

1 Introduction

Spermatogonial stem cells (SSCs) are unique in that, upon mitosis, the progeny cells can undertake one of two alternative fates to either retain SSC fate (= self-renewal) or lose SSC fate to become committed to the spermatogenic pathway as progenitor spermatogonia (= differentiation). This retention or transition in cell fate is dependent upon differential epigenetic programming that predis-

Jon M. Oatley and Brian P. Hermann (eds.), *Spermatogonial Stem Cells: Methods and Protocols*, Methods in Molecular Biology, vol. 2656, https://doi.org/10.1007/978-1-0716-3139-3_6,

poses differential gene expression in the progeny of SSC mitosis. Because epigenetic programming is manifest in vivo as a multiparametric mechanism, it is ultimately important to view these parameters as an integrated code or program. Indeed, this concept has spawned its own field now known as "multiomics" [1, 2]. Current methodologies typically facilitate assessment of each distinct epigenetic parameter individually, with deduction of the integrated multiparametric profile then being accomplished by combining results from these individual assessments. Below we describe methodologies for analyzing DNA sequence datasets indicative of genome-wide patterns of each individual epigenetic parameter. In addition, we describe methodology for combining individual datasets representative of genome-wide patterns of different epigenetic parameters to deduce the integrated, multiparametric profile that regulates fate and function within each cell type. Taken together, these assays form the basis of the approach termed "epigenomics" [3, 4].

2 Materials

2.1 Hardware

It is recommended to set up an environment in Mac OS or Linux hardware and to analyze next-generation sequence (NGS) data using a high-performance computing environment or server. Typically, more than 12 cores with more than 64GB memory can deal with these analyses. Before analyzing NGS data, sufficient computing capacity and data storage space should be confirmed.

2.2 Software

For Mac OS, xcode and Command line tools can be downloaded from the Mac App store. Linux OS is free open-access software and always comes with libraries in build. If neither is available, virtual system and docker container in Windows OS are alternatives.

Using conda manage packages—Typically, all relevant software can be accessed on github, and most can be installed through conda. Primary analysis includes demultiplexing fastq files and alignment of reads. Secondary analysis includes peak calling, identification of differentially programmed regions, motif analysis, and visualization of data. Primary analysis usually requires command line tools, and secondary analysis usually requires command line tools, R or Python. Software packages applicable to analysis of data from each individual epigenomic profiling assay are summarized in Table 1.

Table 1
Software packages for analysis of epigenomics data

Epigenomics assay	Software package
WGBS	*Bismark, bsseq, methylKit*
ChIP-seq	Bowtie, subread/Rsubread, QuasR, MACS2, Homer, BEDTools, ChIPpeakAnno, ChIPseeker
ATAC-seq	Bowtie, subread/Rsubread, QuasR, MACS2, Homer, BEDTools, ChIPpeakAnno, ChIPseeker
HiC-seq	Bwa, HiCUP
Integration	ChromHMM, epigenomix, The UCSC genome browser,
Motif analysis	MEME, Homer

3 Methods

3.1 Analysis of WGBS Data

3.1.1 Prepare Genome for Bismark

```
bismark_genome_preparation [options] <path_to_genome_folder>
```

A typical command for generating a human reference genome could look like this:

```
bismark_genome_preparation --path_to_aligner /usr/bin/bow-
tie2/ --verbose /data/genomes/homo_sapiens/GRCh38/
```

A typical terminal screenshot is shown in Fig. 1, below.

3.1.2 Align Reads to Reference Genome

Typical commands for reads to a reference genome could look like these:

```
USAGE: bismark [options] --genome <genome_folder> {-1 <mates1>
-2 <mates2> | <singles>} bismark --genome /data/genomes/homo_-
sapiens/GRCh38/ sample.fastq.gz
```

A typical terminal screenshot is shown in Fig. 2, below.

3.1.3 Bismark BAM/SAM Output (Default)

By default, Bismark generates SAM output for all alignment modes. Please note that reported quality values are encoded in Sanger format (Phred 33 scale), even if the input is in Phred64.

1. QNAME (seq-ID).
2. FLAG (this flag tries to take the strand a bisulfite read originated from into account (this is different from ordinary DNA alignment flags!))
3. RNAME (chromosome).

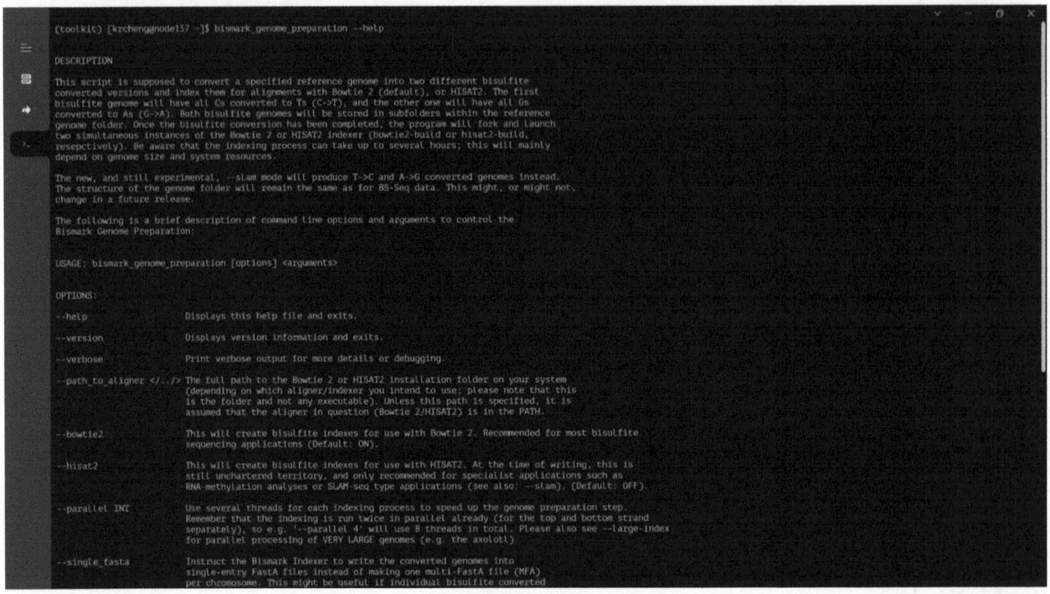

Fig. 1 Screenshot of terminal image showing reference genome

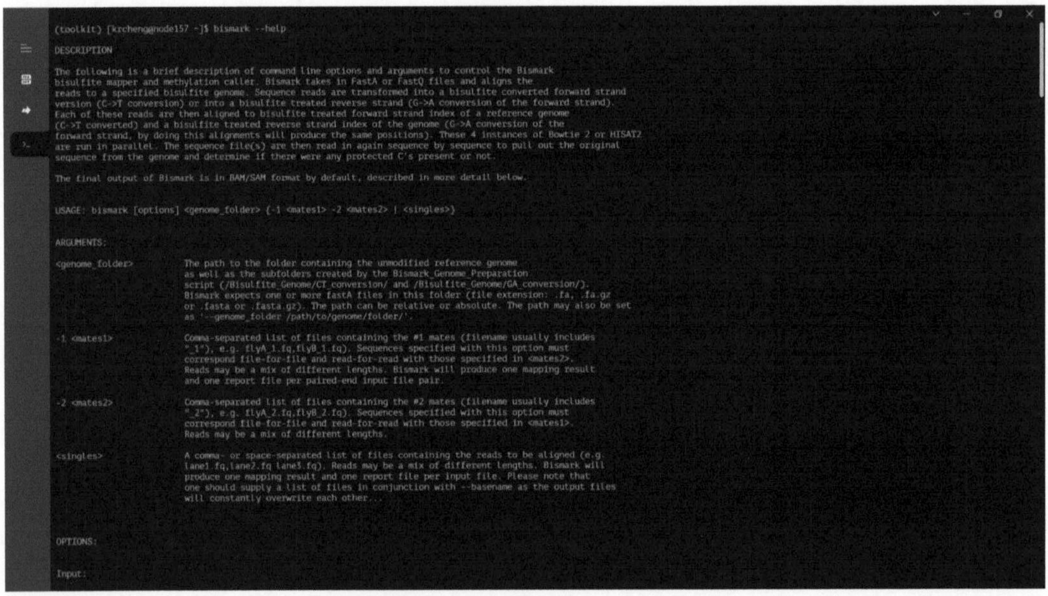

Fig. 2 Screenshot of terminal image showing a typical single-end analysis

4. POS (start position).
5. MAPQ (calculated for Bowtie 2 and HISAT2).
6. CIGAR.
7. RNEXT.
8. PNEXT.
9. TLEN.

```
Final Alignment report
======================
Sequence pairs analysed in total:   296359
Number of paired-end alignments with a unique best hit: 183561
Mapping efficiency: 61.9%
Sequence pairs with no alignments under any condition:  87783
Sequence pairs did not map uniquely:    25015
Sequence pairs which were discarded because genomic sequence could not be extracted:    1

Number of sequence pairs with unique best (first) alignment came from the bowtie output:
CT/GA/CT:   40923   ((converted) top strand)
GA/CT/CT:   51273   (complementary to (converted) top strand)
GA/CT/GA:   50700   (complementary to (converted) bottom strand)
CT/GA/GA:   40664   ((converted) bottom strand)

Final Cytosine Methylation Report
=================================
Total number of C's analysed:   6348472

Total methylated C's in CpG context:    293129
Total methylated C's in CHG context:    34814
Total methylated C's in CHH context:    53442
Total methylated C's in Unknown context:    1198

Total unmethylated C's in CpG context:  78485
Total unmethylated C's in CHG context:  1461257
Total unmethylated C's in CHH context:  4427345
Total unmethylated C's in Unknown context:  10203

C methylated in CpG context:    78.9%
C methylated in CHG context:    2.3%
C methylated in CHH context:    1.2%
C methylated in unknown context (CN or CHN):    10.5%


Bismark completed in 0d 0h 2m 4s
```

Fig. 3 A final bismark report using a snippet of sequence data

10. SEQ.
11. QUAL (Phred33 scale).
12. NM-tag (edit distance to the reference).
13. MD-tag (base-by-base mismatches to the reference) (14) XM-tag (methylation call string).
14. XR-tag (read conversion state for the alignment) (16) XG-tag (genome conversion state for the alignment).

The mate read of paired-end alignments is written out as an additional separate line in the same format.

After finishing Alignment in Bismark, a sample bismark report using a snippet of sequence data is shown in Fig. 3, below.

3.2 Analysis of ChIP-seq Data

Almost all sequencing is carried out on Illumina platforms. Illumina has provided pipelines to convert basic raw data to fastq files. Here, we will start from these fastq files.

3.2.1 Alignment

Among many available software options for alignment, one of the most popular is Bowtie2. Bowtie2 is a classic sequence comparison software package. There are many free tutorials explaining how Bowtie2 carries out sequence alignment. Many current software packages are based on R language for sequence comparison, and

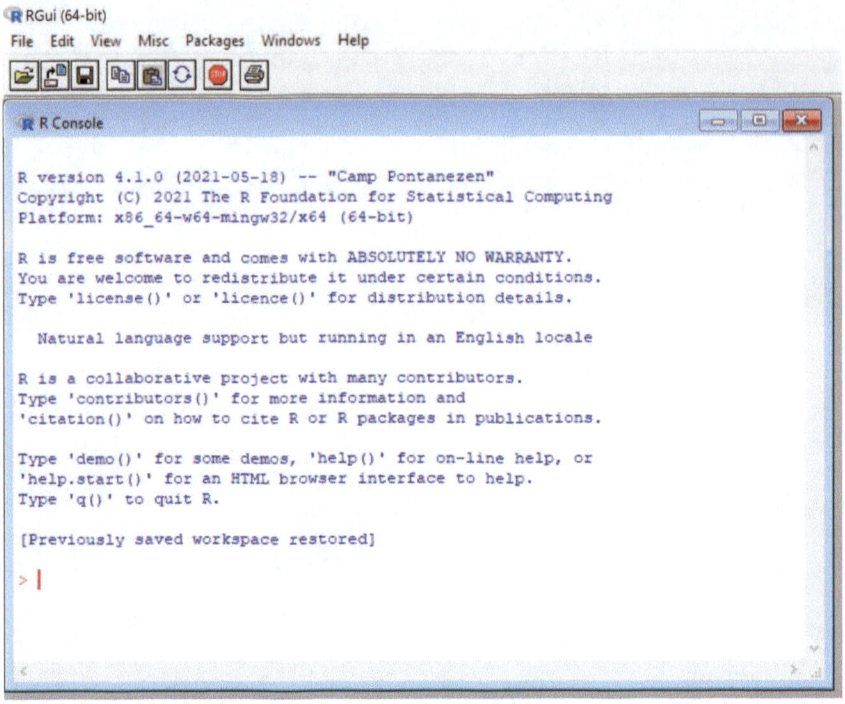

Fig. 4 An example R Console image

the speed and efficiency of these packages are very good. Importantly, no background knowledge of the terminal is required, because the downstream analysis is still in R, rendering this package very convenient and user-friendly. Here, we briefly introduce the use of R to quickly analyze original sequence data.

Download R for your operating system from https://www.r-project.org/.

For Windows OS, Rtools is needed. For linux, please check the dependencies. For Mac OS, xcode, commandlinetools and brew are needed to install certain packages. A typical R console image is shown in Fig. 4.

Alignment of mouse ChIP-seq data to mm10 reference genome. A typical alignment screenshot is shown in Fig. 5.

3.2.2 Peak Calling

Alignment will yield bam files. For analysis of ChIP-seq data, a primary objective is to identify enrichment loci. These loci are sites enriched for histone modifications in promoters or enhancers, typically near transcription factor binding sites. In ChIP-seq data, these sites are typically enriched with more reads than background resulting in these sites showing higher levels of reads which appear as peaks. A typical peak calling screenshot is shown in Fig. 6.

MACS is a classic peak calling software as described below:

```
# snippet 1
library(Rsubread)
setwd('/krcheng/data/multispecies')
# Build an index for reference genome.
buildindex(basename="Rsubread_index",reference=ref)
mouse_read_1 <- 'SRR9109870_1.fastq.gz' # left side
mouse_read_2 <- 'SRR9109870_2.fastq.gz' # right side
mouse_align_pe <- align(index = "/krcheng/Genome/mouse/Rsubread_index",
readfile1 = mouse_read_1,
readfile2 = mouse_read_2,
annot.inbuilt = "mm10",
nthreads = 30)

q()
```

Fig. 5 An example alignment screenshot image

```
## for narrow peaks:
macs2 callpeak -t IP.bam -c Input.bam -n test -p 0.01 --nomodel --extsize
fragment_length --keep-dup all -g mm

## for borad regions:
macs2 callpeak -t IP.bam -c Input.bam -n test --broad -p 0.01 --nomodel --
extsize fragment_length --keep-dup all -g mm
```

Fig. 6 An example peak calling screenshot image

3.2.3 Peak Annotation

Identification of genomic coordinates of peaks indicates the location of peaks within the reference genome. Other analyses can provide information regarding association of multiple ChIP peaks with transcription start site (TSS) regions. Many R packages can be used for these analyses, including Chipseekr or ChIPpeakAnno, as examples. A typical peak annotation screenshot is shown in Fig. 7.

3.2.4 Motif Analysis

Motif Enrichment Analysis (MEA) is a method to predict which DNA-binding transcription factors most likely control the transcription of a set of genes by detecting enrichment of known binding motifs in regulatory regions, such as promoters, enhancers, and open chromatin regions specifically associated with the set of genes in question. Here, we briefly introduce using homer to do Motif Enrichment Analysis. A typical MEA screenshot is shown in Fig. 8.

3.3 Analysis of ATAC-seq Data

Methods for analysis of ATAC-seq data are essentially identical to those described above under Subheading 3.2 for analysis of ChIP-seq data, and so will not be repeated here.

3.4 Analysis of Hi-C Data

There are many software packages that can analyze *Hi-C* data. Here, we present a brief introduction to one of these, HiCUP [5].

```
# snippet 2
BiocManager::install("ChIPpeakAnno")
BiocManager::install("ChIPseeker")
## loading packages
library(ChIPseeker)
library(TxDb.Hsapiens.UCSC.hg19.knownGene)
library(clusterProfiler)
library(ChIPpeakAnno)
##
txdb <- TxDb.Hsapiens.UCSC.hg19.knownGene
## import the MACS output
macs <- system.file("extdata", "MACS_peaks.xls", package="ChIPpeakAnno") #
this should be the output from MACS, usually it's a bed files or list of
bed files.
macsOutput <- toGRanges(macs, format="MACS")
## annotate the peaks with precompiled ensembl annotation
data(TSS.human.GRCh38)
macs.anno <- annotatePeakInBatch(macsOutput,
AnnotationData=TSS.human.GRCh38)
## add gene symbols
library(org.Hs.eg.db)
macs.anno <- addGeneIDs(annotatedPeak=macs.anno,
                        orgAnn="org.Hs.eg.db",
                        IDs2Add="symbol")

if(interactive()){## annotate the peaks with UCSC annotation
    library(GenomicFeatures)
    library(TxDb.Hsapiens.UCSC.hg38.knownGene)
    ucsc.hg38.knownGene <- genes(TxDb.Hsapiens.UCSC.hg38.knownGene)
    macs.anno <- annotatePeakInBatch(macsOutput,
                                     AnnotationData=ucsc.hg38.knownGene)
    macs.anno <- addGeneIDs(annotatedPeak=macs.anno,
                            orgAnn="org.Hs.eg.db",
                            feature_id_type="entrez_id",
                            IDs2Add="symbol")}
```

Fig. 7 An example peak calling screenshot image

```
## homer
# install human dataset for homer
perl
/project/ksasakilab/yard/apps/miniconda3/envs/toolkit/share/homer/.//configureHo
mer.pl -install hg38
perl
/project/ksasakilab/yard/apps/miniconda3/envs/toolkit/share/homer/.//configureHo
mer.pl -install human-o
perl
/project/ksasakilab/yard/apps/miniconda3/envs/toolkit/share/homer/.//configureHo
mer.pl -install human-p
# find motifs using bed files
findMotifsGenome.pl promoters_down.bed hg38 down -p 10
findMotifsGenome.pl promoters_down.bed hg38 down -p 10
# find motifs enrichment among a list of genes'promoters.
findMotifs.pl genes_down_ID.txt human homer/down -p 10
findMotifs.pl genes_up_ID.txt human homer/up -p 10
```

Fig. 8 An example Motif Enrichment Analysis screenshot image

Requirements

HiCUP should work on most Linux-based operating systems. It requires a working version of Perl and uses Bowtie or Bowtie2 to perform the mapping. Full functionality requires R (tested with version 3.1.2) and SAM tools (version 0.1.18 or later). Memory requirements depend on the size of the input files, but as a rough guide allocating 2Gb of RAM per file processed simultaneously (defined by –threads argument) should suffice.

Installation

HiCUP is written in Perl and executed from the command line. To install HiCUP, download the hicup_v0.X.Y.tar.gz file and extract all files by typing:

```
tar -xvzf hicup_v0.X.Y.tar.gz
```

Check after extracting that the Perl scripts are executable by all, which can be achieved with the command:

```
chmod a+x [files]
```

Running HiCUP

1. To create Aligner Indices HiCUP uses the aligner Bowtie or Bowtie2 to map sequences to a reference genome, requiring the construction of genome index files. These indices MUST be constructed from the same reference genome files as those used by the HiCUP Digester script.

   On the command line enter "bowtie-build" (or bowtie2-build) to construct the indices, followed by a comma-separated list of the sequence files, and then a space followed by the name of the output indices:

```
## download reference genome from UCSC or NCBI, then
create genome index.
bowtie-build 1.fa,2.fa,...,MT.fa Mus_muscu-
lus_GRCm38
bowtie2-build 1.fa,2.fa,...,MT.fa Mus_muscu-
lus_GRCm38
```

   Refer to the Bowtie or Bowtie2 manuals for further guidance.

2. Create a digested reference genome. To filter out common experimental artifacts, HiCUP requires input of the positions at which restriction enzyme(s) were used to digest the genome. The script HiCUP Digester creates this reference genome digest file. The example below performs an in silico HindIII digest of all DNA sequences contained within the files in the current working directory suffixed with ".fa." The digest output file will be labeled as the genome "Human_GRCh37." Provide the full path to HiCUP Digester or the sequence files to be digested if they are not in the current working directory.

   Execute the script:

```
hicup_digester --genome Mus_musculus_GRCm38 --re1
A^AGCTT,HindIII *.fa
```

The argument "–re1" specifies the restriction enzyme used to digest the genome (a caret symbol "^" is used to denote the restriction enzyme cut site, and a comma separates the DNA sequence from the restriction enzyme name). The argument "–genome" is for specifying the name of the genome to be digested, it is NOT used to specify the path to the genome aligner indices.

Hi-C Protocol Variations: some Hi-C protocols may use two restriction enzymes at this stage (i.e., the creation of the initial Hi-C interaction). To specify two enzymes use the nomenclature: –re1 A^GATCT,BglII:A^AGCTT,HindIII.

Run the HiCUP Pipeline.

Create an example HiCUP configuration file in your current working directory:

```
hicup --example
```

Use a text editor to edit the configuration file as required, such as in the example shown below in Fig. 9:

Rename the configuration file if desired.

Enter the following text in the command line to run the whole HiCUP pipeline using the parameters specified in the configuration file:

```
hicup --config [Configuration Filename]
```

The –config flag is used to specify the configuration filename. Remember to provide the full path to the HiCUP script and/or the configuration file if they are not in the current working directory.

Please Note: HiCUP attempts to intelligently name files as the pipeline proceeds, so avoid using HiCUP input files with identical names prior to the filename extension. For example, the files "sample.fa" and "sample.fastq" will produce files with identical names as the pipeline progresses. This problem can be overcome by renaming one of the files to "sample2.fa." To minimize inconvenience, HiCUP will immediately produce a warning message and not run if the input filenames are too similar.

Output

The pipeline produces paired-read BAM files representing the filtered di-tags. HiCUP also generates an HTML summary report for each sample and a text file summarizing every sample processed. Summary text files and SVG-format charts are also created at each step along the pipeline.

The "Conversion" folder within the main HiCUP directory contains Perl scripts to convert HiCUP BAM/SAM files into a format compatible with other analysis tools. Executing one of these files with the command line argument—help prints instructions on how to use the conversion script.

```
#Directory to which output files should be
#written (optional parameter)
#Set to current working directory by default
Outdir:

#Number of threads to use
Threads: 1

#Suppress progress updates (0: off, 1: on)
#Quiet:0

#Retain intermediate pipeline files (0: off, 1: on)
Keep:0

#Compress outputfiles (0: off, 1: on)
Zip:1

#Path to the alignment program (Bowtie or Bowtie2)
#Remember to include the executable Bowtie/Bowtie2 filename.
#Note: ensure you specify the correct aligner i.e.
#Bowtie when using Bowtie indices, or Bowtie2 when using Bowtie2 indices.
#In the example below Bowtie2 is specified.
Bowtie2: /usr/local/bowtie2/bowtie2

#Path to the reference genome indices
#Remember to include the basename of the genome indices
Index: /data/public/Genomes/Mouse/NCBIM38/Mus_musculus

#Path to the genome digest file produced by hicup_digester
Digest: Digest_Mouse_genome_HindIII_None_12-32-06_17-02-2012.txt

#FASTQ format (valid formats: 'Sanger', 'Solexa_Illumina_1.0',

#'Illumina_1.3' or 'Illumina_1.5'). If not specified, HiCUP will
#try to determine the format automatically by analysing one of
#the FASTQ files. All input FASTQ will assumed to be in that
#format.
Format: Sanger

#Maximum di-tag length (optional parameter)
Longest: 700

#Minimum di-tag length (optional parameter)
Shortest: 50

#FASTQ files to be analysed, placing paired files on adjacent lines
s_1_1_sequence.txt
s_1_2_sequence.txt

s_2_1_sequence.txt
s_2_2_sequence.txt

s_3_1_sequence.txt.gz
s_3_2_sequence.txt.gz
```

Fig. 9 An example of a HiCUP configuration file

3.5 Integration of Transcriptomic + Epigenomic Profiling Data Using ChromHMM

Integration of genome-wide patterns of transcriptome data plus multiple different epigenomic programming parameters can be achieved using a variety of software packages, one of the most commonly used is Chromatin State Discovery and Characterization based on use of a multivariate hidden Markov model (ChromHMM) as described below.

3.5.1 Installation

1. Install Java 1.5 or later if not already installed.
2. Unzip the file ChromHMM.zip.
3. From the command line, convert the ChromHMM.zip file into the now-unzipped ChromHMM subdirectory containing the ChromHMM.jar file.

   Note: ChrommHMM can be installed in conda environment. First, install conda, and in the terminal, run the following code.

```
conda install -c bioconda chromhmm
```

3.5.2 Binarization

1. Prepare a tab-delimited text file specifying the design file for the cell-mark-aligned read files according to whether there is a single-cell type (option A) or multiple cell types that will be treated independently (option B) or concatenated (option C), or multiple cell types for which the features will be stacked to provide a single-genome annotation (option D).

   (i) Option A: Single-cell type,

   Format a single .txt file in the form:

```
cell1 mark1 cell1_mark1_file
cell1 mark2 cell1_mark2_file
```

   (ii) Option B: Multiple cell types being treated independently,

   Format a single .txt file for each cell type in the form:

```
File 1:
cell1 mark1 cell1_mark1_file
cell1 mark2 cell1_mark2_file
File 2:
cell2 mark1 cell2_mark1_file
cell2 mark2 cell2_mark2_file
```

   (iii) Option C: Multiple cell types being concatenated,

   Format a single .txt file in the form:

```
cell1 mark1 cell1_mark1_file
cell1 mark2 cell1_mark2_file
cell2 mark1 cell2_mark1_file
cell2 mark2 cell2_mark2_file
```

   (iv) Option D: Multiple cell types with stacking features.

   Format a single .txt file in the form:

```
genome cell1_mark1 cell1_mark1_file
genome cell1_mark2 cell1_mark2_file
genome cell2_mark1 cell2_mark1_file
genome cell2_mark2 cell2_mark2_file
```

2. If a mark is specified only in a subset of the cell types, then it will be treated as missing in the remaining cell types (encoded with "2") in the binarized files.
3. If there are multiple files for the same cell and mark combination, then each can be specified and their reads will be pooled, or alternatively, the files can be combined outside of ChromHMM.
4. The .bed or .bam files supplied to BinarizeBed or BinarizeBam should normally contain the locations of aligned reads. ChromHMM also allows binarizing data on the basis of already-called peaks, as opposed to aligned reads. In this case, the "-peaks" flag needs to be specified when executing the BinarizeBam or BinarizeBed commands. Otherwise, the output will not be meaningful. If you are supplying already-called peaks to ChromHMM, it is not recommended to use broad peak calls. ChromHMM already considers spatial information, such that if you are supplying broad peaks, then ChromHMM could specify states indicative of a combination of marks that do not actually co-occur at the resolution at which this analysis operates.
5. If input control data such as whole-cell extract or IgG ChIP data are available, consider using them either as an additional feature (option A) or adjusting the binarization threshold locally based on the relative level of the control reads (option B). Option A is recommended if the control data are relatively flat, except in specific positions (often associated with artifacts), or if the control data were not sequenced deeply. Option B is recommended if the control data show peaks in regions other than artifact-associated regions and are sequenced sufficiently deeply to estimate accurate levels of the local background.

(i) Option A: Treat control data as an additional feature,

Format the cell-mark-aligned read file(s) to have the form:

```
cell1 mark1 cell1_mark1_file
cell1 mark2 cell1_mark2_file
cell1 control cell1_control_file
```

(ii) Option B: Use control data to adjust the binarization threshold locally.

Format the cell-mark-aligned read file(s) to have the form:

```
cell1 mark1 cell1_mark1_file cell1_control_file
cell1 mark2 cell1_mark2_file cell1_control_file
```

The control file is the fourth column. Mark-specific control files can also be specified.

6. Make sure the genome assembly corresponding to the data has a chromosome-length file in the CHROMSIZES directory. Chromosome-length files for the assemblies hg18, hg19, hg38, mm9, mm10, rn5, rn6, danRer7, danRer10, dm3, dm6, ce6, and ce10 are currently provided in the CHROMSIZES directory. If the desired assembly is already included, nothing further needs to be done at this step.

   If the desired assembly is not found in the directory, then it must be prepared and placed in the CHROMSIZES directory. The format is a two-column format in which the first column is the chromosome name and the second is the chromosome length. Such information for a few additional assemblies is available from the UCSC Genome Browser. Linux and Mac users can obtain such files directly by first downloading the fetchChromSizes script from the corresponding directory for their system here: http://hgdownload.cse.ucsc.edu/admin/exe/. At the command prompt, type ./fetchChromSizes assembly > assembly.txt.

7. Execute the command to binarize the data.

   (i) Option A Execute BinarizeBed,
   Enter the command:

```
java -mx4000M -jar ChromHMM.jar BinarizeBed chrlengthfile
inputdir cellmarkfiletable outputdir
```

- chrlengthfile contains the chromosome-length file, which can be downloaded from UCSC.
- inputdir is the directory with the aligned reads.
- cellmarkfiletable is the design file described.
- outputdir is the directory in which the binarized data should be written. For each cell type and chromosome.
- "-mx4000M" specifies the amount of memory given to Java and should be adjusted as needed for this and other commands and depends on your own system.

   (ii) Option B Execute BinarizeBam.
   Enter the command:

```
java -mx4000M -jar ChromHMM.jar BinarizeBam chrlengthfile
inputdir cellmarkfiletable outputdir
```

The parameters for BinarizeBam are the same as those listed above for BinarizeBed.

3.5.3 File Preparation for Enrichment Analyses

1. The coordinate files should be in .bed format. This step can also be done after executing the LearnModel command with enrichments then computed by running OverlapEnrichment.
2. Anchor files for external position annotation. This file should be a tab-delimited text file in the subfolder of the corresponding assembly's directory, named ANCHORFILES. The format of this file should be a tab-delimited text file in which the first column is the chromosome, the second column is the coordinate, and, optionally, the third column is the strand ("+" or "−"). This step can also be done after executing LearnModel with enrichments then computed by running NeighborhoodEnrichment, although the neighborhood enrichment analysis would not be automatically included in the report generated by the LearnModel command.

3.5.4 Model Learning

1. To start model learning and have automatic enrichments computed after the model is learned, execute the LearnModel command:

```
java -mx4000M -jar ChromHMM.jar LearnModel inputdir outputdir
numstates assembly
```

In this command, "inputdir" specifies the directory containing the binarized data, which will often be the output directory from previous Steps. "outputdir" is the directory in which the output files from the LearnModel command should be placed. "numstates" specifies the number of states. "assembly" specifies the genome assembly for the input.

Repeat this step for modeling with different numbers of states. If a computer cluster is available, then model learning can proceed in parallel.

3.5.5 Post-Model-Learning Analysis

1. The individual steps in this section are optional; users can choose any or all steps depending on how they wish to use and analyze the learned models. The full set of options is discussed in the ChromHMM manual.

   To compare models with different numbers of states at the emission parameter level, which can complement comparisons at the enrichment level, use the CompareModels command to generate a heatmap comparison by entering the following command:

```
java –mx4000M –jar ChromHMM.jar CompareModels referencemodel
comparedir outputprefix
```

2. To compute overlap enrichments in addition to those that were already computed after running the LearnModel command, use the OverlapEnrichment command by entering:

```
java -mx4000M -jar ChromHMM.jar OverlapEnrichment inputsegment
inputcoorddir outfileprefix
```

3. To compute neighborhood enrichments other than those that were already computed after applying the LearnModel command, use the NeighborhoodEnrichment command by entering:

```
java -mx4000M -jar ChromHMM.jar NeighborhoodEnrichment input-
segment anchorpositions outfileprefix
```

4. Reorder the states of the model to a different order by entering:

```
java -mx4000M -jar ChromHMM.jar Reorder –o stateorderingfile
inputmodel outputdir
```

5. Generate a segmentation corresponding to the reordered states. To generate a segmentation for data not used in learning an existing model, but in the same format, apply the MakeSegmentation command by entering:

```
java -mx4000M -jar ChromHMM.jar MakeSegmentation modelfile
inputdir outputdir
```

6. To change state colors in the browser files, use the MakeBrowserFiles command by entering:

```
java -mx4000M -jar ChromHMM.jar MakeBrowserFiles –c color-
mappingfile segmentfile segmentationname outputfileprefix
```

Acknowledgments

This work was supported by NIH grants P50 HD98593, U01 DA054179, and R01 HD78679 and by generous gifts from the Robert J. Kleberg, Jr. and Helen C. Kleberg Foundation and the Nancy Smith Hurd Foundation.

References

1. Guo F, Lin L, Jingyun L, Wu X, Hu B, Zhu P, Wen L, Tang F (2017) Single-cell multi-omics sequencing of mouse early embryos and embryonic stem cells. Cell Res 27:165–183
2. Argelaguet R, Clark SJ, Mohammed H, Stapel LC, Reik W et al (2019) Multi-omics profiling of mouse gastrulation at single-cell resolution. Nature 576:487–491
3. ENCODE Project Consortium (2012) An integrated encyclopedia of DNA elements in the human genome. Nature 489:57–74

4. Roadmap Epigenomics Consortium, Kundaje A, Meuleman W, Ernst J, Bilenky M, Yen A, Heravi-Moussavi A, Kheradpour P, Zhang Z, Wang J et al (2015) Integrative analysis of 111 reference human epigenomes. Nature 518: 317–330

5. Wingett S et al (2015) HiCUP: pipeline for mapping and processing Hi-C data. F1000Research 4:1310. https://doi.org/10.12688/f1000research.7334.1

Chapter 7

Primary Cultures of Spermatogonia and Testis Cells

Fan Yang, Jiachen Sun, and Xin Wu

Abstract

Spermatogonial stem cells (SSCs) maintain adult spermatogenesis in mammals by undergoing self-renewal and differentiation into spermatozoa. In order to study the biology of SSCs as related to spermatogenesis, an in vitro, long-term expansion system of SSCs constitutes an ideal tool. In this chapter, we describe a robust culture system for mouse and rat SSCs in vitro. In the presence of GDNF, GFRα1, and bFGF, SSCs maintained on STO feeder layers with serum-free medium continuously proliferate for over 6 months. Complete spermatogenesis in infertile recipient mice can be attained following transplantation of the cultured mouse and rat SSCs. Using the in vitro SSC culture systems, elucidation of stem cell biology can be advanced that significantly advances our understanding of spermatogenesis and male fertility.

Key words Spermatogonial stem cell, Spermatogonia, Culture, Spermatogenesis, Testis, Developmental biology

1 Introduction

The testis is the only male reproductive organ that produces sperm to deliver the genetic information for next generation. There are multiple types of testis cells, including spermatogonial stem cells (SSCs), spermatogonia, spermatocytes, spermatids, spermatozoa, and somatic cells (e.g., Sertoli cells, Leydig cells, and Myoid cells). A robust in vitro culture system of testicular cells has utility in furthering our knowledge of the biology of the male reproductive system. SSCs are the founders of normal spermatogenesis and male fertility. These cells either self-renew to produce two stem cells or differentiate into later stages of germ cells [1, 2]. As the only stem cells that contribute genetic information to the next generation, SSCs possess unique advantages that make them a powerful tool for genetic modification and generation of transgenic animals [3, 4]. Due to the rarity of SSCs in testes and a lack of specific markers to accurately identify them, large-scale and unlimited propagation of SSCs in vitro can be a powerful tool for studying

Jon M. Oatley and Brian P. Hermann (eds.), *Spermatogonial Stem Cells: Methods and Protocols*, Methods in Molecular Biology, vol. 2656, https://doi.org/10.1007/978-1-0716-3139-3_7,

molecular mechanisms underlying SSCs self-renewal and differentiation [5].

A successful SSC culture system consists of two critical components. One is the enrichment of SSCs at the beginning of culture, and the other is to mimic an in vitro SSC niche environment similar to that in vivo, including preparation of culture medium, feeder layers, and growth factors. Given that a low number of SSCs (0.03%) reside in the adult testes [6]—and that testicular somatic cells may stimulate SSC differentiation or outcompete SSCs so as to inhibit their proliferation at the beginning of culture [7, 8]—it is important to remove as many somatic cells as possible to achieve a starting population of cells that is enriched for SSCs. There are multiple ways to enrich SSCs from the testis, including differential plating [9], fluorescence-activated cell sorting (FACS) [10], magnetic-activated cell sorting (MACS) [11], and the use of cryptorchid animals [12]. Investigators have previously demonstrated via FACS and transplantation that enrichment of mouse SSCs was found in the THY-1$^+$ testicular cell population. [13]. MACS using THY-1 antibody is also more flexible and rapid than other methods, thus making it a preferable option for SSC enrichment in the mouse testis [11]. In addition, EpCAM$^+$ rat testicular cell populations enriched by MACS show augmented colony numbers compared with unsorted cell populations following transplantation [14].

The second key factor for SSC culture is the creation of an in vitro SSC niche environment similar to that in vivo, including feeder layers, culture medium, and growth factors. The first long-term maintenance of mouse SSCs in vitro was reported in 1998 [15], where mouse SSCs could be maintained for 4 months and using SIM mouse embryo-derived thioguanine and ouabain-resistant (STO) cells as a feeder layer lengthened SSC survival. Subsequently, mouse embryonic fibroblasts (MEFs) were used as a feeder layer in the first successful long-term expansion of mouse SSCs [16]. Glial cell-derived neurotrophic factor (*Gdnf*)-overexpressing STO feeder layers also promoted undifferentiated spermatogonia proliferation in culture without the supplementation of GDNF as growth factor. However, transfection with *Gdnf*-overexpressing plasmid accelerated the aging of STO feeder cells [17]. There have also been some studies that focused on the application of Sertoli cell lines in the long-term culture of SSCs [18–20]. Collectively, these studies suggested that an ideal feeder cell population will be extremely helpful in producing a successful SSC culture system in vitro.

Serum has been used to culture many different cell types. StemPro supplemented with 1% fetal bovine serum (FBS) and other components was used to culture gonocytes successfully [16], and FBS can also promote cell binding to gelatin-coated dishes to enrich a mouse SSC population [21]. However, the components of serum are undefined and vary between different batches. Also, inclusion of FBS in SSC culture medium promotes growth of somatic cells at the start of culture which can outcompete the SSC population, and several studies have shown that FBS exerts a negative effect on the growth of mouse and rat SSCs in vitro [14, 22]. A serum- and feeder-free culture medium obviates the disadvantage of FBS and allows us to define the extrinsic factors that regulate self-renewal and differentiation of SSCs. Investigators have over many years attempted to fabricate a serum- and feeder-free culture system [21, 23–25]. For example, an anchorage-independent culture condition was developed for SSC expansion in vitro, and fertile offspring were generated following transplantation [23]. However, nonadherent SSCs grow more slowly relative to cells in a conventional culture system containing StemPro, FBS, and MEFs. In 2014, an improved serum- and feeder-free culture system was established using laminin-coated plates and IMDM medium supplemented with knockout serum replacement. Furthermore, SSC amplification in IMDM serum-free medium was improved relative to conventional culture conditions [25].

Many growth factors have been tested in cultures of SSCs in vitro—including GDNF [11, 16], basic fibroblast growth factor (bFGF) [16, 26], epidermal growth factor (EGF) [11, 27], leukemia inhibitory factor (LIF) [11, 28], stem cell factor (SCF) [11, 29], colony-stimulating factor-1 (CSF-1) [30], insulin-like growth factor-1 (IGF-1) [11, 31], and FGF9 [32]. GDNF has been shown to be critical in regulating SSC self-renewal, and its overexpression caused the accumulation of undifferentiated spermatogonia in the mouse testis [33]. Additionally, STO feeders overexpressing *Gdnf* were shown to stimulate amplification of undifferentiated spermatogonia in culture [17]. bFGF, CSF-1, and IGF-1 also showed positive effects on SSC activity [22, 30, 34]. In a long-term SSC culture system, a cocktail of GDNF, bFGF, and/or GFRα1 was used to induce in vitro expansion of mouse, hamster, rat, rabbit, and tree shrew SSCs [7, 14, 19, 22, 35, 36]. However, culture of SSCs from domestic animals could only be maintained in vitro for a short time [27, 37–40]. Using bovine fetal fibroblasts as feeders, with defined serum-free medium supplemented with GDNF, FGF2, and LIF, cattle undifferentiated spermatogonia could be maintained in vitro for 2 months [41]. Although several studies have shown long-term expansion of human spermatogonia in vitro [42–46], demonstrating the presence of human SSCs in culture is challenging due to the lack of functional identification of stem cell activity [47–50].

In this chapter, we present methodology for establishing cultures of mouse and rat SSCs that are based on previous reports [7], with some modifications. Overall, both culture methods include selection of the SSC-enriched THY-1$^+$ or EpCAM$^+$ testis population using MACS, plating of the cells onto the STO feeder layers, and maintenance in MEM-α based serum-free medium supplemented with GDNF, GFRα1, and bFGF. This system allows robust expansion of SSCs for up to 6 months in vitro from mice with DBA, C57, ROSA26-LacZ, and *W*ᵛ backgrounds [22, 51–55], and rats with a MTLacZ background [14].

2 Materials

2.1 Preparation of Feeder Layers

1. Mitomycin C stock solution (500 μg/mL): Dissolve 2 mg of mitomycin C (Sigma) in 4 mL of DPBS and store it in the dark at 4 °C.
2. Medium for STO cell culture (STO medium): Add 2.5 mL of 10,000 U/mL penicillin–streptomycin, 5 mL of 200 mM L-glutamine (Gibco), 3.6 μL of 14.3 M 2-mercaptoethanol (Sigma), and 35 mL of FBS (Sigma) in a 500 mL storage bottle. Make up to 500 mL with DMEM (Gibco). Mix the components well and filter through a 0.2-μm rapid-flow bottle top filter and store at 4 °C. Place FBS in a 56 °C water bath for 30 min before using.
3. HBSS.
4. 0.25% trypsin–EDTA (Gibco).
5. DMSO.
6. Tissue culture dishes, 100 × 20 mm.
7. Cryogenic vials.
8. Gelatin, from porcine skin (Sigma).
9. Tissue culture plates, 12-well.
10. STO cells (SNL76/7 cells, ATCC).
11. Mr. Frosty™ freezing container.

2.2 Preparation of Testicular Single-Cell Suspension

1. Dissection instruments: including the fine forceps, ophthalmic surgery scissors, and dissecting microscope.
2. Tissue culture dishes, 35 × 10 mm.
3. Collagenase solution: Dissolve 10 mg of collagenase (Sigma) in 10 mL of HBSS, and filter through a 0.22-μm filter unit.
4. DNase solution: Dissolve 70 mg of DNase (Sigma) in 10 mL of HBSS, and filter through a 0.22-μm filter unit. Store at 4 °C.
5. 15 mL and 50 mL polypropylene conical tubes.
6. PBS-S: Weigh 0.5 g of D-glucose, transfer it to a 500 mL bottle, and add 50 mL of 10 × DPBS, 5 mL of 1 M HEPES (Gibco), 5 mL of FBS, 5 mL of 100 mM sodium pyruvate (Gibco), 2.5 mL of 10,000 U/mL penicillin-streptomycin, and 2 mL of 0.5 M EDTA (Invitrogen). Make up to 500 mL

with distilled water and mix them well. Filter through a 0.2-μm Rapid-flow bottle top filter and store at 4 °C.

7. 30% Percoll: Add 25 mL of 10 × DPBS, 2.5 mL of FBS, 1.25 mL of 10,000 U/mL penicillin–streptomycin, and 75 mL of Percoll (GE Healthcare, 17-0891-01) to a 250 mL storage bottle. Make up to 250 mL with distilled water and mix well. Filter through a 0.2-μm Rapid-flow bottle top filter and store at 4 °C.
8. 5 mL polystyrene round-bottom tube.

2.3 SSC Enrichment

1. CD90.2 Microbeads, mouse (Miltenyi).
2. Anti-rat EpCAM antibody (Abcam).
3. Anti-mouse IgG microbeads (Miltenyi).
4. MS column (Miltenyi).
5. Thirty-μm Pre-Separation Filters (Miltenyi).
6. MiniMACS™ Separator (Miltenyi).
7. MACS MultiStand (Miltenyi).

2.4 Culture of SSCs

1. Free fatty acid mixture (FFA): All FFAs are dissolved in ethanol. To make 1 mL of 100 meq/L FFA mixture, add 5.6 μL of 1 M linolenic acid (Sigma), 13.4 μL of 1 M oleic acid (Sigma), 2.8 μL of 1 M palmitoleic acid (Sigma), 35.6 μL of 1 M linoleic acid (Sigma), 31.0 μL of 1 M palmitic acid (Sigma), 76.9 μL of 151 mM stearic acid (Sigma), and 834.7 μL of ethanol. Make 50 μL/0.5 mL tube aliquots, purge with N_2 or CO_2, seal with parafilm, and store at −20 °C [7].
2. Mouse serum-free medium (mSFM): Weigh 0.5 g of BSA (*see* **Note 1**) and transfer it to a 250 mL storage bottle. Add 0.25 mL of 10 mg/mL transferrin (dissolved in DPBS; Sigma), 1.25 mL of 10,000 U/mL penicillin–streptomycin, 2.5 mL of 200 mM L-glutamine, 25 μL of 3×10^{-4} M Na_2SeO_3 (dissolved in distilled water; Sigma), 125 μL of 10 mg/mL insulin (dissolved in 0.01 N HCl; Sigma), 19 μL of FFA, 150 μL of 100 mM putrescine (dissolved in DPBS; Sigma), 125 μL of 100 mM 2-mercaptoethanol, and 2.5 mL of 1 M HEPES. Make up to 250 mL with MEM Alpha (Gibco). Leave the solution in the refrigerator overnight, filter through a 0.2-μm Rapid-flow bottle top filter, and store at 4 °C [7].
3. Rat serum-free medium (rSFM): Weigh 1.5 g of BSA (*see* **Note 1**) and 25 mg of transferrin, and transfer to a 250 mL storage bottle—to which we add 1.25 mL of 10,000 U/mL penicillin–streptomycin, 2.5 mL of 200 mM L-glutamine, 50 μL of 3×10^{-4} M Na_2SeO_3, 0.625 mL of 10 mg/mL insulin, 38 μL of FFA, 0.3 mL of 100 mM putrescine, 0.25 mL of

100 mM 2-mercaptoethanol, and 2.5 mL of 1 M HEPES. Make up to 250 mL with MEM Alpha. Add 37.5 mL of distilled water, leave in the refrigerator overnight (*see* **Note 2**), filter through a 0.2-μm Rapid-flow bottle top filter, and store at 4 °C [7].

4. Growth factor stock solution: (1) GDNF (20 μg/mL): Dissolve 50 μg of GDNF (R&D) in 2.5 mL of 0.1% BSA (dissolved in DPBS), aliquot 500 μL/vial, and store at −80 °C. (2) GFRα1 (100 μg/mL): Dissolve 100 μg of GFRα1 (R&D) in 1 mL of 0.1% BSA, aliquot 500 μL/vial, and store at −80 °C. (3) bFGF (1 μg/mL): Dissolve 10 μg of bFGF (Corning) in 10 mL of 0.1% BSA, aliquot 500 μL/vial, and store at −20 °C.
5. 0.01% trypsin–EDTA: Mix 0.4 mL of 0.25% trypsin–EDTA with 9.6 mL of HBSS and store at 4 °C.

3 Methods

3.1 Preparation of Feeder Layers

1. Plate STO cells into 10-cm dishes, and initiate mitomycin C treatment when cells reach 80–90% confluency (*see* **Note 3**).
2. Remove the medium in dishes, add 6 mL of STO medium containing 10 μg/mL mitomycin C, and incubate dishes at 37 °C for 4 h. Shake the dishes gently every hour to make sure the cells near the edge of dish can be exposed to mitomycin C fully.
3. Discard medium containing mitomycin C and wash dishes 3 times with 10 mL of HBSS.
4. Digest STO cells with 2 mL of trypsin–EDTA for 5 min at 37 °C. Add 3 mL of STO medium to stop digestion and aliquot cell suspension into a 50 mL conical tube. Wash each dish with 5 mL of STO medium and aliquot medium into a 50 mL conical tube.
5. Centrifuge cells at 600 × g for 5 min at 4 °C and resuspend them with STO medium at 6×10^6 cells/mL. Then, add same volume of STO medium containing 20% DMSO dropwise and mix well.
6. Add 1 mL of cell suspension (3×10^6 cells/mL) into each cryogenic vial. Put vials in freezing container and store it at −80 °C (*see* **Note 4**).
7. To prepare STO feeder plates, add 1 mL of 0.1% gelatin into each well of 12-well plates and incubate plates at 37 °C at least 1 h prior to use.
8. Take 1 vial mitomycin C-treated STO cells (3×10^6 cells/mL) from the freezer, transfer it to a 37 °C water bath, and shake the vial gently until the frozen material liquifies. Deposit cells into a

50 mL conical tube and add 4 mL of STO medium dropwise. Centrifuge cells at 600 × g for 5 min at 4 °C.

9. Discard supernatant, add 20 mL of STO medium to resuspend the cells at a concentration of 1.5×10^5 cells/mL (*see* **Note 5**). Add 1 mL of cells to each well of 12-well plates coated with gelatin, and transfer plates to a 5% CO_2 incubator at 37 °C.

3.2 Isolation of Mouse SSCs

3.2.1 Preparation of Single-Cell Suspensions from Mouse Testis

1. Instruments are sterilized by autoclaving. Prepare 2 dishes (35 × 10 mm) with 2 mL HBSS and place them on ice.
2. Add 4.5 mL of 0.25% trypsin–EDTA and 0.5 mL of 7 mg/mL DNase to a 15 mL conical tube, and transfer the tube to a 37 °C water bath.
3. Collect 2–4 testes from mouse pup testes at 6–8 days postpartum (dpp) and place them in a 35 mm dish containing 2 mL of HBSS. Remove the tunica under a dissecting microscope and shift the testes to the next dish.
4. Bring testes to a hood, and deliver them into a 15 mL conical tube containing 4.5 mL of trypsin–EDTA and 0.5 mL of DNase.
5. Use 1 mL pipette to gently aspirate and move the testicular tissues up and down for 2 min, incubate them in a 37 °C water bath for 5 min, pipet them again for 2 min, and re-incubate the tissues in a water bath for another 3 min.
6. Add 0.7 mL of FBS and 1 mL of DNase and pipet testicular cells to be a single-cell suspension (*see* **Note 6**). Use 40-μm cell strainer to filter the cell suspension and then wash the cell strainer with 5 mL of HBSS.
7. Centrifuge cells at 600 × g for 5 min at 4 °C, resuspend them in 10 mL of PBS-S, and count them (*see* **Note 7**).
8. Add 2 mL of 30% Percoll into each 15 mL conical tube and slowly overlay a 2–5 mL cellular suspension on the Percoll. Do not lay more than 2×10^7 testicular cells on 2 mL of 30% Percoll.
9. Centrifuge the tubes at 600 × g for 7 min at 4 °C and slowly transport the tubes to the hood without disturbing the interface between the Percoll layer and the cell suspension.
10. Remove dead cells and cellular debris and fragments at the interface between the PBS-S and Percoll, discard both the cellular suspension and Percoll, resuspend the cell pellet with 2 mL of PBS-S, and transfer the suspension to a 5 mL polystyrene round-bottom tube. Then count the cells (*see* **Note 8**).

3.2.2 Enrichment of Mouse SSCs

1. Following the use of Percoll, centrifuge the cells at 600 × g for 5 min at 4 °C, remove supernatant, and resuspend cells in 90 μL of PBS-S.
2. Incubate the cells with 10 μL of mouse CD90.2 microbeads for 15 min at 4 °C (*see* **Note 9**) and tap the bottom of the tube once during the incubation period.
3. Add 2 mL of PBS-S to the tube and centrifuge it at 600 × g for 5 min at 4 °C.
4. Discard supernatant and resuspend the pellet in 0.5 mL of PBS-S.
5. Sort THY-1^+ cells magnetically according to the manufacturer's protocol, place a MS column in the magnetic field of a MiniMACS™ Separator, and rinse the column with 0.5 mL of PBS-S.
6. The cell suspension is applied to a column. Wash column 3 times with 0.5 mL of PBS-S.
7. Remove column from the separator and place it in a 5 mL round-bottom tube, and add 1 mL of mSFM to the column. Immediately flush out the magnetically labeled THY-1^+ cells by firmly pushing the plunger into the column (*see* **Note 10**).
8. After eluting THY-1^+ cells, rinse them twice with 1 mL of mSFM, and count the cells.

3.3 Isolation of Rat SSCs

3.3.1 Preparation of Rat Testis Single-Cell Suspensions

1. The testis preparation is performed as described in the above section for mice. This protocol is designed for 2–4 rat pup testes at 8–12 dpp.
2. Bring testes to the hood, place them in a 15 mL conical tube containing 10 mL of 1 mg/mL collagenase, and incubate them in a 37 °C water bath for 6–8 min. Invert the tube several times every 2–3 min (*see* **Note 11**).
3. Centrifuge the tube at 600 × g for 1 min at 4 °C, remove supernatant and add 10 mL of cold HBSS, and invert tube several times to loosen fragments of the seminiferous tubules.
4. Centrifuge the tube at 600 × g for 1 min at 4 °C, remove supernatant, and add 6 mL of warmed trypsin–EDTA and 1 mL of 7 mg/mL DNase.
5. Use 1 mL pipette to gently pipet the testicular tissues up and down for 2 min, and then incubate them in a 37 °C water bath for 5 min, pipet them again for 2 min, and incubate them in a water bath for 3 min.
6. Add 0.9 mL of FBS and 1 mL of DNase, and pipet the testicular cells so as to create a single-cell suspension (*see* **Note 6**). Use 40-μm cell strainer to filter the cell suspension, and subsequently wash the cell strainer with 5 mL HBSS.

7. Centrifuge cells at 600 × g for 5 min at 4 °C, resuspend cells in 10 mL of cold PBS-S and count them.
8. Perform the Percoll procedure as described in the above section for mice. Resuspend the cell pellet with 5 mL of PBS-S and count the cells (*see* **Note 8**).

3.3.2 Enrichment of Rat SSCs

1. Following the Percoll, centrifuge cells at 600 × g for 5 min at 4 °C, remove the supernatant, and resuspend the cells in 1 mL of PBS-S.
2. Cells are incubated with anti-rat EpCAM antibody (1:200 dilution) (*see* **Note 12**) for 15 min at 4 °C, and tap the bottom of the 15 mL conical tube once during the incubation period.
3. Add 1–2 mL of PBS-S per 10^7 cells and centrifuge at 600 × g for 5 min at 4 °C.
4. Discard supernatant and add 80 μL of PBS-S per 10^7 cells to resuspend. Then add 20 μL of anti-mouse IgG microbeads per 10^7 cells. Mix them well and incubate for 15 min at 4 °C in the dark, tap the bottom of tube once during the incubation period.
5. Add 1–2 mL of PBS-S per 10^7 cells, and centrifuge at 600 × g for 5 min at 4 °C.
6. Resuspend cells in 1 mL of PBS-S. Do not aliquot more than 10^8 cells into 0.5 mL of PBS-S.
7. Sort EpCAM$^+$ cells magnetically using 2 MS columns with 30-μm Pre-Separation Filters according to the manufacturer's protocol (*see* **Note 13**), and add 500 μL of cell suspension to each column.
8. After eluting EpCAM$^+$ cells, rinse cells twice with 2 mL of rSFM and count the cells.

3.4 Culture of SSCs

3.4.1 Culture of Mouse SSCs

1. Resuspend THY-1$^+$ cells in warmed mSFM containing 20 ng/mL GDNF, 150 ng/mL GFRα1, and 1 ng/mL bFGF.
2. Discard STO medium from the 12-well plates with the STO feeder, and wash twice with 1 mL of HBSS per well.
3. Add 1.5 mL of THY-1$^+$ cells (1–2 × 10^5 cells/well) to each well of 12-well plates containing the STO feeder layer.
4. Incubate cells in a humidified atmosphere of 5% CO_2 in compressed air or 5% O_2/5% CO_2/90% N_2 at 37 °C (*see* **Note 14**), and culture medium should be changed every 2–3 days (Fig. 1a, b).
5. Perform the first subculture after an initial seeding of 10–14 days at a ratio of 1:2 (*see* **Note 15**). After the culture medium is discarded, add 0.5 mL of warmed 0.25% trypsin–

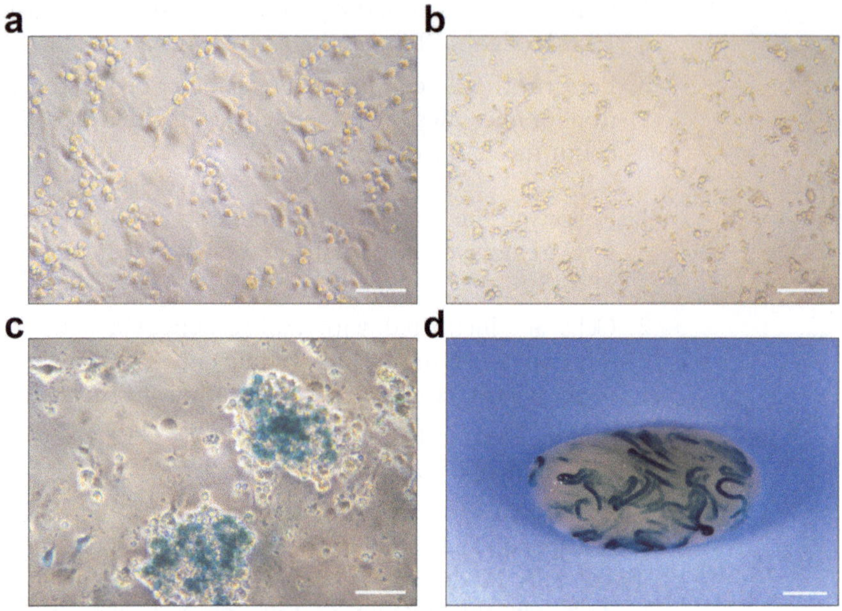

Fig. 1 Morphology of mouse SSCs in culture and colonies following transplantation. (**a**) Representative morphology of THY-1^+ mouse testicular cells following a 4-hour initial seeding onto a STO feeder layer. C57/BL6 mouse pup testes at 6–8 dpp were collected and digested to produce a single-cell suspension. THY-1^+ testicular cells were isolated by THY-1 MACS and plated onto a STO feeder layer. Scale bar, 50 μm. (**b**) Representative morphology of THY-1^+ mouse testicular cells following a 1-day initial seeding onto a STO feeder layer. Scale bar, 100 μm. (**c**) Representative morphology of ROSA mouse SSCs stained with X-gal solution following a 3-month culture. Photomicrograph depicts germ-cell clumps. Scale bar, 50 μm. (**d**) Colony formation from a recipient testis that was transplanted with cultured ROSA mouse SSCs, followed by X-gal staining. Scale bar, 1 mm

EDTA per well of a 12-well plate, and incubate cells at 37 °C for 5 min (*see* **Note 16**).

6. Add 0.1 mL of FBS to stop digestion and collect the cells into a 15 mL conical tube. Then add 1.4 mL of mSFM.
7. Centrifuge at 600 × g for 5 min at 4 °C.
8. Remove the supernatant and rinse again with 2 mL of mSFM.
9. Resuspend cells in mSFM containing GNDF, GFRα1, and bFGF, and plate cells on a new STO monolayer (*see* **Note 17**).
10. When the germ cells expand consistently, passage them every 4–7 days thereafter at a ratio of 1:2 to 1:4 (Fig. 1c) (*see* **Note 22**).
11. The SSC population in culture can be identified functionally using transplantation. Blue colonies can be found in the transplanted testes following injection of ROSA mouse germ cells (Fig. 1d).

3.4.2 Culture of Rat SSCs

1. Resuspend EpCAM$^+$ cells in warmed rSFM containing 20 ng/mL GDNF, 75 ng/mL GFRα1, and 1 ng/mL bFGF.
2. Discard STO medium from the 12-well plates with the STO feeder and wash twice with 2 mL of HBSS per well.
3. Add 1.5 mL of EpCAM$^+$ cells (1–2 × 10^5 cells/well) to each well of the 12-well plates containing the STO feeder. Cells are maintained in an atmosphere of 5% O_2/5% CO_2/90% N_2 at 37 °C, and the culture medium is changed every 2–3 days (*see* **Note 18**).
4. Perform the first subculture after an initial seeding of 6–8 days at a ratio of 1:1 or 1:2. Gently blow off the cell clumps using a 1 mL pipette and collect them in a 15 mL conical tube (*see* **Note 19**).
5. Centrifuge at 600 × g for 5 min at 4 °C, remove the supernatant, and add 100 μL of pre-warmed 0.01% trypsin–EDTA. Place the tube in a 37 °C water bath and continuously shake it for 1 min (*see* **Note 20**). Then, add 100 μL of FBS to stop the digestion and add 1.8 mL of rSFM.
6. Centrifuge at 600 × g for 5 min at 4 °C, remove supernatant, and again rinse with 2 mL of rSFM.
7. Cells are resuspended in rSFM containing GNDF, GFRα1, and bFGF and plated on a new STO monolayer (*see* **Note 21**).
8. When germ cells begin to expand consistently, passage cells every 10–14 days thereafter at a ratio of 1:2 (*see* **Note 22**).

4 Notes

1. There are batch effects of BSA on mouse and rat SSC cultures. BSA (Sigma, A1470, Lot # SLBG4990V) is used for establishing and maintaining mouse germ cells. We use BSA (Calbio, 12660, Lot # B73082) to establish the rat germ-cell lines. BSA from Millipore (81-003-N, Lot # 744) is used to maintain established rat germ-cell lines. We suggest testing different batches of BSA from MP biochemicals (810661), Sigma (A3803 and A1470), and Millipore (12660 and 81-003-N).
2. Add 15% (vol/vol) distilled water to the medium to reduce osmolarity. Reduction of medium osmolarity helps to prevent cellular differentiation and to promote proliferation. Leave medium overnight without filtering to guarantee that all components are dissolved well in the medium.
3. STO cells are to be passaged rapidly once they become confluent, as contact inhibition may occur when cells reach confluency. We usually use STO cells at low passage numbers (<14).

4. Cell storage in liquid nitrogen is better than in a −80 °C freezer, and cells that remain in liquid nitrogen for longer than 1 year are not recommended for use.
5. For 12-well plates, the density of plating STO cells is 1.5–2 × 10^5 cells/well. We prefer to use STO feeder layers 2 days after seeding and recommend usage within 4 days) but STO plates can be used up to 8 days after seeding.
6. We add DNase and continue to pipette for approximately 1 min. If the cell suspension is still viscous, we add more DNase as necessary.
7. Recovered testicular cell number is approximately 1–1.3 × 10^6 cells/testis at 5 dpp and 1.7–2 × 10^6 cells/testis at 8 dpp. Cell viability is normally higher than 85%.
8. We suggest being careful when overlaying the cells on Percoll. Disturbing the interface between cells and Percoll will affect cellular viability and the recovery rate. Our cell recovery and viability rates after the Percoll procedure are normally 40–70% and>90%, respectively.
9. The capacity of CD90.2 microbeads is 10 μL of beads per 1 × 10^7 total cells.
10. A second THY-1^+ cell sorting using MACS might be helpful in improving the purity of SSCs. Using a more-enriched population of THY-1^+ cells allows seeding at a higher density of cells (2–3 × 10^5 cells/well) initially, which might be useful in the establishment of cell lines.
11. After 6–8 min of incubation, fragments of seminiferous tubules can be observed. If fragments remain large, we digest them further until it is difficult to visualize them.
12. We usually add at least 5 μL of EpCAM antibody to perform the incubation. A 1:200 dilution is the lowest we use, but a higher concentration of antibody can be utilized. If different brands of EpCAM antibody are used, the dilution may vary according to instructions.
13. A pre-separation filter is recommended as it can remove cell aggregates and large particles, and this will guarantee optimal magnetic cell binding.
14. Low O_2 atmosphere (5% or 10%) has positive effect on establishment and long-term culture of mouse SSCs. In vitro culture of SSCs from W^v/W^v mice only could be established in a 10% O_2 condition instead of 21% [55]. A 10% O_2 atmosphere is also beneficial to the regenerative integrity of mouse SSCs in the long-term culture [56].
15. Germ-cell clumps will be observed 4–5 days after initial seeding. We find that maintaining germ cells longer (>10 days) in

the first 2–3 passages will achieve larger clumps, which is an important factor in establishing cell lines successfully.

16. We suggest using trypsin digestion to subculture cells most of the time. If somatic cells grow so well that they inhibit germ-cell proliferation, we gently blow off clumps and digest with 0.01% trypsin–EDTA as a better choice for passaging cells. However, note that physical damage occurs by blowing off clumps.
17. In the first several passages, the ratio of subculture is usually 1:1 or 1:2 and depends upon germ cell and somatic cell numbers. One needs to consider the possibility that the somatic cells will overgrow the germ cells or that there may be a gradual loss of germ cells due to a low seeding number. After approximately 1 month of maintenance, a gradual increase in germ-cell clumps will be observed. When the cell line is established, subculture can be performed by digesting all cells with 0.25% trypsin–EDTA at a ratio of 1:2 to 1:4.
18. Germ-cell clumps are observed 4–5 days after initial seeding. An atmosphere of 5% O_2/5% CO_2/90% N_2 is better than humidified 5% CO_2 in air for maintaining rat germ-cell culture.
19. Because of the contamination of testicular somatic cells, we do not recommend digesting all cells directly to subculture. Rather, we gently blow off germ-cell clumps to remove somatic cells and maintain the clumps.
20. Because trypsin–EDTA damages cells, we use diluted (1:25) trypsin–EDTA to digest germ-cell clumps when performing subculture. Digestion time may also be longer than 1 min if the cell clumps are large. Avoid hard pipetting of cell clumps during the process of digestion as it will cause more physical damage to the clumps. Single-cell suspension is not required after digestion, as both small clumps and single cells are acceptable.
21. After a second passage of rat germ cells, we transport them to a 24-well plate. In our experience, 24-well plates are better than 12-well plates for maintaining cell clumps. Subculture is to be conducted every 14 days, with the ratio of subculture at 2:1 or 3:1 because of a gradual decrease in cell clumps after passaging. After a few weeks of maintenance, we observe a gradual increase in germ-cell clumps. When the cell line is established, subculture can be performed by digesting all cells with 0.25% trypsin–EDTA at a ratio of 1:2.
22. Make sure the temperature, humidity, and the gas composition are accurate. These factors are often overlooked in culture, and each has been shown to have an effect. Temperature should be checked with a thermometer that has been certified accurate, since laboratory thermometers can vary ±1°. Maintenance of

humidity is critical and should be maintained as close to 100% as possible, by including supplementary open trays of water when necessary. Particularly, measurement of CO_2 concentration in commercial incubators is important since it is often incorrect. The effect of CO_2 on pH of media may take a long period of time, and the time necessary depends on the ratio of media surface area to media volume.

Acknowledgments

We thank Dr. Ralph Brinster and Dr. Eoin Whelan for helpful advice. This work was supported by the National Key R&D Program of China (Grant No. 2018YFC1003302) and the National Natural Science Foundation of China (Grant No. 31872844 and 32070831).

References

1. de Rooij DG, Russell LD (2000) All you wanted to know about spermatozoon but were afraid to ask. J Androl 21:776–798
2. De Rooij DG, Griswold MD (2012) Questions about spermatogonia posed and answered since 2000. J Androl 33(6):1085–1095. https://doi.org/10.2164/jandrol.112.016832
3. Brinster RL (2002) Germline stem cell transplantation and transgenesis. Science 296(5576):2174–2176. https://doi.org/10.1126/science.1071607
4. Brinster RL (2007) Male germline stem cells: from mice to men. Science 316(5823):404–405. https://doi.org/10.1126/science.1137741
5. Oatley JM, Avarbock MR, Telaranta AI, Fearon DT, Brinster RL (2006) Identifying genes important for spermatogonial stem cell self-renewal and survival. Proc Natl Acad Sci U S A 103(25):9524–9529. https://doi.org/10.1073/pnas.0603332103
6. Tegelenbosch RA, de Rooij DG (1993) A quantitative study of spermatogonial multiplication and stem cell renewal in the C3H/101 F1 hybrid mouse. Mutat Res 290(2):193–200. https://doi.org/10.1016/0027-5107(93)90159-d
7. Kubota H, Brinster RL (2008) Culture of rodent spermatogonial stem cells, male germline stem cells of the postnatal animal. Methods Cell Biol 86:59–84. https://doi.org/10.1016/S0091-679X(08)00004-6
8. Nagano M, Ryu BY, Brinster CJ, Avarbock MR, Brinster RL (2003) Maintenance of mouse male germ line stem cells in vitro. Biol Reprod 68(6):2207–2214. https://doi.org/10.1095/biolreprod.102.014050
9. Shinohara T, Avarbock MR, Brinster RL (1999) beta1- and alpha6-integrin are surface markers on mouse spermatogonial stem cells. Proc Natl Acad Sci U S A 96(10):5504–5509. https://doi.org/10.1073/pnas.96.10.5504
10. Shinohara T, Orwig KE, Avarbock MR, Brinster RL (2000) Spermatogonial stem cell enrichment by multiparameter selection of mouse testis cells. Proc Natl Acad Sci U S A 97(15):8346–8351. https://doi.org/10.1073/pnas.97.15.8346
11. Kubota H, Avarbock MR, Brinster RL (2004) Culture conditions and single growth factors affect fate determination of mouse spermatogonial stem cells. Biol Reprod 71(3):722–731. https://doi.org/10.1095/biolreprod.104.029207
12. Shinohara T, Avarbock MR, Brinster RL (2000) Functional analysis of spermatogonial stem cells in steel and cryptorchid infertile mouse models. Dev Biol 220(2):401–411. https://doi.org/10.1006/dbio.2000.9655
13. Kubota H, Avarbock MR, Brinster RL (2003) Spermatogonial stem cells share some, but not all, phenotypic and functional characteristics with other stem cells. Proc Natl Acad Sci U S A 100(11):6487–6492. https://doi.org/10.1073/pnas.0631767100

14. Ryu BY, Kubota H, Avarbock MR, Brinster RL (2005) Conservation of spermatogonial stem cell self-renewal signaling between mouse and rat. Proc Natl Acad Sci U S A 102(40): 14302–14307. https://doi.org/10.1073/pnas.0506970102
15. Nagano M, Avarbock MR, Leonida EB, Brinster CJ, Brinster RL (1998) Culture of mouse spermatogonial stem cells. Tissue Cell 30(4): 389–397. https://doi.org/10.1016/s0040-8166(98)80053-0
16. Kanatsu-Shinohara M, Ogonuki N, Inoue K, Miki H, Ogura A, Toyokuni S, Shinohara T (2003) Long-term proliferation in culture and germline transmission of mouse male germline stem cells. Biol Reprod 69(2):612–616. https://doi.org/10.1095/biolreprod.103.017012
17. Wei X, Jia Y, Xue Y, Geng L, Wang M, Li L, Wang M, Zhang X, Wu X (2016) GDNF-expressing STO feeder layer supports the long-term propagation of undifferentiated mouse spermatogonia with stem cell properties. Sci Rep 6:36779. https://doi.org/10.1038/srep36779
18. Guo Y, Hai Y, Gong Y, Li Z, He Z (2014) Characterization, isolation, and culture of mouse and human spermatogonial stem cells. J Cell Physiol 229(4):407–413. https://doi.org/10.1002/jcp.24471
19. Li CH, Yan LZ, Ban WZ, Tu Q, Wu Y, Wang L, Bi R, Ji S, Ma YH, Nie WH, Lv LB, Yao YG, Zhao XD, Zheng P (2017) Long-term propagation of tree shrew spermatogonial stem cells in culture and successful generation of transgenic offspring. Cell Res 27(2):241–252. https://doi.org/10.1038/cr.2016.156
20. Hamra FK, Chapman KM, Nguyen DM, Williams-Stephens AA, Hammer RE, Garbers DL (2005) Self renewal, expansion, and transfection of rat spermatogonial stem cells in culture. Proc Natl Acad Sci U S A 102(48): 17430–17435. https://doi.org/10.1073/pnas.0508780102
21. Kanatsu-Shinohara M, Miki H, Inoue K, Ogonuki N, Toyokuni S, Ogura A, Shinohara T (2005) Long-term culture of mouse male germline stem cells under serum-or feeder-free conditions. Biol Reprod 72(4):985–991. https://doi.org/10.1095/biolreprod.104.036400
22. Kubota H, Avarbock MR, Brinster RL (2004) Growth factors essential for self-renewal and expansion of mouse spermatogonial stem cells. Proc Natl Acad Sci U S A 101(47): 16489–16494. https://doi.org/10.1073/pnas.0407063101
23. Kanatsu-Shinohara M, Inoue K, Lee J, Miki H, Ogonuki N, Toyokuni S, Ogura A, Shinohara T (2006) Anchorage-independent growth of mouse male germline stem cells in vitro. Biol Reprod 74(3):522–529. https://doi.org/10.1095/biolreprod.105.046441
24. Kanatsu-Shinohara M, Inoue K, Ogonuki N, Morimoto H, Ogura A, Shinohara T (2011) Serum- and feeder-free culture of mouse germline stem cells. Biol Reprod 84(1):97–105. https://doi.org/10.1095/biolreprod.110.086462
25. Kanatsu-Shinohara M, Ogonuki N, Matoba S, Morimoto H, Ogura A, Shinohara T (2014) Improved serum- and feeder-free culture of mouse germline stem cells. Biol Reprod 91(4):88. https://doi.org/10.1095/biolreprod.114.122317
26. Ishii K, Kanatsu-Shinohara M, Toyokuni S, Shinohara T (2012) FGF2 mediates mouse spermatogonial stem cell self-renewal via upregulation of Etv5 and Bcl6b through MAP2K1 activation. Development 139(10):1734–1743. https://doi.org/10.1242/dev.076539
27. Aponte PM, Soda T, Teerds KJ, Mizrak SC, van de Kant HJ, de Rooij DG (2008) Propagation of bovine spermatogonial stem cells in vitro. Reproduction 136(5):543–557. https://doi.org/10.1530/REP-07-0419
28. Sharma A, Shah SM, Saini N, Mehta P, Kumar BSB, Dua D, Singh MK, Singla SK, Palta P, Manik RS, Chauhan MS (2019) Optimization of serum-free culture conditions for propagation of Putative Buffalo (Bubalus bubalis) spermatogonial stem cells. Cell Reprogram 21(1): 1–10. https://doi.org/10.1089/cell.2018.0018
29. Dolci S, Pellegrini M, Di Agostino S, Geremia R, Rossi P (2001) Signaling through extracellular signal-regulated kinase is required for spermatogonial proliferative response to stem cell factor. J Biol Chem 276(43): 40225–40233. https://doi.org/10.1074/jbc.M105143200
30. Oatley JM, Oatley MJ, Avarbock MR, Tobias JW, Brinster RL (2009) Colony stimulating factor 1 is an extrinsic stimulator of mouse spermatogonial stem cell self-renewal. Development 136(7):1191–1199. https://doi.org/10.1242/dev.032243
31. Huang YH, Chin CC, Ho HN, Chou CK, Shen CN, Kuo HC, Wu TJ, Wu YC, Hung YC, Chang CC, Ling TY (2009) Pluripotency of mouse spermatogonial stem cells maintained by IGF-1- dependent pathway. FASEB J 23(7): 2076–2087. https://doi.org/10.1096/fj.08-121939

32. Yang F, Whelan EC, Guan X, Deng B, Wang S, Sun J, Avarbock MR, Wu X, Brinster RL (2020) FGF9 promotes mouse spermatogonial stem cell proliferation mediated by p38 MAPK signalling. Cell Prolif 54. https://doi.org/10.1111/cpr.12933
33. Meng X, Lindahl M, Hyvonen ME, Parvinen M, de Rooij DG, Hess MW, Raatikainen-Ahokas A, Sainio K, Rauvala H, Lakso M, Pichel JG, Westphal H, Saarma M, Sariola H (2000) Regulation of cell fate decision of undifferentiated spermatogonia by GDNF. Science 287(5457):1489–1493. https://doi.org/10.1126/science.287.5457.1489
34. Takashima S, Kanatsu-Shinohara M, Tanaka T, Morimoto H, Inoue K, Ogonuki N, Jijiwa M, Takahashi M, Ogura A, Shinohara T (2015) Functional differences between GDNF-dependent and FGF2-dependent mouse spermatogonial stem cell self-renewal. Stem Cell Reports 4(3):489–502. https://doi.org/10.1016/j.stemcr.2015.01.010
35. Kubota H, Wu X, Goodyear SM, Avarbock MR, Brinster RL (2011) Glial cell line-derived neurotrophic factor and endothelial cells promote self-renewal of rabbit germ cells with spermatogonial stem cell properties. FASEB J 25(8):2604–2614. https://doi.org/10.1096/fj.10-175802
36. Kanatsu-Shinohara M, Muneto T, Lee J, Takenaka M, Chuma S, Nakatsuji N, Horiuchi T, Shinohara T (2008) Long-term culture of male germline stem cells from hamster testes. Biol Reprod 78(4):611–617. https://doi.org/10.1095/biolreprod.107.065615
37. Kuijk EW, Colenbrander B, Roelen BA (2009) The effects of growth factors on in vitro-cultured porcine testicular cells. Reproduction 138(4):721–731. https://doi.org/10.1530/REP-09-0138
38. Zhu H, Liu C, Li M, Sun J, Song W, Hua J (2013) Optimization of the conditions of isolation and culture of dairy goat male germline stem cells (mGSC). Anim Reprod Sci 137(1–2):45–52. https://doi.org/10.1016/j.anireprosci.2012.12.005
39. Kadam PH, Kala S, Agrawal H, Singh KP, Singh MK, Chauhan MS, Palta P, Singla SK, Manik RS (2013) Effects of glial cell line-derived neurotrophic factor, fibroblast growth factor 2 and epidermal growth factor on proliferation and the expression of some genes in buffalo (Bubalus bubalis) spermatogonial cells. Reprod Fertil Dev 25(8):1149–1157. https://doi.org/10.1071/RD12330
40. Zheng Y, Tian X, Zhang Y, Qin J, An J, Zeng W (2013) In vitro propagation of male germline stem cells from piglets. J Assist Reprod Genet 30(7):945–952. https://doi.org/10.1007/s10815-013-0031-0
41. Oatley MJ, Kaucher AV, Yang QE, Waqas MS, Oatley JM (2016) Conditions for long-term culture of cattle undifferentiated spermatogonia. Biol Reprod 95(1):14. https://doi.org/10.1095/biolreprod.116.139832
42. Sadri-Ardekani H, Mizrak SC, van Daalen SK, Korver CM, Roepers-Gajadien HL, Koruji M, Hovingh S, de Reijke TM, de la Rosette JJ, van der Veen F, de Rooij DG, Repping S, van Pelt AM (2009) Propagation of human spermatogonial stem cells in vitro. JAMA 302(19):2127–2134. https://doi.org/10.1001/jama.2009.1689
43. He Z, Kokkinaki M, Jiang J, Dobrinski I, Dym M (2010) Isolation, characterization, and culture of human spermatogonia. Biol Reprod 82(2):363–372. https://doi.org/10.1095/biolreprod.109.078550
44. Kokkinaki M, Djourabtchi A, Golestaneh N (2011) Long-term culture of human SSEA-4 positive Spermatogonial Stem Cells (SSCs). J Stem Cell Res Ther 2(2). https://doi.org/10.4172/2157-7633.S2-003
45. Sadri-Ardekani H, Akhondi MA, van der Veen F, Repping S, van Pelt AM (2011) In vitro propagation of human prepubertal spermatogonial stem cells. JAMA 305(23):2416–2418. https://doi.org/10.1001/jama.2011.791
46. Hou JM, Niu MH, Liu LH, Zhu ZJ, Wang XB, Sun M, Yuan QQ, Yang S, Zeng WX, Liu Y, Li Z, He ZP (2015) Establishment and characterization of human germline stem cell line with unlimited proliferation potentials and no tumor formation. Sci Rep-Uk 5. doi:ARTN 16922. https://doi.org/10.1038/srep16922
47. Kossack N, Terwort N, Wistuba J, Ehmcke J, Schlatt S, Scholer H, Kliesch S, Gromoll J (2013) A combined approach facilitates the reliable detection of human spermatogonia in vitro. Hum Reprod 28(11):3012–3025. https://doi.org/10.1093/humrep/det336
48. Langenstroth D, Kossack N, Westernstroer B, Wistuba J, Behr R, Gromoll J, Schlatt S (2014) Separation of somatic and germ cells is required to establish primate spermatogonial cultures. Hum Reprod 29(9):2018–2031. https://doi.org/10.1093/humrep/deu157
49. Zheng Y, Thomas A, Schmidt CM, Dann CT (2014) Quantitative detection of human spermatogonia for optimization of spermatogonial stem cell culture. Hum Reprod 29(11):

2497–2511. https://doi.org/10.1093/humrep/deu232

50. Medrano JV, Rombaut C, Simon C, Pellicer A, Goossens E (2016) Human spermatogonial stem cells display limited proliferation in vitro under mouse spermatogonial stem cell culture conditions. Fertil Steril 106(6):1539–1549. e1538. https://doi.org/10.1016/j.fertnstert.2016.07.1065
51. Li L, Wang M, Wang M, Wu X, Geng L, Xue Y, Wei X, Jia Y, Wu X (2016) A long non-coding RNA interacts with Gfra1 and maintains survival of mouse spermatogonial stem cells. Cell Death Dis 7:e2140. https://doi.org/10.1038/cddis.2016.24
52. Wang M, Guo Y, Wang M, Zhou T, Xue Y, Du G, Wei X, Wang J, Qi L, Zhang H, Li L, Ye L, Guo X, Wu X (2017) The glial cell-derived neurotrophic factor (GDNF)-responsive phosphoprotein landscape identifies raptor phosphorylation required for Spermatogonial progenitor cell proliferation. Mol Cell Proteomics 16(6):982–997. https://doi.org/10.1074/mcp.M116.065797
53. Wang M, Yu L, Wang S, Yang F, Wang M, Li L, Wu X (2020) LIN28A binds to meiotic gene transcripts and modulates their translation in male germ cells. J Cell Sci 133(12). https://doi.org/10.1242/jcs.242701
54. Du G, Wang X, Luo M, Xu W, Zhou T, Wang M, Yu L, Li L, Cai L, Wang PJ, Zhong Li J, Oatley JM, Wu X (2020) mRBPome capture identifies the RNA-binding protein TRIM71, an essential regulator of spermatogonial differentiation. Development 147(8). https://doi.org/10.1242/dev.184655
55. Kubota H, Avarbock MR, Schmidt JA, Brinster RL (2009) Spermatogonial stem cells derived from infertile Wv/Wv mice self-renew in vitro and generate progeny following transplantation. Biol Reprod 81(2):293–301. https://doi.org/10.1095/biolreprod.109.075960
56. Helsel AR, Oatley MJ, Oatley JM (2017) Glycolysis-optimized conditions enhance maintenance of regenerative integrity in mouse Spermatogonial stem cells during long-term culture. Stem Cell Reports 8(5): 1430–1441. https://doi.org/10.1016/j.stemcr.2017.03.004

Chapter 8

Reconstitution of Human Prospermatogonial Development from Human-Induced Pluripotent Stem Cells

Yasunari Seita, Young Sun Hwang, and Kotaro Sasaki

Abstract

There is a scarcity of information regarding the molecular mechanisms underlying human germ cell development due to limitations in obtaining the relevant materials. Reconstitution of human germ cell development from pluripotent stem cells in vitro would provide critical insight into the etiology of various reproductive conditions and disorders, including infertility.

Recently, we reported the in vitro reconstitution of human prospermatogonial development from human-induced pluripotent stem cells through human primordial germ cell (PGC)-like cells (hPGCLCs) using long-term cultured xenogeneic reconstituted testes. Here, we describe a method to generate M-prospermatogonia-like cells (MLCs) and T1-prospermatogonia-like cells (T1LCs), which closely resemble M- and T1-prospermatogonia present in second-trimester human fetal testes in vivo.

Key words Human primordial germ cell-like cells, In vitro gametogenesis, Prospermatogonia, T1-prospermatogonia-like cells

1 Introduction

Among germ cells, which include oocytes and spermatozoa and their unique and common progenitors, primordial germ cells (PGCs) are programmed for transmission of genetic and epigenetic information to subsequent generations. The germline is established as PGCs undergo a complex cascade of developmental processes that ultimately result in the formation of either spermatozoa or oocytes, depending on the sex-specific niche signals in the gonads. In mice, male-specific germ cell development begins when PGCs colonize the fetal testis and immediately differentiate into prospermatogonia (also known as gonocytes), which divide several times as M-prospermatogonia (M) and then become mitotically quiescent, as T1-prospermatogonia (T1) [1]. Immediately after birth, T1 resumes proliferation and differentiates into T2-prospermatogonia (T2), which are considered the immediate

Jon M. Oatley and Brian P. Hermann (eds.), *Spermatogonial Stem Cells: Methods and Protocols*, Methods in Molecular Biology, vol. 2656, https://doi.org/10.1007/978-1-0716-3139-3_8,

precursors of spermatogonia/spermatogonial stem cells, a founding population that establishes ongoing spermatogenesis [1].

Infertility is a major clinical problem affecting approximately 10% of couples worldwide. Among all infertility cases, approximately 50% are attributed to male infertility, of which a certain fraction is due to the permanent loss of functional spermatogenesis because of errors that occurred during fetal or postnatal testis development [2, 3]. The diagnosis and treatment of disorders originating from defects in germ cells are currently limited by the lack of understanding of human fetal germ cell development. Direct characterization of human fetal germ cells in vivo has been hampered by both ethical and technical constrains. A reliable in vitro reconstitution method that accurately recapitulates the fetal phases of human male germ cell development would enable scalable expansion, manipulation, and visualization of developing male germ cells, thereby allowing investigation of the molecular mechanisms underlying human male germ cell development.

Human-induced pluripotent stem cells (hiPSCs) are the somatic cell-derived pluripotent stem cells, which were initially independently established by the Yamanaka and Thomson laboratories [4, 5]. These cells represent versatile tools to potentially regenerate patient-derived cells, including germ cells [6–8]. In mice, both functional spermatozoa [9, 10] and oocytes [11, 12] have been successfully generated from iPSCs; however, completion of spermatogenesis requires transplantation of PGCLCs or GSCLCs (germ line stem cell-like cells) into mouse seminiferous tubules [9, 10]. For human cells, we have previously established in vitro methods to induce hiPSCs into human PGC-like cells (hPGCLCs) that resemble pre-migratory stage human PGCs [7] (Fig. 1). More recently, Yamashiro et al. successfully induced oogonia/gonocyte-like cells from hPGCLCs using xenogeneic reconstituted ovaries (xrOvaries) [8]. Using this method, hiPSC-derived hPGCLCs were subsequently cultured with female mouse gonadal somatic cells (mGSOs) for approximately 4 months to establish oogonia/gonocyte-like cells. We recently reported the analogous method in males, in which hiPSC-derived hPGCLCs were further induced to differentiate into M-prospermatogonia-like cells (MLCs) and T1-prospermatogonia-like cells (T1LCs) through prolonged culture of xenogeneic reconstituted testes (xrTestes) formed from a mixture of hPGCLCs and mouse fetal testicular somatic cells [13]. Transcriptomes of MLCs and T1LCs resemble those of M- and T1-prospermatogonia present in human fetal testes during the second trimester [8]. In this chapter, we describe the detailed protocols for the stepwise induction of MLCs and T1LCs from hiPSCs using xrTestis [13]. The endpoint determination of the cellular state is based on marker gene expression and transcriptomes, and functional determination that a germ cell state has been induced is not possible at this time.

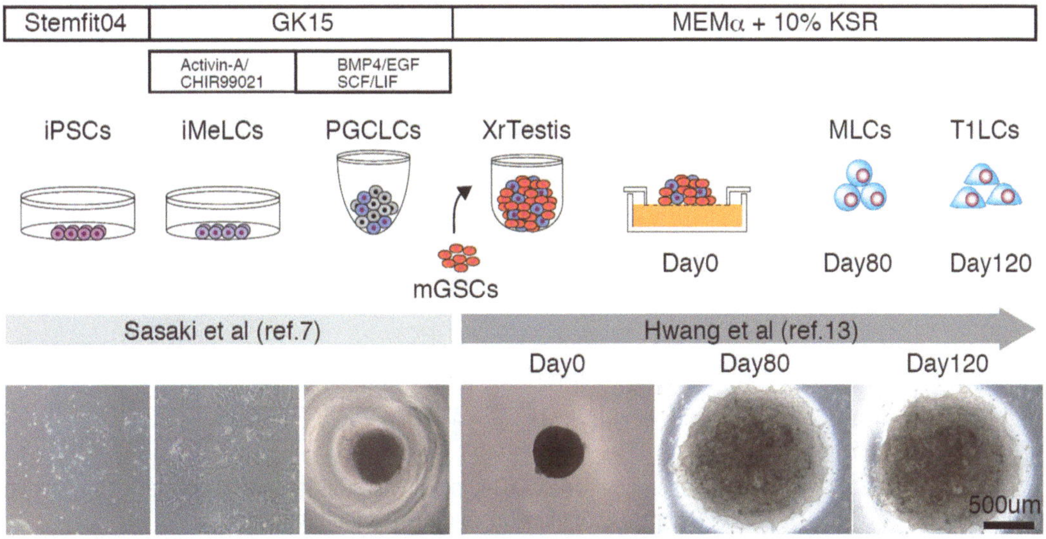

Fig. 1 Scheme for the reconstitution of human prospermatogonial development from human-induced pluripotent stem cells. Top row shows basal medium and cytokines for the xrTestes culture. Middle row shows scheme for the xrTestes culture. Bottom row shows images captured under inverted microscope for each stage

2 Materials

2.1 Animals

Mice (pregnant female ICR mice at 12.5 d post-coitum (d.p.c.) for the isolation of gonadal somatic cells.

2.2 Human iPSCs (hiPSCs)

Male iPSCs bearing fluorescent reporter alleles: *TFAP2C-2A-EGFP* (AG); *DDX4(hVH)-2A-tdTomato* (VT); *PIWIL4-2A-ECFP* (PC) (AGVTPC 9A13 hiPSCs) (Fig. 2) are used in this protocol (*see* **Notes 1** and **2**).

2.3 Equipment

1. Stereomicroscope for mouse embryo handling.
2. Inverted microscope.
3. CO_2 incubator. Standard setting: 37 °C, 5% CO_2, and 95% air, humidified by a pan containing sterile distilled water.
4. Isopropanol cryo chamber.
5. Liquid nitrogen storage tank.
6. Table-top centrifuge.
7. 37 °C water bath.
8. Magnetic cell separator.
9. Cell culture hood.
10. Vortex.
11. FACS sorter.

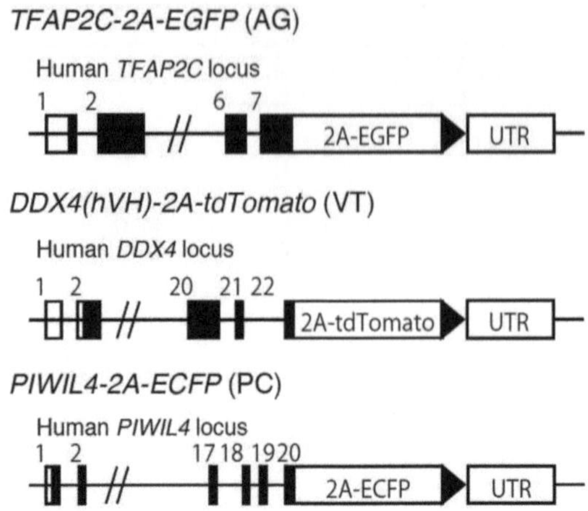

Fig. 2 Generation of AGVTPC-knock-in hiPSCs. Targeting schemes of the human *TFAP2C* (top), *DDX4* (middle), and *PIWIL4* (bottom) loci

2.4 Tools

1. Glass capillary.
2. Glass cutter.
3. Hemocytometer.
4. Micropipettes.
5. Sterile plastic tips.
6. 5 mL FACS tubes.
7. Petri dishes (35 mm, 60 mm, and 100 mm).
8. 1.5 and 15 mL centrifuge tubes.
9. 6-, 12-, and 24-well tissue culture plates.
10. CytoOne cell scraper (USA Scientific).
11. 96-well V-bottom, low-binding culture plates (Greiner Bio-one, #651970).
12. 96-well Lipidure-coated U-bottom plates (Amsbio, #AMS. LCP-A-U96).
13. Cell strainer, 70 μm (Fisher, #22363548).
14. MACS MS column (Miltenyi Biotec, #130-042-201).
15. Surgical scissors, microforceps, and surgical forceps.
16. Tungsten needles [14].
17. Collagen-coated PTFE Transwell membrane (Corning, #3496).
18. Cryomold (Sakura Finetek USA, #4565).

2.5 Reagents (See Note 3)

1. Collection buffer: DMEM +10% FBS.
2. Wash buffer: DMEM/F12 (Gibco, #11320033) + penicillin–streptomycin (Gibco, #15140122) + 0.1% bovine serum albumin (BSA, Gibco, #15260037).
3. Dissociation buffer: 1 mg/mL hyaluronidase type IV (Sigma, #H4272) + 5 U dispase (Corning, #354235) + 5 U DNase (Promega, #M6101) in wash buffer.
4. 0.05% trypsin/EDTA.
5. 0.1% trypsin/EDTA.
6. Phosphate-buffered saline (PBS).
7. MACS buffer: PBS + 2 mM EDTA +0.5% BSA.
8. Cell Banker Type I (Takara, ZNQ CB011).
9. iPS culture medium: StemFit04 + 50 ng/mL FGF2 (*see* **Note 4**).
10. iPS medium + Y: iPS medium +10 μM Y-27632 (Tocris Bioscience, #1254).
11. GK15: GMEM +15% (v/v) KnockOut Serum Replacement (KSR) + 2 mM L-glutamine +100 U/ml Penicillin/streptomycin + 0.1 mM β-mercaptoethanol (β-ME) + 1 mM sodium pyruvate + 0.1 mM nonessential amino acid.
12. GK15 + Y; GK15 + 10 μM Y-27632.
13. iMeLC medium: GK15 + 50 ng/mL activin-A (R&D systems, #338-AC) + 3 μM CHIR-99021 (Tocris Bioscience, #4423/10) + 10 μM Y-27632.
14. hPGCLC induction medium: GK15 + 200 ng/mL BMP4 + 10 ng/mL human LIF (Millipore, #LIF1005) + 100 ng/mL human SCF (R&D Systems, #255-SC-010) + 50 ng/mL human EGF + 10 μM Y-27632.
15. xrTestis medium: MEM-α + 10% KSR + 1× penicillin/streptomycin +55 μM β-mercaptoethanol.
16. 0.5 × Triple select: mix 1× Triple select and 0.5 mM EDTA/PBS(−) solution at a 1:1 ratio.
17. 0.1% human plasma fibronectin solution (Millipore, FC010).
18. Fetal bovine serum (FBS) (Corning).
19. FACS buffer: PBS + 0.1% BSA.
20. Anti-SSEA1 microbeads (Miltenyi Biotec, 130-094-530).
21. BV412-conjugated anti-human ITGA6 antibody (Biolegend, #313623).
22. APC-conjugated anti-human EPCAM antibody (Biolegend, #324207).
23. Cellotion (Amsbio).

24. PBS-T: 0.1% Tween-20 in PBS.
25. 4% PFA: 4% paraformaldehyde in PBS.
26. OCT compound.

3 Methods

3.1 Preparation of Mouse Gonadal Somatic Cells (mGSC)

1. Dissect out gonads and mesonephros from E12.5 male embryos (Fig. 3a).
2. Isolate gonads from mesonephros using a tungsten needle and carefully transfer them into collection buffer using wide-bore P1000 tips (Fig. 3b).
3. Add 500 μL PBS to each well of a 60 mm dish and transfer the gonads using wide-bore P1000 tips.
4. Homogenize the gonads using a tungsten needle (Fig. 3c).
5. Transfer the PBS-gonad homogenate to a 1.5 mL centrifuge tube.
6. Centrifuge the tube at 250 × g for 5 min at room temperature (RT) and discard the supernatant.
7. Add 1 mL dissociation buffer and incubate at 37 °C for 15 min (invert-mix the tube contents every 5 min).
8. Centrifuge the tube at 250 × g for 5 min at RT and discard the supernatant.
9. Add 1 mL PBS, centrifuge the tube at 250 × g for 5 min at RT, and discard the supernatant.
10. Add 500 μL 0.05% trypsin and incubate at 37 °C for 10 min (Fig. 3d).
11. Mix the samples using a P1000 micropipette until fully dissociated.
12. Add 500 μL collection buffer and filter the cell suspension through a cell strainer into a clean 15 mL centrifuge tube.
13. Rinse the tube with 1 mL collection buffer and filter through a cell strainer into the same tube.
14. Centrifuge the tube at 250 × g for 5 min at RT, discard the supernatant, and tap the tube to loosen the pellet.
15. Add 80 μL MACS buffer and 20 μL anti-SSEA1 microbeads to the tube and incubate on ice for 20 min (Fig. 3e).
16. Add 5 mL MACS buffer, centrifuge at 250 × g for 5 min at RT, discard the supernatant, and tap the tube to loosen the pellet.
17. Resuspend the pellet in 500 μL MACS buffer.
18. Attach a MACS MS column to a MiniMACS separator.

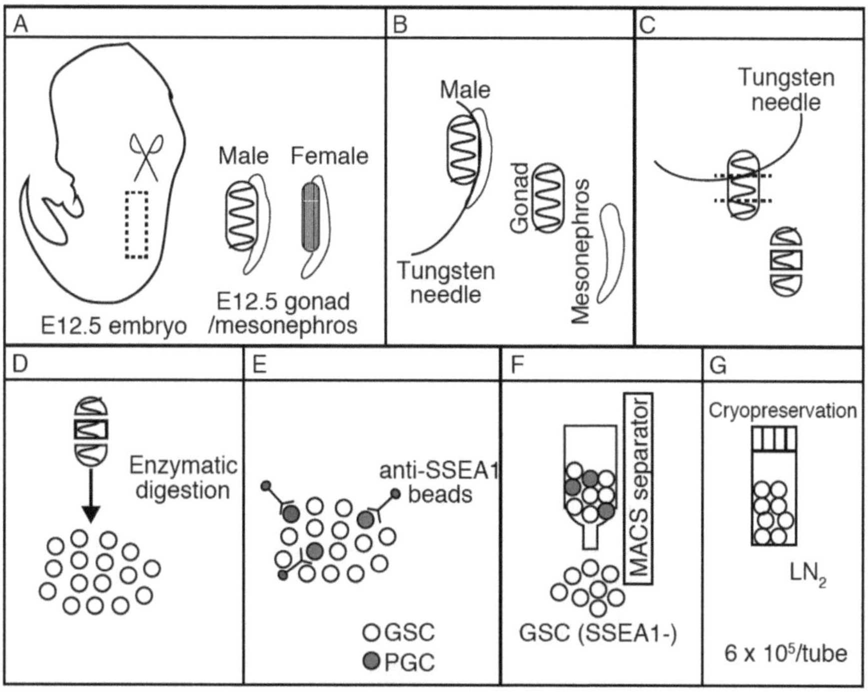

Fig. 3 mGSC collection from E12.5 embryos. (**a**) Dissection of male gonads and mesonephros from E12.5 mouse embryos. (**b**) Separation of the gonads and mesonephros using tungsten needle. (**c**) Cutting the gonads into several pieces with tungsten needle. (**d**) Enzymatic dissociation of the gonads. (**e**) Incubation of the dissociated gonadal cells with anti-SSEA1 antibodies conjugated with magnetic beads. (**f**) Purification of mGSC by magnetic separation. mGSCs should be in the flow through (**g**). Cryopreservation of mGSCs at 6×10^5 cells/tube

19. Wash the MACS column with 500 μL MACS buffer. Immediately proceed to **step 20**.
20. Add the cell suspension to the MACS MS column (Fig. 3f) and collect the flow-through.
21. Rinse the MACS MS column 3 times with 500 μL MACS buffer and collect the flow through.
22. Resuspend the cells and count.
23. Centrifuge the cells at 250 × g for 5 min at RT, discard the supernatant, and tap the tube to loosen the pellet.
24. Add Cell Banker Type I (200 μL/ 6×10^5 cells).
25. Dispense the cell suspension into cryotubes (200 μL/tube) (Fig. 3g).
26. Transfer the cryotubes to a freezing container and maintain at −80 °C overnight.
27. Transfer the samples to LN_2 and store until use.

3.2 hPGCLC Induction

Day 0

1. Coat a 12-well plate with human plasma fibronectin (10 μL fibronectin/600 μL PBS/well). Incubate at 37 °C for at least 2 h (overnight is acceptable).
2. Wash hiPSCs on a 6-well plate with PBS (1 mL/well) and discard the PBS.
3. Add 350 μL 0.5 × triple select solution and incubate at 37 °C for ~12 min (*see* **Note 5**).
4. Discard the triple select solution and wash with PBS (2 mL/well).
5. Discard the PBS, add 1 mL iPS medium + Y, and collect the cells using a cell scraper.
6. Gently mix 6 times using a P1000 micropipette and collect the cell suspension in a 15 mL tube.
7. Centrifuge at 160 × g for 5 min at RT, discard the supernatant, and tap the tube to loosen the pellet.
8. Resuspend the cells in 200 μL GK15 + Y and count.
9. Dispense the cells into a tube containing iMeLC induction medium to yield a 1.5–2.0 × 10^5/mL cell suspension.
10. Add 300 μL/well GK15 to the fibronectin-coated 24-well plate and remove all the medium.
11. Immediately, dispense 1 mL cell suspension in iMeLC induction medium into the fibronectin-coated 24-well plate.
12. Culture the cells at 37 °C for 31–48 h in a CO_2 incubator (*see* **Note 6**).

Day 2

1. Discard the medium and wash with 500 μL PBS.
2. Discard the PBS, add 200 μL 0.5 × triple select solution, and incubate at 37 °C for 12 min. Gently swirl the plate every 2 min.
3. Add 800 μL GK15 + Y to each well, mix 10 times using a P1000 micropipette, and collect the cell suspension in a 15 mL tube.
4. Centrifuge the tube at 160 × g for 5 min at RT, remove the supernatant, and tap the tube to loosen the pellet.
5. Resuspend the cells in 200 μL GK15 + Y and count.
6. Transfer the cells to a tube containing PGCLC induction medium to yield a concentration of 3.5 × 10^4 cells/mL.
7. Aliquot 100 μL into each well of a 96-well low-binding V-bottom plate.
8. Culture the cells at 37 °C for ~120 h in a CO_2 incubator.

Day 7

1. Harvest the aggregates into a 15 mL tube using a P1000 micropipette (*see* **Note 7**).
2. Centrifuge the tube at 160 × g for 3 min at RT and remove the supernatant.
3. Add 1 mL PBS and centrifuge the tube at 160 × g for 3 min at RT.
4. Discard the supernatant and add 150 μL pre-warmed 0.05% trypsin/EDTA.
5. Incubate the aggregates at 37 °C for 20 min with shaking at 450 rpm. Gentle tap every 3 min.
6. To quench the reaction, add 100 μL FBS and mix 15 times using a P200 micropipette.
7. Add 250 μL FACS buffer and filter the cells through a cell strainer.
8. For reporter hiPSC lines, proceed to sorting (Subheading 3.4), and for non-reporter lines, proceed to cell surface marker staining (Subheading 3.3).

3.3 Cell Surface Marker Staining

1. Centrifuge the tube at 160 × g for 5 min at RT and remove the supernatant.
2. Resuspend the cells in 50 μL FACS buffer.
3. Add 5 μL BV412-conjugated anti-human ITGA6 and APC-conjugated anti-human EPCAM.
4. Mix using a vortex for 2 s and incubate for 30 min at RT in the dark.
5. Add 1 mL FACS buffer and centrifuge the tube at 160 × g for 5 min at RT.
6. Remove the supernatant and resuspend the cells in 500 μL FACS buffer.

3.4 Sorting

1. Add 500 μL Cellotion to a 1.5 mL centrifuge tube.
2. Reporter lines: FACS-sort AG (TFAP2C-EGFP) single-positive population; non-reporter lines: FACS-sort ITGA6/EPCAM double-positive population in the tube (Prepared **step 1**).
3. Centrifuge the tube at 500 × g for 10 min at 4 °C and discard the supernatant.
4. Add 100 μL xrTestis medium + Y per 5 × 10^3 hPGCLCs.

3.5 xrTestis

Day 0

1. Thaw 1 vial (6×10^5) mGSCs in a 37 °C water bath for 25 s (6×10^4 cells per xrTestis).
2. Transfer the cells to a 15 mL tube containing 5 mL xrTestis medium + Y.
3. Centrifuge the tube at 160 × g for 5 min at RT and discard the supernatant.
4. Resuspend the cells in 1 mL xrTestis medium.
5. Add 5×10^4 hPGCLCs to the tube (**step 4**) (5×10^3 cells per xrTestis).
6. Mix 10 times using a P200 micropipette and collect the cell suspension in a 15 mL tube.
7. Aliquot 100 μL to each well of a 96-well low-binding V-bottom plate.
8. Culture the cells at 37 °C for 48 h in a CO_2 incubator.

Day 2

1. Add 400 μL xrTestis medium to 1 well of the 24-well plate and insert a collagen-coated PTFE Transwell membrane.
2. Add 1 mL PBS to the surrounding wells and equilibrate the plate at 37 °C for 1 h.
3. Transfer up to 3 aggregates formed in the 96-well plate into the collagen-coated PTFE Transwell membrane using glass capillaries (Fig. 3) (*see* **Note 7**).
4. Adjust the volume of medium in the wells to avoid the aggregates being completely soaked in medium (*see* **Note 8**).

Day 5

1. Remove 190 μL medium and replace with 200 μL xrTestis medium every 3 days.
2. Harvest xrTestes for FACS analysis/sorting or fixation after ~80 (MLCs) or ~120 days (T1LCs) of culture.

3.6 Harvest of xrTestis for FACS Analysis/Sorting

1. Wash the Transwell membranes with 1 mL PBS.
2. Clip out xrTestes with a minimum amount of membrane attached using sterile ophthalmic scissors.
3. Place xrTestes with attached membranes into a 15 mL tube containing 1 mL pre-warmed 0.1% trypsin/EDTA.
4. Incubate at 37 °C for 15 min, gentle tapping every 3 min.
5. To quench the reaction, add 100 μL FBS and mix 15 times using a P1000 micropipette (*see* **Note 9**).
6. Remove the membranes and add 5 mL PBS.
7. Centrifuge the tube at 160 × g for 5 min at RT and discard the supernatant.

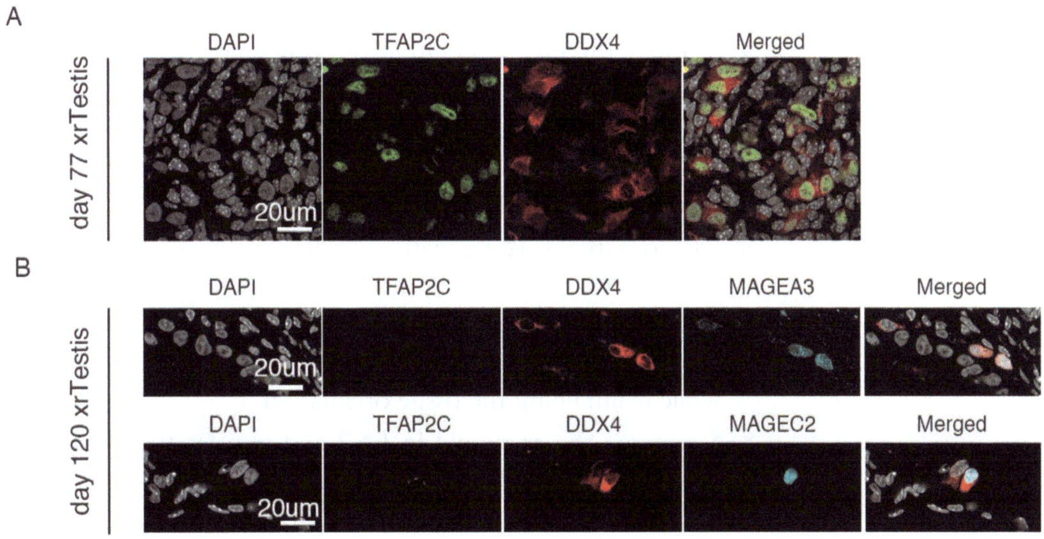

Fig. 4 Generation of MLCs and T1LCs. (**a**) IF images of day 77 xrTestis for their expression of key germ cell markers (green: TFAP2C-EGFP[AG]; red: DDX4-tdTomato[VT]) (**b**) IF images of day 120 xrTestis for their expression of key germ cell markers. Top (green: TFAP2C-EGFP [AG]; red: DDX4-tdTomato [VT]; cyan: MAGEA3) Bottom (green: TFAP2C-EGFP [AG]; red: DDX4-tdTomato [VT]; cyan: MAGEC2)

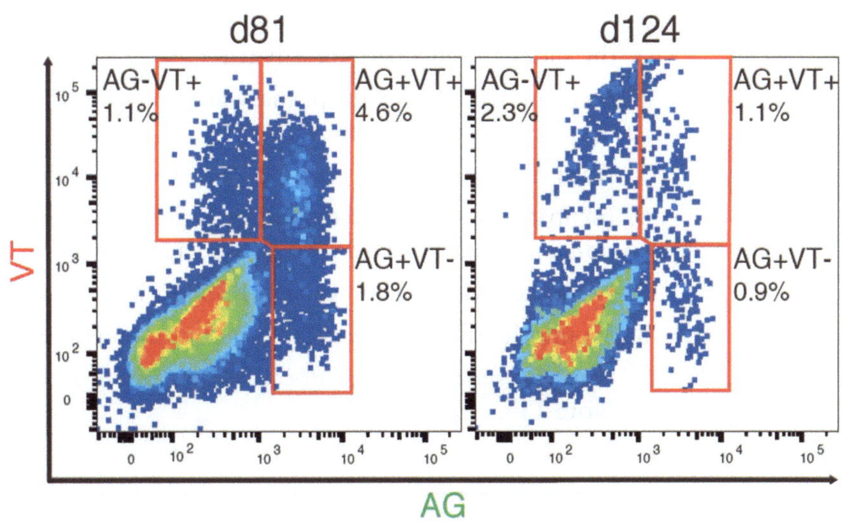

Fig. 5 FACS analysis for AG-VT expression of day 81 and day 124 xrTestes. AG: TFAP2C-EGFP, VT: DDX4-tdTomato

8. Add 250 μL FACS buffer and filter the cell suspension through a cell strainer.
9. Confirm the presence of three populations by FACS (if AG-VT reporter hiPSCs are used): $AG^{+}VT^{-}$, $AG^{+}VT^{+}$, and $AG^{-}VT^{+}$ cells (*see* **Note 10**) (Figs. 4a and 5). These cells can be sorted for downstream analyses.

3.7 Fixation of xrTestis for IF Analyses

1. Wash the Transwell membrane with 1 mL PBS.
2. Transfer the membrane to a well containing 1 mL prechilled (4 °C) 2% PFA and incubate on ice for 3 h.
3. Transfer the membrane to a well containing 1 mL prechilled PBS-T and incubate on ice for 10 min.
4. Repeat **steps 4** and **5** two times.
5. Clip out the xrTestes with a minimum amount of residual membrane.
6. Gently and thoroughly soak xrTestes in OCT compound.
7. Transfer xrTestes to a cryomold filled with OCT compound while maintaining the tissue orientation.
8. Submerge the cryomold in LN2 and transfer to −80 °C for long-term storage (up to 2 years).
9. Make a 10-μm-thick cryosection using a cryostat.
10. Place the section on a positively charged glass microscope slide.
11. Air-dry the slides and store at −80 °C until use.

3.8 Immunofluorescence Analysis

1. Air-dry the slides for 10 min.
2. Wash the slides three times with PBS.
3. Apply blocking solution (5% normal goat serum in PBS-T) to the slides and incubate for 1 h at RT.
4. Incubate the slides with primary antibodies in blocking solution for 2 h at RT (Table 1).
5. Wash the slides four times with PBS (15 min each).
6. Incubate the slides with secondary antibodies in blocking solution and 1 μg/mL DAPI for 50 min at RT.
7. Wash the slides four times with PBS (15 min each).
8. Mount the slides in Vectashield mounting medium.
9. Scan the slides using confocal laser scanning microscopy (*see* **Note 11**).

4 Notes

1. Fluorescence reporters are useful to visualize and sort MLCs and T1LCs but may not be essential for the differentiation and detection of MLCs/T1LCs.
2. We strongly recommend checking the karyotype by G-banding prior to use in experiments. The detailed protocol to establish AVGTPC hiPSCs has been described in previous publications [7, 13].

Table 1
List of antibodies

Antibodies	Concentration	Supplier	Cat.#
Antibodies for sorting			
BV 421™ conjugated anti-CD49f	1:10	Biolegend	313623
APC-conjugated anti-CD326	1:10	Biolegend	324207
Antibodies for MLCs			
Mouse anti-TFAP2C	1:200	Santa Cruz Biotechnology	sc-12762
Goat anti-NANOG	1:200	R&D Systems	AF1997
Rabbit anti-OCT3/4	1:200	Abcam	ab19857
Rabbit anti-DDX4	1:200	Abcam	ab13840
Goat anti-DDX4	1:200	R&D Systems	AF2030
Rabbit anti-DAZL	1:200	Abcam	ab34139
Antibodies for T1LCs			
Mouse anti-TFAP2C	1:200	Santa Cruz Biotechnology	sc-12762
Rabbit anti-DDX4	1:200	Abcam	ab13840
Goat anti-DDX4	1:200	R&D Systems	AF2030
Mouse anti-MAGEA3	1:200	Millipore	MABC1150
Rabbit anti-MAGEC2	1:500	Abcam	ab209667

3. We have validated that reagents from the listed manufacturer work in this protocol, but others may also be suitable and should be tested for efficacy.
4. Dispense 40 mL each and store at −80 °C. Add FGF2 after thawing and use within 1 week.
5. Longer incubation up to ~15 min is recommended for certain cell lines. Periodically check the cells under an inverted microscope and incubate until the cells become rounded and separate from each other but remain attached to the plate.
6. Optimal duration for iMeLC induction may be different among cell lines; therefore, the duration of iMeLC induction that allows maximal hPGCLC induction should be selected for subsequent experiments.
7. Place the aggregate in the center of the membrane. The edge of the membrane should be easily soaked in medium.
8. Remove the extra medium from the membrane; the medium needs to be at the same level as the bottom of the Transwell.
9. Cells will detach from the membranes.

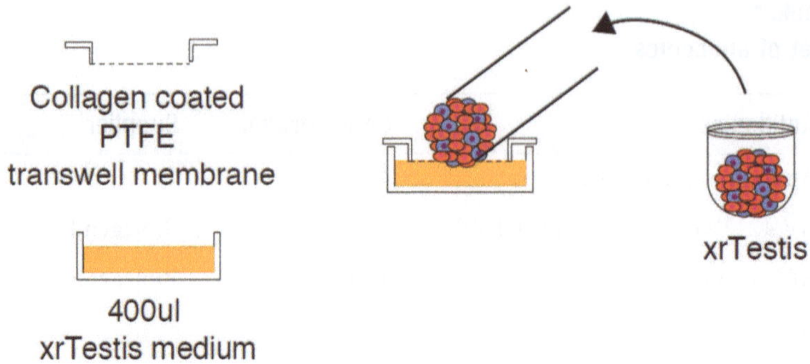

Fig. 6 Air–liquid interface culture of xrTestes. Preparation of a collagen-coated PTEF transwell membrane on 1 well of a 12-well plate by submerging with 400 ul of xrTestis medium. Subsequently, xrTestes are transferred to the Transwell membrane using a glass pipette

Table 2
Markers to detect MLCs and T1LCs

MLCs and T1LC detection antibodies				
	MAGEA3	**MAGEC2**	**DDX4**	**TFAP2C**
MLCs	+	–	++	++
T1LCs	+++	+++	+++	–

– negative
+ weakly positive
++ positive
+++ strongly positive

10. Day 80 xrTestes contain $AG^{+}VT^{+}$ and $AG^{-}VT^{+}$ cells, which usually consist of ~60% and ~10% of hPGCLC-derived germ cells, respectively (Figs. 5 and 6). Day 120 xrTestes contain more AG-VT+ cells. $AG^{+}VT^{+}$ and $AG^{-}VT^{+}$ cells are enriched for those representing MLCs or TC/T1LCs, respectively, based on transcriptome and IF analyses [13] . Also note that a fraction (~10–20%) of $AG^{+}VT^{+}$ and $AG^{-}VT^{+}$ cells express PC.
11. Markers should be used in combination to detect MLCs and T1LCs by IF. Please see Table 2.

References

1. McCarrey JR (2013) Toward a more precise and informative nomenclature describing fetal and neonatal male germ cells in rodents. Biol Reprod 89(2):47. https://doi.org/10.1095/biolreprod.113.110502
2. Hamada AJ, Esteves SC, Agarwal A (2013) A comprehensive review of genetics and genetic testing in azoospermia. Clinics (Sao Paulo) 68 Suppl 1:39–60. https://doi.org/10.6061/clinics/2013(sup01)06

3. Poongothai J, Gopenath TS, Manonayaki S (2009) Genetics of human male infertility. Singap Med J 50(4):336–347
4. Takahashi K, Tanabe K, Ohnuki M, Narita M, Ichisaka T, Tomoda K et al (2007) Induction of pluripotent stem cells from adult human fibroblasts by defined factors. Cell 131(5): 861–872. https://doi.org/10.1016/j.cell.2007.11.019
5. Yu J, Vodyanik MA, Smuga-Otto K, Antosiewicz-Bourget J, Frane JL, Tian S et al (2007) Induced pluripotent stem cell lines derived from human somatic cells. Science 318(5858):1917–1920. https://doi.org/10.1126/science.1151526
6. Irie N, Weinberger L, Tang WW, Kobayashi T, Viukov S, Manor YS et al (2015) SOX17 is a critical specifier of human primordial germ cell fate. Cell 160(1–2):253–268. https://doi.org/10.1016/j.cell.2014.12.013
7. Sasaki K, Yokobayashi S, Nakamura T, Okamoto I, Yabuta Y, Kurimoto K et al (2015) Robust in vitro induction of human germ cell fate from pluripotent stem cells. Cell Stem Cell 17(2):178–194. https://doi.org/10.1016/j.stem.2015.06.014
8. Yamashiro C, Sasaki K, Yabuta Y, Kojima Y, Nakamura T, Okamoto I et al (2018) Generation of human oogonia from induced pluripotent stem cells in vitro. Science 362(6412): 356–360. https://doi.org/10.1126/science.aat1674
9. Hayashi K, Ohta H, Kurimoto K, Aramaki S, Saitou M (2011) Reconstitution of the mouse germ cell specification pathway in culture by pluripotent stem cells. Cell 146(4):519–532. https://doi.org/10.1016/j.cell.2011.06.052
10. Ishikura Y, Yabuta Y, Ohta H, Hayashi K, Nakamura T, Okamoto I et al (2016) In vitro derivation and propagation of spermatogonial stem cell activity from mouse pluripotent stem cells. Cell Rep 17(10):2789–2804. https://doi.org/10.1016/j.celrep.2016.11.026
11. Hayashi K, Ogushi S, Kurimoto K, Shimamoto S, Ohta H, Saitou M (2012) Offspring from oocytes derived from in vitro primordial germ cell-like cells in mice. Science 338(6109):971–975. https://doi.org/10.1126/science.1226889
12. Hikabe O, Hamazaki N, Nagamatsu G, Obata Y, Hirao Y, Hamada N et al (2016) Reconstitution in vitro of the entire cycle of the mouse female germ line. Nature 539(7628):299–303. https://doi.org/10.1038/nature20104
13. Hwang YS, Suzuki S, Seita Y, Ito J, Sakata Y, Aso H et al (2020) Reconstitution of prospermatogonial specification in vitro from human induced pluripotent stem cells. Nat Commun 11(1):5656. https://doi.org/10.1038/s41467-020-19350-3
14. Brady J (1965) A simple technique for making very fine, durable dissecting needles by sharpening tungsten wire electrolytically. Bull World Health Organ 32(1):143–144

Chapter 9

RNA Interference as a Method of Gene Knockdown in Cultured Spermatogonia

Connor Cason and Tessa Lord

Abstract

Maintenance and self-renewal of the spermatogonial stem cell (SSC) population in the testis are dictated by the expression of a unique suite of genes. In manipulating gene expression through loss-of-function approaches, we can identify important regulatory mechanisms that dictate spermatogonial fate decisions. One such approach is RNA interference (RNAi), which uses natural cellular responses to small interfering RNAs to decrease levels of a targeted transcript. RNAi is performed in primary cultures of undifferentiated spermatogonia, and can be paired with techniques such as spermatogonial transplantation to assess the functional consequences of downregulated expression of the target gene on stem cell maintenance. This approach provides an alternative or complementary strategy to the generation of knockout mouse lines / cell lines. Here, we describe the methodology of RNAi in undifferentiated spermatogonia, and outline its inherent advantages and disadvantages over other technologies in the study of gene regulation in these cells.

Key words RNAi-mediated knockdown, siRNAs, Cultured undifferentiated spermatogonia, Genetic manipulation, Cell fate decisions

1 Introduction

1.1 Spermatogonial Stem Cell Biology

Spermatogenesis is fueled by a small pool of undifferentiated spermatogonia, which are charged with the production of millions of spermatozoa daily. The optimal activity of spermatogonial stem cells (SSCs) in particular is paramount to ensure that this proliferative process continues throughout the reproductive life span of the adult male. SSCs undergo self-renewal to maintain the stem cell reservoir in the testis, while also giving rise to transit-amplifying progenitor spermatogonia that are poised for differentiation. Every 8.6 days in the cycle of the seminiferous epithelium, a pulse of retinoic acid initiates the transition of a majority of progenitor spermatogonia into a differentiating pathway, terminally committing these cells to form spermatozoa [1].

There are several models that aim to characterize the dynamics of the SSC population and its relationship with progenitor and

Jon M. Oatley and Brian P. Hermann (eds.), *Spermatogonial Stem Cells: Methods and Protocols*, Methods in Molecular Biology, vol. 2656, https://doi.org/10.1007/978-1-0716-3139-3_9,

differentiating spermatogonial subtypes. In the mouse, these include the "traditional" and "revised" A_{single} models, as well as the fragmentation model: all of which have been described and compared in detail previously [2, 3]. Briefly, the A_{single} models propose that SSCs primarily exist as "single" cells (A_{single}) along the basement membrane of the seminiferous tubule, whereas progenitor and differentiating spermatogonia remain interconnected by cytoplasmic bridges to form pairs (A_{paired}), or chains of 4, 8 or 16 spermatogonia ($A_{aligned}$) [2, 4, 5]. These models posit that there is little capacity for reversion of progenitor spermatogonia back into a stem cell state, with the possible exception of the earliest transitory stages during nonsteady state spermatogenesis [6]. Contrastingly, the fragmentation model proposes that replenishment of the SSC pool is supplemented via the fragmentation of individual spermatogonia from A_{paired} and short $A_{aligned}$ chains [7–9], with this abscission event reigniting the capacity of the spermatogonium for self-renewal [7]. The conflicting nature of these models is, in part, attributed to discrepancies in the genetic markers that have been used to identify and trace the activities of the "SSC" pool. In the case of the revised A_{single} model, hypotheses were formed using Inhibitor of DNA binding 4 (ID4) as an SSC marker, with observations that high levels of ID4-eGFP expression in a transgenic reporter mouse line were restricted to A_{single} spermatogonia [10], complementing original morphology-based observations by Huckins and Oakberg [4, 5]. Alternatively, the fragmentation model is based on GFRA1 (GDNF Family Receptor Alpha 1) expression being exclusive to the SSC pool [7], although the specificity of this marker has been contested based on putative expression in both SSC and progenitor populations [2, 11]). Thus, inconsistencies in our understanding of the germline stem cell population and how we can identify and trace spermatogonial sub-populations highlight the requirement for continual investigation into gene signatures that represent SSC and progenitor pools, and the characterization of these genes in regulating spermatogonial fate decisions. With the advent of single-cell RNA sequencing (scRNA-seq) technology, the identification of SSC and progenitor gene signatures is becoming more feasible; however, confirmation of the role of these genes is reliant on downstream loss-of-function studies.

In this chapter, we discuss the utility of RNA interference (RNAi) techniques to explore gene function in undifferentiated spermatogonia. RNAi uses small, noncoding RNAs to disrupt the production of a specific protein for a short window of time via degradation of the target transcript or impairment of translation. This technique allows the functional response of spermatogonia to be assessed in the wake of reduced gene expression, without the complexity and time commitment associated with knockout techniques. Here we will detail the current methods and tools used for RNAi in undifferentiated spermatogonia, and the merits and limitations of this approach when compared to alternate loss-of-function techniques.

1.2 The Physiology of RNAi

Known as post-transcriptional gene silencing, quelling or co-suppression, the term RNAi was originally coined by Fire et al. (1998) as the phenomenon of sequence-specific gene interference that occurs after the introduction of double-stranded RNA (dsRNA) into a cell [12]. Fire et al. demonstrated in *C. Elegans*, that long dsRNAs disrupted gene expression to a higher degree than individual sense and antisense strands alone. When observed within a mammalian model, long dsRNAs were found to instigate a nonspecific shutdown of protein production due to initiation of a physiological reaction involving interferon synthesis [13]; however, this was overcome in a landmark study by Elbashir et al. [14] which showed that dsRNAs of 21–23 nucleotides in length (termed small interfering RNAs or siRNAs) could rapidly and specifically shut-down gene expression in mammalian cells.

Thus, RNAi can be harnessed as an experimental tool to instigate knockdown of expression of specific genes within the mammalian cell, including in spermatogonia, via the introduction of exogenous small RNAs. The success of this strategy is based on exploitation of enzymes that regulate endogenous RNAi pathways that co-ordinate the synthesis and function of microRNAs (miRNAs), piwi-interacting RNAs (piRNAs), and tRNA fragments (tRFs) in the mammalian cell. Endogenous RNAi is a multistep process which can be effectively broken into three main stages: (1) processing of long dsRNA molecules by the RNAse III enzyme DICER, (2) formation of the RNA-Induced Silencing complex (RISC), and (3) gene silencing guided by sequence-specific base pairing that stimulates endonuclease activity / transcript degradation, or alternatively, silencing by translational repression when base pair complementarity with the seed region is low.

The activity of endogenous RNAi pathways within germ cells, and the requirement of these pathways for normal spermatogenesis, has been demonstrated previously via the production of DICER conditional knockout mouse lines. In these mouse models, male infertility ensues following DICER ablation in early stage germ cells, i.e., in prospermatogonia [15] or spermatogonia [16]. Additionally, a number of miRNAs have been shown to be involved in regulating SSC self-renewal [17, 18], with inhibition of their function resulting in diminishment of the SSC population, as assessed by spermatogonial transplantation experiments. These studies highlight the activity of endogenous RNAi enzymes in the spermatogonial population, and thus, the amenability of these cells to the introduction of exogenous small RNAs (siRNAs or short hairpin RNAs (shRNAs)) for experimental purposes. Below, we will describe a methodology for RNAi in undifferentiated cultures of spermatogonia that can be used to achieve optimal knockdown of a gene of interest, and will explore potential analyses postknockdown to elucidate effects on the SSC population.

2 Materials

2.1 siRNA or shRNA Selection

RNAi has been successfully performed in undifferentiated spermatogonia using siRNAs [19–24] and shRNA vectors [25, 26], and selection between these two molecules determines the downstream methodology. RNAi using shRNA has the benefit of creating long-lasting knockdown effects, as the shRNA integrates into the endogenous miRNA pathway of the cell and will be continuously synthesized. As the efficiency of plasmid transfection using lipid-mediated delivery in spermatogonia is less than 10% [27], RNAi using shRNA is generally achieved using viral transduction [25, 26], which is covered in detail in Chapter 10 and will not be discussed further here. As an alternative to shRNA, siRNAs can be utilized to elicit short-term knockdown of target transcripts within the cell. Unlike shRNAs, siRNAs exert their activity independent of DICER, and instead are directly incorporated into the RISC complex upon their entry into the cell. The benefit of siRNA-mediated RNAi is that these 19–25 nucleotide duplexes are amenable to lipid-mediated delivery, which produces a rapid and cost-effective result. Although short acting, with a majority of siRNA duplexes being destroyed by 48 h post-transfection [28], several studies have demonstrated the utility of performing multiple siRNA transfections over time within cultures of undifferentiated spermatogonia (for example, once per week over a 3 week period, as described in Oatley et al. [29]), to instigate a more severe phenotype. Because of its short-term effects, higher concentrations of siRNA need to be introduced into the cell to achieve sufficient knockdown of the target transcript when compared to shRNA. As such, siRNA experiments in particular have the potential to instigate more pronounced off-target effects [28], a limitation that will be discussed in detail shortly.

In selecting siRNAs for transfection, researchers are provided with the option of single siRNAs that target one specific sequence within the transcript, or alternatively, a "pool" of siRNAs, in which multiple duplexes targeting different sequences within the same transcript are present in equimolar quantities. Validated siRNAs targeted to >19,000 mouse transcripts are available in both of these formats from reputable companies such as Horizon Discoveries. Successful knockdown of gene expression in spermatogonia has been achieved using single siRNAs [21, 29] and pooled siRNAs [30–32]. However, recent research has suggested that the likelihood of successful knockdown of the target transcript is significantly increased with the use of pooled siRNAs, which have been shown to exert an increased hit frequency and generate a more severe phenotype than individual siRNA duplexes [33]. Finally, in addition to selecting siRNA(s) to target the transcript of interest, a nontargeting or "scrambled" siRNA / siRNA pool is also required to act as a negative control in RNAi experiments, providing a baseline to exclude the influence of the transfection process itself on the cell in downstream analyses. Again, control siRNAs are available for purchase from companies such as Horizon Discovery.

Given the capacity to order prevalidated siRNAs from reputable companies, we have not focused on the methodology of designing siRNAs here. However, parameters that maximize the likelihood of successful knockdown by siRNA include: (1) homology of the siRNA to the mRNA target with minimal homology to unintended targets, (2) ensuring that siRNA homology is to the coding region of the gene, (3) keeping G-C content of the siRNA duplex between 36% and 52%, (4) including three or more A/U bases in the final five bases of the 5′ antisense end to lower internal stability and favor RISC formation, and (5) ensuring that there are no inverted repeats in the structure that could cause self-annealing (described in detail in Reynolds et al. [34] and Naito et al. [35]).

2.2 Reagents

1. Nontargeting siRNA / siRNA pool, and targeted siRNA / siRNA pool (e.g., Horizon Discoveries).
2. OptiMEM.
3. Lipofectamine 3000.
4. HBSS.
5. Trypsin-EDTA (0.25%).
6. Fetal Bovine Serum (FBS).
7. Mouse serum-free culture medium: 0.2% Bovine serum albumin (BSA), 5 μg/mL insulin, 10 μg/mL iron-saturated transferrin, 7.6 μeq/L free fatty acids, 3×10^{-8} M H_2SeO_3, 50 μM 2-mercaptoethanol, 10 mM HEPES, 60 μM putrescine, 2 mM glutamine, and penicillin/streptomycin in minimum essential medium alpha (*see* **Note 1**).
8. Recombinant human GDNF and recombinant human bFGF (FGF2) protein.
9. Fluorescent oligo "siGLO" (Optional, for protocol optimization).

3 Methods

3.1 Preparation of Spermatogonia for RNAi

RNAi is performed using undifferentiated spermatogonia from primary culture. Cultures should be established and maintained as described in Chapter 7. In order to retrieve pure populations of spermatogonia for transfection, colonies must be dislodged from

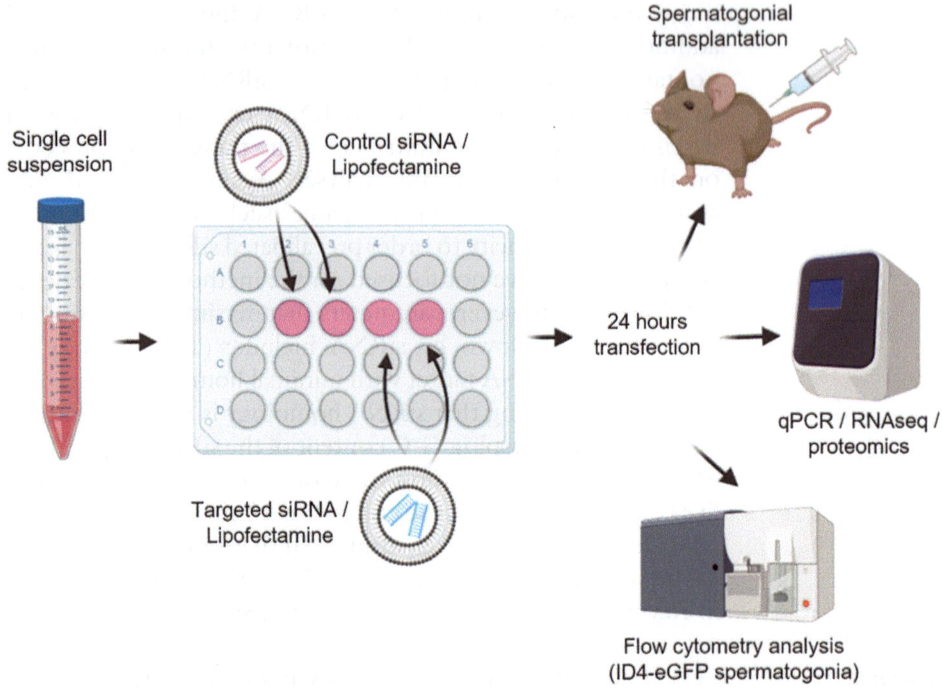

Fig. 1 Overview of RNAi in undifferentiated spermatogonia. Undifferentiated spermatogonia are dislodged from feeder cells in primary culture, and are trypsinized to create a single-cell suspension. Cells are transfected in feeder-free conditions with the selected targeting siRNA, alongside a nontargeting control, using Lipofectamine. From 16–24 h post-transfection, cells can be subjected to RNA / protein extraction to assess knockdown efficiency or dysregulated gene expression. Alternatively, alterations to SSC content can be assessed by performing spermatogonial transplantation using both control and targeted cell populations, or, if spermatogonia have been taken from a transgenic reporter mouse line (i.e., ID4-eGFP), levels of reporter expression can be assessed via flow cytometry. (This figure was created in part using BioRender.com)

the feeder cell layer via gentle pipetting, then digested to form a single-cell suspension using trypsin (represented diagrammatically in Fig. 1).

1. For cultures in a 24-well plate format (in 1 mL of culture media): an additional 1 mL of HBSS is added to each well. To dislodge spermatogonia from the feeder cell layer, a P1000 is used to gently pipette around the entire area of the well, being sure not to scrape the bottom / disrupt the feeder cells.
2. Once all spermatogonia have been dislodged from the feeder layer (check under microscope), cells are transferred into a 15- or 50 mL conical tube.
3. After cell collection, each well of the culture dish is washed with 1 mL HBSS, and this volume of HBSS is transferred to the cell suspension tube.

4. To create a single-cell suspension, one volume of trypsin-EDTA is added to the cells, followed by a 5 min incubation in a 37 °C water bath.
5. To quench trypsin activity, the tube is removed from the water bath and FBS is added at 10% of the total volume. A serological pipette is then used to break up any remaining clumps.
6. Cells are centrifuged at 600 g for 7 min.
7. Supernatant is removed and cells are washed in mouse serum-free culture media. If cells were collected from 5 wells in **step 1**, then they will be washed in 5 mL of media here, etc.
8. Cells are counted using a hemocytometer, then centrifuged at 600 g for 7 min.
9. Cells are resuspended in mouse serum-free culture media (no growth factor supplementation required for this step), and transferred to an empty 24-well plate (i.e., on plastic—no feeder cells) at an equal cell concentration per well. For transfection, 100,000–200,000 cells per well is recommended, in 1 mL of mouse serum-free culture media (*see* **Note 2**).
10. Cells are then placed in an incubator (37 °C, 5% CO_2) until transfection reagents have been prepared.

3.2 Preparation of Transfection Reagents, Transfection, and Collection of Spermatogonia for Downstream Analyses

Efficient transfection of spermatogonia can be achieved using lipid-mediated delivery of siRNAs. We have achieved a transfection efficiency of >95% using Lipofectamine 3000, as assessed by cellular uptake of the fluorescent oligo "siGLO" (Horizon Discovery) (Fig. 2, replicated from Lord et al. [36], with permission).

1. For *each well* of culture intended for transfection, the following reagents are prepared:
   (a) siRNA solutions: 100 μL OptiMEM +75 pmol siRNA (we recommend keeping 50 pmol/μL stock solutions of siRNA, meaning that 1.5 μL of siRNA is utilized for each well to be transfected. (Separate tubes will be required for control and targeted siRNA dilutions).
   (b) Lipofectamine solution: 100 μL OptiMEM +2 μL Lipofectamine 3000.
2. Each solution is mixed gently and incubated at room temperature for 5 min.
3. 100 μL of Lipofectamine solution (b) is added to every 100 μL of siRNA solution (a). This "transfection mixture" is incubated at room temperature for 20 min.
4. 200 μL of transfection mixture is added to each culture well containing isolated spermatogonia.

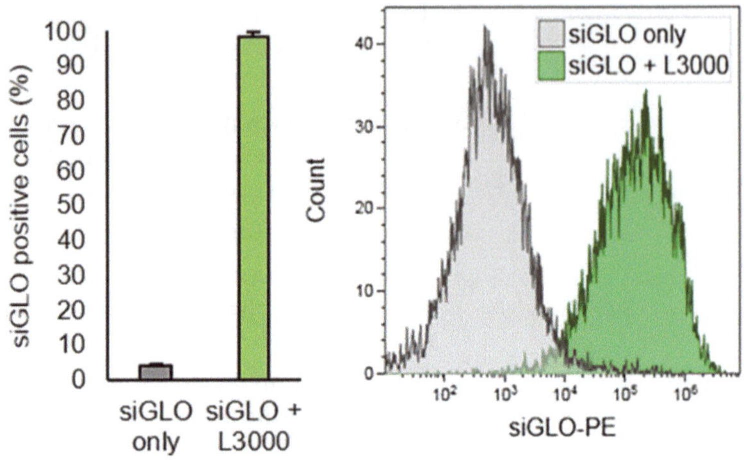

Fig. 2 Transfection efficiency of undifferentiated spermatogonia using Lipofectamine 3000. Over 95% of undifferentiated spermatogonia uptake the fluorescent oligo "siGLO" using the Lipofectamine 3000 transfection regime described in this chapter. (Figure taken with permission from Lord et al. [36])

5. Spermatogonia are transfected overnight (~16–24 h) at 37 °C, 5% CO_2.

The next day, transfected spermatogonia are collected as follows:

6. Cells are collected and placed in a 15 mL conical tube. Be aware that spermatogonia weakly adhere to plastic, so gentle pipetting is required to dislodge them from the culture plate, as described in Subheading 3.1, **step 1**.
7. Cells are centrifuged at 600 g for 7 min.
8. If assessing knockdown efficiency: supernatant is removed and the cell pellet is used for RNA extraction.
9. If further culture is desired: cells are washed in mouse serum-free culture media, then resuspended in 1 mL of culture media containing growth factors (GDNF, bFGF: *see* **Note 1**), for each well to be plated. Spermatogonia are then added onto a fresh feeder cell monolayer, and are maintained in standard culture conditions (Chapter 7).

3.3 Assessing Knockdown Efficiency

In order to assess knockdown efficiency, RNA extraction should be performed at ~16–24 h post-transfection, and a comparison of transcript levels generated between control and targeted cells using quantitative RT-PCR. A knockdown efficiency of >80% is considered ideal and should be achievable for most targets using the optimized transfection regime described above. A comparison of target protein levels can also be conducted using standard immunoblotting techniques. Depending on protein stability, a significant decrease in protein expression may be evident at 24 h postknockdown, however in some cases, will not be appreciable until 96 h [37]. For targets that are resistant to knockdown, troubleshooting could involve the confirmation of effective transfection using a fluorescent oligo such as siGLO, as well as performing knockdown of the desired target alongside transfection with siRNA(s) that have previously been shown to be effective (such as *Rb1* siRNA [19]). If knockdown efficiency remains low, it may be pertinent to try an alternate siRNA or pool of siRNAs for the mRNA target. Alternatively, to maximize transcript knockdown, transfection can be repeated a second time at 16–24 h to further stimulate the degradation of target mRNA, or, transfection can be repeated multiple times over several weeks to increase the severity of the phenotype, as was conducted in Oatley et al. [29].

3.4 Assessing Altered SSC Function in Response to RNAi

There are several approaches that can be taken to assess the biological effects of knockdown of target gene expression in the undifferentiated spermatogonia population (Fig. 1). In order to quantify effects on SSC maintenance specifically, spermatogonial transplantation is the "gold-standard" approach [11, 38] (protocol described in Chapter 5). Indeed, the quantification of donor-derived colonies of spermatogenesis following transplantation of RNAi-treated spermatogonia has formed the basis for defining self-renewal activity of a number of key genes, including *Id4* [29], *Rb1* [19], and *Etv5* [22]. Beyond this, the assessment of patchy versus complete colonies of spermatogenesis following spermatogonial transplantation [39] may provide an indication of defects in progenitor formation / expansion, or spermatogonial differentiation.

Given that spermatogonial transplantation is somewhat laborious and time-intensive, alternative approaches have been explored to characterize disruptions to SSC and progenitor populations following RNAi. One such approach has been to perform RNAi in primary cultures derived from the *Id4-eGfp* mouse line, and assess fluctuations in SSC and progenitor numbers based on levels of GFP fluorescence [30, 32, 36]. Using limiting dilution spermatogonial transplantation, as well as standard spermatogonial transplantation approaches, it has been demonstrated that ID4-eGFP^{Bright} spermatogonia represent a perceivably pure population of SSCs, while ID4-eGFP^{Dim} spermatogonia possess the characteristics of progenitors [10]. Thus, it stands to reason that monitoring the composition of these populations using flow cytometry provides a rapid readout that reflects disruptions to individual sub-populations in the undifferentiated spermatogonia pool. The validity of this approach has been demonstrated in Fon Tacer et al. [30], where knockdown of different *Mage-a* isoforms resulted in a 20–40% reduction in the eGFP^{Bright} population by day 6 post-knockdown, along with a corresponding increase in the eGFP^{Dim} population. This was validated by spermatogonial transplantation analyses that demonstrated a 40% reduction in SSC content following targeting of the *Mage-a* family for knockdown. Similarly, we have demonstrated that knockdown of the nucleosome remodeling protein CHD4 instigates a ~50% loss of ID4-eGFP^{Bright} spermatogonia in culture, which equates to a ~40% reduction in the number of donor-derived colonies of spermatogenesis following spermatogonial transplantation (Cafe et al. [32]). Importantly, these approaches have paved the way for large-scale siRNA screening in undifferentiated spermatogonia (Lord et al. [36]), by providing a high-throughput readout in the form of flow-cytometry analysis.

In addition to determining the effects of RNAi on the distribution of spermatogonia among different sub-populations, it is evident that one may also want to determine how knockdown of the target gene effects downstream gene expression (Fig. 1). This can of course be achieved via qRT-PCR analysis, comparing expression levels of genes-of-interest in knockdown spermatogonia and control spermatogonia that were transfected with nontargeting siRNA. Alternatively, "discovery" approaches such as RNAseq, scRNA-seq (Cafe et al. [32]), and proteomics can be used to produce a more complete picture of dysregulated gene expression.

3.5 In Vivo Approaches for RNAi in Spermatogonia

The focus of this book chapter has been RNAi in primary cultures of undifferentiated spermatogonia, with the potential for downstream in vivo analyses upon spermatogonial transplantation. It should be noted, however, that it is possible to perform RNAi on spermatogonia in vivo by microinjecting siRNA or shRNA into the seminiferous tubules via the rete or efferent duct. Unfortunately, these techniques have generally been shown to have low efficiency, and do not target the spermatogonia population, or even the germ cell population, with any specificity. Indeed, it has been demonstrated that microinjection of shRNAs into the adult mouse testis, followed by electroporation, primarily results in transfection of a sub-population of Sertoli cells (~40%), with less than 1% of germ cells being transfected [40]. Improved success rates for in vivo germ cell transfection have been reported by targeting the prepubertal testis with shRNA [41]; however, the transfection efficiency for spermatogonia remained under 10%. Finally, microinjection of siRNAs into the prepubertal mouse testis has also been reported, targeting spermatogonia, Sertoli cells, and stromal cells; however, the effects were transient and the phenotype difficult to assess [42]. Regardless of the efficiency of transfection techniques for RNAi of germ cells in vivo, the primary limitation of such experiments would be the inability to interpret the phenotype being produced due to the wide array of germ and somatic cells being targeted. Thus, the utility of this approach is limited.

3.6 Considerations When Adopting RNAi Technology: Off-Target Effects

One limitation of RNAi technology is the potential for "off-target" effects [43]. While the process of transfection itself can instigate gene expression changes within the cell that are independent of specific mRNA targeting, the adoption of a nontargeting siRNA control in comparative experiments should largely negate this issue. Thus, off-targets are most commonly the consequence of sequence complementarity between the siRNA and a nontarget transcript. Perhaps surprisingly, the overall identity of the siRNA is not the determining factor when considering propensity for off-targets. Instead, it is the homology of the siRNA seed region on the antisense strand with the 3′ untranslated region (UTR) of nontarget transcripts [44]. As such, siRNA duplexes that have low seed complement frequencies across the 3′ UTR transcriptome induce significantly less off-target effects than duplexes with more abundant seed complement frequencies [45].

With this knowledge, siRNA technologies have now been developed to significantly diminish off-target effects via modification of the seed region of the siRNA guide strand. This has included the generation of siRNAs that have low seed frequencies [45], but also siRNAs that have been chemically modified to further prevent pairing with off-target transcripts. For example, a 29-O-methyl ribosyl modification at position 2 (seed region) in the guide strand has been shown to achieve a 66% reduction in the silencing of off-target transcripts [46]. Finally, off-target effects can also be minimized by incorporating several duplexes (generally 3–4) that have homology to different regions of the mRNA transcript (i.e., an siRNA pool), rather than transfecting with a higher concentration of a single species of siRNA. Given that off-targets are unlikely to be identical between siRNA duplexes, this approach dilutes the incidence of gene silencing of unintended targets [47, 48]. The authors regularly use ON-TARGET plus SMART pool siRNA (Horizon Discovery), which adopts all of the aforementioned strategies / siRNA modifications to minimize off-target effects in RNAi experiments using undifferentiated spermatogonia.

3.7 RNAi Versus Knockout Techniques: Competing or Complementary Approaches?

Recent progress in the realm of gene editing technology, particularly the development of CRISPR Cas9 approaches, means it is easier and more cost-effective than ever before to knockout gene expression in cell lines in vitro, or generate knockout mouse lines. As such, it has been a subject of debate as to whether there is still a place for RNAi in conducting loss-of-function experiments [49]. Calls for RNAi to be replaced by knockout experiments have been previously ignited by publications that show a lack of harmony between knockdown and knockout phenotypes [50, 51], with initial criticisms suggesting that this must be attributed to off-target effects of RNAi reagents. However, more recent investigation in this area has demonstrated that the lack of phenocopy between these experimental models may instead be a consequence of "genetic compensation" caused by knockout approaches that are not instigated by RNAi. For example, in a zebra fish model, it has been demonstrated that transcriptional adaption in functionally related gene pathways can occur in knockout mutants (but not knockdown mutants), thus explaining alterations in the phenotype ([52], and reviewed in El-Brolosy and Stainier [53]). Additionally, it has been noted that several knockdown experiments that are used as an example of off-target effects causing a lack of phenocopy are lacking the appropriate controls, or have used concentrations of reagent that exceed the recommended dose [49].

In the context of comparing knockdown and knockout approaches in primary cultures of undifferentiated spermatogonia, RNAi has significant benefits in terms of cost- and time-effectiveness. As detailed in this chapter, RNAi experiments can be completed in 24 hours, provide downstream results in a matter of days, have a > 90% transfection efficiency, require only a small number of cells, and are amenable to large-scale high-throughput approaches. Contrastingly, although CRISPR Cas9 gene editing of undifferentiated spermatogonia in culture has been successfully carried out [54], the methodology associated is significantly more laborious. Specifically, following transfection of spermatogonia with a plasmid containing Cas9, a fluorescent reporter tag, and a guide RNA; the small number of spermatogonia that have been successfully transfected are isolated using flow sorting. Given that each targeted cell likely carries a different iteration of editing to the target gene, the cells cannot be pooled, so a single-cell plating methodology must be adopted. Thus, to establish robust spermatogonial cultures that are clonal for a single gene edit takes many weeks of in vitro expansion. Additionally, these clones will not necessarily experience ablation of expression of the target gene, for instance, in the circumstance of some in-frame deletions. It should also be noted that the potential for off-target effects is not restricted to RNAi, with CRISPR Cas9 technology having its own set of concerns with off-targets [55].

In comparing in vitro RNAi strategies with the generation of knockout mouse lines to decipher the role of a gene in regulating SSC function: we consider these to be complementary approaches, albeit not required to go hand-in-hand. In circumstances where the gene being investigated is not germline specific, or is an essential gene that causes lethality in a knockout model, the benefits of performing RNAi in SSCs in vitro are obvious. This is particularly the case when considering the expense and time required to generate conditional knockout mouse lines. Thus, results obtained in RNAi experiments paired with spermatogonial transplantation or transgenic reporter analysis could provide the impetus to progress with conditional knockout studies, or alternatively, negate the need for knockout studies altogether if results are interpreted with caution and the correct controls are used. To address concerns that phenotypes observed in the SSC population following RNAi may not phenocopy that of SSCs in a knockout mouse line, it should be noted that numerous instances exist where self-renewal defects caused by loss of gene function were firstly reported using siRNA studies, only to be validated in knockout mice. Such examples include: *Id4* [29], *Rb1* [19], *Etv5* [22], and *Bcl6b* [22], which, when targeted by RNAi, caused a significant reduction in SSC number as assessed by spermatogonial transplantation. Accordingly, gene-edited mouse models with varying degrees of infertility were generated [10, 22, 29, 56], with impairments to spermatogenesis being attributed to dysfunction in SSC activities. In alignment with this, one could argue that quantitative assessment of self-renewal defects in the SSC population is only possible using RNAi strategies paired with spermatogonia transplantation, with knockout mouse lines only being confirmatory to these findings.

To exclude RNAi experiments in favor of "knockout" techniques would significantly slow progress in the field, potentially creating a situation in which loss-of-function studies could only be performed by laboratories that have the considerable resources and finances required to generate knockout mouse lines. Thus, given the demonstrated capacity for RNAi experiments (particularly when paired with spermatogonial transplantation) to accurately predict the phenotype of SSCs in a number of knockout models, we believe that RNAi remains a valuable technique for germline stem cell biology, so long as the appropriate reagents and controls are adopted.

4 Notes

1. For prolonged culture post-RNAi, mouse serum-free culture media should be supplemented with 20 ng/ml GDNF and 1 ng/mL bFGF.
2. At least two wells will be required: one for targeted knockdown, and one for transfection with nontargeting siRNA (control). However, duplicate wells may be desired, depending on what downstream experiments are planned and thus the number of cells required.

Acknowledgments

T.L's research is supported by an NHMRC Ideas Grant (APP1181024), an ARC Discovery Early Career Research Award (DE220100032) and a Bob and Terry Kennedy HMRI Infertility Grant. The authors gratefully acknowledge the support of the University of Newcastle, Priority Research Center for Reproductive Science.

References

1. Tegelenbosch RAJ, de Rooij DG (1993) A quantitative study of spermatogonial multiplication and stem cell renewal in the C3H/101 Fl hybrid mouse. Mutat Res - Fundam Mol Mech Mutagen 290(2):193–200. https://doi.org/10.1016/0027-5107(93)90159-D
2. Lord T, Oatley JM (2017) A revised A_{single} model to explain stem cell dynamics in the mouse male germline. Reproduction 154: R55–R64. https://doi.org/10.1530/REP-17-0034
3. de Rooij DG (2017) The nature and dynamics of spermatogonial stem cells. Development 144(17):3022–3030. https://doi.org/10.1242/dev.146571
4. Oakberg EF (1971) Spermatogonial stem-cell renewal in the mouse. Anat Rec 169:515. https://doi.org/10.1002/ar.1091690305
5. Huckins C (1971) The spermatogonial stem cell population in adult rats. I. Their morphology, proliferation and maturation. Anat Rec 169:533. https://doi.org/10.1002/ar.1091690306
6. Zhang T, Oatley J, Bardwell VJ, Zarkower D (2016) DMRT1 is required for mouse spermatogonial stem cell maintenance and replenishment. PLoS Genet 12(9):e1006293. https://doi.org/10.1371/journal.pgen.1006293
7. Hara K, Nakagawa T, Enomoto H, Suzuki M, Yamamoto M, Simons Benjamin D, Yoshida S (2014) Mouse spermatogenic stem cells continually interconvert between equipotent singly isolated and syncytial states. Cell Stem Cell 14(5):658–672. https://doi.org/10.1016/j.stem.2014.01.019
8. Nakagawa T, Nabeshima Y, Yoshida S (2007) Functional identification of the actual and potential stem cell compartments in mouse spermatogenesis. Dev Cell 12:195. https://doi.org/10.1016/j.devcel.2007.01.002
9. Nakagawa T, Sharma M, Nabeshima Y, Braun RE, Yoshida S (2010) Functional hierarchy and reversibility within the murine spermatogenic stem cell compartment. Science 328(5974): 62–67. https://doi.org/10.1126/science.1182868
10. Helsel AR, Yang Q-E, Oatley MJ, Lord T, Sablitzky F, Oatley JM (2017) ID4 levels dictate the stem cell state in mouse spermatogonia. Development 144(4):624–634. https://doi.org/10.1242/dev.146928
11. Lord T, Oatley JM (2018) Functional assessment of spermatogonial stem cell purity in experimental cell populations. Stem Cell Res 29:129–133. https://doi.org/10.1016/j.scr.2018.03.016
12. Fire A, Xu S, Montgomery MK, Kostas SA, Driver SE, Mello CC (1998) Potent and specific genetic interference by double-stranded RNA in Caenorhabditis elegans. Nature 391(6669):806–811. https://doi.org/10.1038/35888
13. Stark GR, Kerr IM, Williams BR, Silverman RH, Schreiber RD (1998) How cells respond to interferons. Annu Rev Biochem 67:227–264. https://doi.org/10.1146/annurev.biochem.67.1.227
14. Elbashir SM, Harborth J, Lendeckel W, Yalcin A, Weber K, Tuschl T (2001) Duplexes of 21-nucleotide RNAs mediate RNA interference in cultured mammalian cells. Nature 411(6836):494–498. https://doi.org/10.1038/35078107
15. Romero Y, Meikar O, Papaioannou MD, Conne B, Grey C, Weier M, Pralong F, De Massy B, Kaessmann H, Vassalli J-D, Kotaja N, Nef S (2011) Dicer1 depletion in male germ cells leads to infertility due to cumulative meiotic and spermiogenic defects. PLoS One 6(10):e25241. https://doi.org/10.1371/journal.pone.0025241
16. Korhonen HM, Meikar O, Yadav RP, Papaioannou MD, Romero Y, Da Ros M, Herrera PL, Toppari J, Nef S, Kotaja N (2011) Dicer is required for haploid Male germ cell differentiation in mice. PLoS One 6(9): e24821. https://doi.org/10.1371/journal.pone.0024821
17. He Z, Jiang J, Kokkinaki M, Tang L, Zeng W, Gallicano I, Dobrinski I, Dym M (2013) MiRNA-20 and MiRNA-106a regulate spermatogonial stem cell renewal at the post-transcriptional level via targeting STAT3 and Ccnd1. Stem Cells 31(10). https://doi.org/10.1002/stem.1474
18. Niu Z, Goodyear SM, Rao S, Wu X, Tobias JW, Avarbock MR, Brinster RL (2011) MicroRNA-21 regulates the self-renewal of mouse spermatogonial stem cells. Proc Natl Acad Sci 108(31):12740–12745. https://doi.org/10.1073/pnas.1109987108
19. Yang Q-E, Gwost I, Oatley MJ, Oatley JM (2013) Retinoblastoma protein (RB1) controls fate determination in stem cells and progenitors of the mouse male germline. Biol Reprod 89(5):113, 111. https://doi.org/10.1095/biolreprod.113.113159
20. Wu X, Goodyear SM, Tobias JW, Avarbock MR, Brinster RL (2011) Spermatogonial stem

cell self-renewal requires ETV5-mediated downstream activation of Brachyury in mice. Biol Reprod 85(6):1114–1123. https://doi.org/10.1095/biolreprod.111.091793

21. He Z, Jiang J, Hofmann M, Dym M (2007) Gfra1 silencing in mouse Spermatogonial stem cells results in their differentiation via the inactivation of RET tyrosine kinase. Biol Reprod 77(4):723–733. https://doi.org/10.1095/biolreprod.107.062513
22. Oatley JM, Avarbock MR, Telaranta AI, Fearon DT, Brinster RL (2006) Identifying genes important for spermatogonial stem cell self-renewal and survival. Proc Natl Acad Sci U S A 103(25):9524–9529. https://doi.org/10.1073/pnas.0603332103
23. Zheng K, Wu X, Kaestner KH, Wang P (2009) The pluripotency factor LIN28 marks undifferentiated spermatogonia in mouse. BMC Dev Biol 9(1):1–11. https://doi.org/10.1186/1471-213x-9-38
24. Lovelace DL, Gao Z, Mutoji K, Song YC, Ruan J, Hermann BP (2016) The regulatory repertoire of PLZF and SALL4 in undifferentiated spermatogonia. Development 143(11):1893–1906. https://doi.org/10.1242/dev.132761
25. Yang Q-E, Kim D, Kaucher A, Oatley MJ, Oatley JM (2013) CXCL12-CXCR4 signaling is required for the maintenance of mouse spermatogonial stem cells. J Cell Sci 126(Pt 4):1009–1020. https://doi.org/10.1242/jcs.119826
26. Kaucher AV, Oatley MJ, Oatley JM (2012) NEUROG3 is a critical downstream effector for STAT3-regulated differentiation of mammalian stem and progenitor spermatogonia. Biol Reprod 86(5):164, 161. https://doi.org/10.1095/biolreprod.111.097386
27. Kanatsu-Shinohara M, Toyokuni S, Shinohara T (2005) Genetic selection of mouse Male germline stem cells in vitro: offspring from single stem Cells1. Biol Reprod 72(1):236–240. https://doi.org/10.1095/biolreprod.104.035659
28. Rao DD, Vorhies JS, Senzer N, Nemunaitis J (2009) siRNA vs. shRNA: similarities and differences. Adv Drug Deliv Rev 61(9):746–759. https://doi.org/10.1016/j.addr.2009.04.004
29. Oatley MJ, Kaucher AV, Racicot KE, Oatley JM (2011) Inhibitor of DNA binding 4 is expressed selectively by single spermatogonia in the male germline and regulates the self-renewal of spermatogonial stem cells in mice. Biol Reprod 85(2):347–356. https://doi.org/10.1095/biolreprod.111.091330
30. Fon Tacer K, Montoya MC, Oatley MJ, Lord T, Oatley JM, Klein J, Ravichandran R, Tillman H, Kim M, Connelly JP, Pruett-Miller SM, Bookout AL, Binshtock E, Kamiński MM, Potts PR (2019) MAGE cancer-testis antigens protect the mammalian germline under environmental stress. J Sci Adv 5(5):eaav4832. https://doi.org/10.1126/sciadv.aav4832
31. Legrand JMD, Chan A-L, La HM, Rossello FJ, Änkö M-L, Fuller-Pace FV, Hobbs RM (2019) DDX5 plays essential transcriptional and post-transcriptional roles in the maintenance and function of spermatogonia. Nat Commun 10(1):2278. https://doi.org/10.1038/s41467-019-09972-7
32. Cafe SL, Skerrett-Byrne DA, De Oliveira CS, Nixon B, Oatley MJ, Oatley JM, Lord T (2021) A regulatory role for CHD4 in maintenance of the spermatogonial stem cell pool. Stem Cell Reports 16(6):1555–1567. https://doi.org/10.1016/j.stemcr.2021.04.003
33. Parsons BD, Schindler A, Evans DH, Foley E (2010) A direct phenotypic comparison of siRNA pools and multiple individual duplexes in a functional assay. PLoS One 4(12):e8471. https://doi.org/10.1371/journal.pone.0008471
34. Reynolds A, Leake D, Boese Q, Scaringe S, Marshall WS, Khvorova A (2004) Rational siRNA design for RNA interference. Nat Biotechnol 22(3):326–330. https://doi.org/10.1038/nbt936
35. Naito Y, Yamada T, Ui-Tei K, Morishita S, Saigo K (2004) siDirect: highly effective, target-specific siRNA design software for mammalian RNA interference. Nucleic Acids Res 32:W124–129. https://doi.org/10.1093/nar/gkh442
36. Lord T, Law NC, Oatley MJ, Miao D, Du G, Oatley JM (2022) A novel high throughput screen to identify candidate molecular networks that regulate spermatogenic stem cell functions. Biol Reprod 106(6):1175–1190. https://doi.org/10.1093/biolre/ioac048
37. Curtis CD, Nardulli AM (2009) Using RNA interference to study protein function. Methods Mol Biol 505:187–204. https://doi.org/10.1007/978-1-60327-575-0_11
38. Helsel AR, Oatley JM (2017) Transplantation as a quantitative assay to study mammalian male germline stem cells. In: Buszczak M (ed) Germline Stem Cells. Springer New York, New York, pp 155–172. https://doi.org/10.1007/978-1-4939-4017-2_12
39. Helsel AR, Oatley MJ, Oatley JM (2017) Glycolysis-optimized conditions enhance

maintenance of regenerative integrity in mouse spermatogonial stem cells during long-term culture. Stem Cell Reports 8(5):1430–1441. https://doi.org/10.1016/j.stemcr.2017.03.004

40. González-González E, López-Casas PP, Del Mazo J (2008) Gene silencing by RNAi in mouse Sertoli cells. Reprod Biol Endocrinol 6:29–29. https://doi.org/10.1186/1477-7827-6-29

41. Shoji M, Chuma S, Yoshida K, Morita T, Nakatsuji N (2005) RNA interference during spermatogenesis in mice. Dev Biol 282(2): 524–534. https://doi.org/10.1016/j.ydbio.2005.03.030

42. Takizawa T, Ishikawa T, Kosuge T, Mizuguchi Y, Sato Y, Koji T, Araki Y, Takizawa T (2012) Gene suppression of mouse testis in vivo using small interfering RNA derived from plasmid vectors. Acta Histochem Cytochem 45(1):77–81. https://doi.org/10.1267/ahc.11024

43. Jackson AL, Bartz SR, Schelter J, Kobayashi SV, Burchard J, Mao M, Li B, Cavet G, Linsley PS (2003) Expression profiling reveals off-target gene regulation by RNAi. Nat Biotechnol 21(6):635–637. https://doi.org/10.1038/nbt831

44. Birmingham A, Anderson EM, Reynolds A, Ilsley-Tyree D, Leake D, Fedorov Y, Baskerville S, Maksimova E, Robinson K, Karpilow J, Marshall WS, Khvorova A (2006) 3′ UTR seed matches, but not overall identity, are associated with RNAi off-targets. Nat Methods 3(3):199–204. https://doi.org/10.1038/nmeth854

45. Anderson EM, Birmingham A, Baskerville S, Reynolds A, Maksimova E, Leake D, Fedorov Y, Karpilow J, Khvorova A (2008) Experimental validation of the importance of seed complement frequency to siRNA specificity. RNA 14(5):853–861. https://doi.org/10.1261/rna.704708

46. Jackson AL, Burchard J, Leake D, Reynolds A, Schelter J, Guo J, Johnson JM, Lim L, Karpilow J, Nichols K, Marshall W, Khvorova A, Linsley PS (2006) Position-specific chemical modification of siRNAs reduces "off-target" transcript silencing. RNA 12(7):1197–1205. https://doi.org/10.1261/rna.30706

47. Hannus M, Beitzinger M, Engelmann JC, Weickert M-T, Spang R, Hannus S, Meister G (2014) siPools: highly complex but accurately defined siRNA pools eliminate off-target effects. Nucleic Acids Res 42(12):8049–8061. https://doi.org/10.1093/nar/gku480

48. Kittler R, Surendranath V, Heninger A-K, Slabicki M, Theis M, Putz G, Franke K, Caldarelli A, Grabner H, Kozak K, Wagner J, Rees E, Korn B, Frenzel C, Sachse C, Sönnichsen B, Guo J, Schelter J, Burchard J, Linsley PS, Jackson AL, Habermann B, Buchholz F (2007) Genome-wide resources of endoribonuclease-prepared short interfering RNAs for specific loss-of-function studies. Nat Methods 4(4):337–344. https://doi.org/10.1038/nmeth1025

49. Zimmer AM, Pan YK, Chandrapalan T, Kwong RWM, Perry SF (2019) Loss-of-function approaches in comparative physiology: is there a future for knockdown experiments in the era of genome editing? J Exper Biol 222(7): jeb175737. https://doi.org/10.1242/jeb.175737

50. Kok FO, Shin M, Ni CW, Gupta A, Grosse AS, van Impel A, Kirchmaier BC, Peterson-Maduro J, Kourkoulis G, Male I, DeSantis DF, Sheppard-Tindell S, Ebarasi L, Betsholtz C, Schulte-Merker S, Wolfe SA, Lawson ND (2015) Reverse genetic screening reveals poor correlation between morpholino-induced and mutant phenotypes in zebrafish. Dev Cell 32(1):97–108. https://doi.org/10.1016/j.devcel.2014.11.018

51. Song H-W, Bettegowda A, Oliver D, Yan W, Phan MH, de Rooij DG, Corbett MA, Wilkinson MF (2015) shRNA off-target effects in vivo: impaired endogenous siRNA expression and spermatogenic defects. PLOS ONE 10(3):e0118549. doi:https://doi.org/10.1371/journal.pone.0118549

52. Rossi A, Kontarakis Z, Gerri C, Nolte H, Hölper S, Krüger M, Stainier DYR (2015) Genetic compensation induced by deleterious mutations but not gene knockdowns. Nature 524(7564):230–233. https://doi.org/10.1038/nature14580

53. El-Brolosy MA, Stainier DYR (2017) Genetic compensation: a phenomenon in search of mechanisms. PLoS Genet 13(7):e1006780. https://doi.org/10.1371/journal.pgen.1006780

54. N'Tumba-Byn T, Yamada M, Seandel M (2020) Loss of tyrosine kinase receptor Ephb2 impairs proliferation and stem cell activity of spermatogonia in culture†. Biol Reprod 102(4):950–962. https://doi.org/10.1093/biolre/ioz222

55. Zhang X-H, Tee LY, Wang X-G, Huang Q-S, Yang S-H (2015) Off-target effects in CRISPR/Cas9-mediated genome engineering. Mol Ther 4:e264. https://doi.org/10.1038/mtna.2015.37

56. Tyagi G, Carnes K, Morrow C, Kostereva NV, Ekman GC, Meling DD, Hostetler C, Griswold M, Murphy KM, Hess RA, Hofmann MC, Cooke PS (2009) Loss of Etv5 decreases proliferation and RET levels in neonatal mouse testicular germ cells and causes an abnormal first wave of spermatogenesis. Biol Reprod 81(2):258–266. https://doi.org/10.1095/biolreprod.108.075200

Chapter 10

Isolation of Undifferentiated Spermatogonia from Adult and Developing Mouse Testes

Vera van Melis, Lorena Roa-de la Cruz, and Brian P. Hermann

Abstract

In the mammalian testis, the mitotic complements of spermatogenic cells are spermatogonia, including spermatogonial stem cells (SSCs) which form the basis of life-long spermatogenesis and male fertility. Thus, investigating spermatogonia and subdivisions thereof is essential to increase our understanding of male germline development and infertility. This protocol describes the isolation of spermatogonia from both adult and developing [postnatal day 6 (P6)] mouse testes. Cell suspensions of the adult mouse testis from the *Id4-Egfp* transgenic mouse line are obtained through a two-step enzymatic digestion and are subjected to Percoll pre-enrichment before spermatogonia are isolated by selecting testis cells that are $CD9^{bright}$ and ID4-EGFP+ through FACS. For P6 mice, the testis is digested using trypsin-DNase, and spermatogonia are isolated by FACS selection of ID4-EGFP+ testis cells. In both cases, nearly pure populations of undifferentiated spermatogonia are obtained that can be further subdivided using additional parameters (e.g., EGFP intensity, cell surface protein immunostaining), and recovered for use in various downstream applications, such as biochemical analyses (e.g., transcriptome/epigenome), functional analyses by SSC transplantation or propagation in vitro.

Key words Spermatogonia, Spermatogenesis, Isolation, Fluorescence-activated cell sorting

1 Introduction

Spermatogenesis takes place within the seminiferous tubules of the testis and leads to the production of sperm [1]. The most primitive cells in the spermatogenic lineage are spermatogonial stem cells (SSCs), a subtype of undifferentiated spermatogonia, which are responsible for maintenance of spermatogenesis throughout the adult lifespan. SSCs alternately self-renew or initiate the process of spermatogenic differentiation by becoming progenitor spermatogonia that mitotically amplify and are poised to respond to a pulse of retinoic acid that will transform these cells into differentiating spermatogonia [2]. Since continual spermatogenesis is founded on undifferentiated spermatogonia (SSCs and progenitor spermatogonia), studying these cells is essential to understanding mammalian

Jon M. Oatley and Brian P. Hermann (eds.), *Spermatogonial Stem Cells: Methods and Protocols*, Methods in Molecular Biology, vol. 2656, https://doi.org/10.1007/978-1-0716-3139-3_10,

male fertility. However, isolating subsets of undifferentiated spermatogonia for biochemical and functional analyses can be difficult due to their relative rarity and the paucity of tools to precisely define the SSC and progenitor subsets [3, 4]. Recent development of a GFP-expressing transgenic mouse line, *Id4-Egfp*, has mitigated both of these challenges by facilitating isolation of SSC-enriched and progenitor-enriched populations of undifferentiated spermatogonia from the developing and adult mouse testis [5–7]. *Id4* is required for spermatogenesis [5, 7–9], transplantable SSC activity was exclusively observed from ID4-EGFP+ spermatogonia in vitro [5], and in vivo, ID4+ cells are largely A_{single} and A_{paired} spermatogonia [5, 7, 9]. More recently, transplantation experiments determined that subdivisions of ID4-EGFP+ spermatogonia exhibited a differing degree of SSC content. Specifically, GFP+ spermatogonia with relatively high levels of EGFP epifluorescence (ID4-$EGFP^{Bright}$) were found to be SSC-enriched, while EGFP+ spermatogonia with low levels of EGFP epifluorescence (ID4-$EGFP^{dim}$) were relatively depleted of SSCs (progenitor-enriched) in both the developing mouse testis (P6–8) [7] and adult mouse testis [6]. Therefore, this tool provides a ready opportunity to select enriched populations of undifferentiated spermatogonia from mice under a variety of conditions (genetic models, drug treatments), to study the stem cell population by transplantation and/or explore the biochemical and molecular underpinnings regulating cell fate.

This protocol describes the procedures to isolate undifferentiated spermatogonia from single-cell suspensions of adult or postnatal day 6 (P6) mouse testis using the *Id4-Egfp* transgenic mouse line [5]. In either case, enzymatic digestion produces a single-cell suspension that is then used for FACS. Isolating these cells from adult testes at high efficiency requires pre-enrichment for spermatogonia using density sedimentation through a Percoll cushion and CD9 immunostaining to eliminate pachytene spermatocyte contamination. Ultimately, these procedures produce highly pure populations of undifferentiated spermatogonia that can be used for various downstream applications, such as transcriptome analysis, cell transplantation or cell culture.

2 Materials

Reagents and materials are needed for isolation of undifferentiated spermatogonia both from adult and P6 testes, unless otherwise stated.

2.1 Mouse Testis Cell Isolation

1. Pipette aid.
2. 10 mL serological pipettes.
3. Micropipettes and tips.

4. 15 and 50 mL conical tubes.
5. Tube racks.
6. Waste container.
7. Timer.
8. 3 cm dish (only for P6 testis isolation).
9. 40 μm (for P6 testis isolation) or 70 μm (for adult testis isolation) basket cell strainer.
10. Hank's Balanced Salt Solution (HBSS).
11. Minimum Essential Medium (MEMα).
12. Fetal Bovine Serum (FBS, defined and heat-inactivated).
13. 0.25% Trypsin/EDTA.
14. Benchtop centrifuge.
15. Microscope (for P6 testis isolation).
16. Analytical balance.
17. Water or bead bath (for adult testis isolation) or culture incubator (for P6 testis isolation).
18. 70% EtOH for disinfection of surfaces.
19. 10 mL syringes and 0.22 μm syringe filters.
20. One pair of spring scissors (autoclaved).
21. Two pairs of fine forceps (autoclaved).
22. Two culture dishes or a well plate for storing the testes after isolation.
23. Dulbecco's Phosphate-Buffered Saline (DPBS, for P6 testis isolation).
24. Collagenase IV (for adult testis isolation).
25. DNase I from bovine pancreas.
26. Hemocytometer.
27. Trypan blue.
28. For isolation of adult testis cells, male mice at least 2 months of age (*see* **Note 1**).
29. For isolation of postnatal day 6 testis cells, 6-day old male mice (*see* **Note 2**).

2.2 Pre-Enrichment of Spermatogonia (Adult Testis Only)

1. Pipette aid.
2. 10 mL serological pipettes.
3. Micropipettes and tips.
4. 15 and 50 mL conical tubes.
5. Tube racks.
6. Waste container.

7. Hemocytometer.
8. Benchtop centrifuge.
9. Dulbecco's Phosphate-Buffered Saline (DPBS).
10. Fetal Bovine Serum (FBS, defined and heat-inactivated).
11. Percoll.
12. Trypan blue.

2.3 Preparation for FACS Isolation of ID4-EGFP+ Undifferentiated Spermatogonia

1. Micropipettes and tips.
2. 1.5 mL microcentrifuge tubes.
3. Tube racks.
4. Waste container.
5. 5 mL (12 × 75 mm) round-bottom plastic tubes suitable for FACS.
6. 35 μm strainer cap 5 mL (12 × 75 mm) round-bottom tubes suitable for FACS.
7. Timer.
8. Benchtop centrifuge.
9. Dulbecco's Phosphate-Buffered Saline (DPBS).
10. Fetal Bovine Serum (FBS, defined and heat-inactivated).
11. Biotin anti-CD9 antibody (Biolegend, for adult testis isolation).
12. Biotin rat IgG2a, κ isotype control antibody (Biolegend, for adult testis isolation).
13. Streptavidin-APC (Biolegend, for adult testis isolation).
14. Propidium Iodide (PI) viability stain.

2.4 FACS Analysis

1. Pipette aid.
2. 10 mL serological pipettes.
3. Micropipettes and tips.
4. Tubes for recovery of sorted cells (configuration FACS machine-specific).
5. Minimum Essential Medium (MEMα).
6. Fetal Bovine Serum (FBS, defined and heat-inactivated).
7. FACS machine.

3 Methods

Once the testes are isolated, all subsequent steps are performed in a biosafety cabinet (laminar flow hood) to maintain a sterile environment for downstream culture applications. If sterile cells are not needed, these steps can be performed on the benchtop. Investigators should always consult and follow their institutional biosafety guidelines.

3.1 Mouse Testis Cell Isolation

3.1.1 Preparation

1. Clean dissecting benchtop and/or microscope with 70% EtOH. Before placing anything in the biosafety cabinet, disinfect with 70% EtOH.
2. Prepare a solution of 10% FBS in MEMα and store in the biosafety cabinet prior to use.
3. Prepare a solution of 10% FBS in DPBS (for P6 testis isolation) and store in the biosafety cabinet prior to use.
4. Prepare a solution of 7 mg/mL DNase I in HBSS. For each adult testis, 0.5 mL DNase I solution is needed. For P6 testes from a typical litter containing 3–6 males, 1.5 mL DNase I solution is needed in total. Mix well and for sterile applications (e.g., culture or transplantation) filter sterilize the solution using a 10 mL syringe and 0.22 μm PES syringe filter (*see* **Note 3**). The solution can be kept at room temperature in the biosafety cabinet until use.
5. Prepare a solution of 1 mg/mL Collagenase IV in HBSS (1 mL Collagenase IV solution per testis, for adult testis isolation). For sterile applications (e.g., culture or transplantation), filter sterilize the solution using a 10 mL syringe and 0.22 μm PES syringe filter. The solution can be kept at room temperature in the biosafety cabinet until use.
6. Transfer **items 1–13** in Subheading 2.1 to the biosafety cabinet.
7. Ensure water or bead bath (for adult testis isolation) or culture incubator (for P6 testis isolation) is set to 37 °C.
8. Add 5 mL HBSS to the two culture dishes or wells in a well plate.
9. Obtain the euthanized animals (adult or P6) and record their information (ID, date of birth, strain, dam/sire).

3.1.2 Procedure for Testis Cell Isolation from Adult Mouse Testes

1. After obtaining the euthanized mice, dissect the testes (without epididymides) using spring scissors and fine forceps and place the testes in HBSS in the well plate or culture dish.
2. Determine and record the wet weight of each testis (x ≈ 100 mg).
3. Remove the seminiferous tubules from one testis by holding the tunica albuginea with forceps and using the closed scissor blades to push the tubules out of a small hole made in the tunica at the opposite pole of the testis. Place the tubules in HBSS in a second dish/well. Repeat for all testes.

4. Move the dish/plate with seminiferous tubules to the biosafety cabinet and transfer the tubules to a 15 mL or 50 mL conical tube using forceps.
5. Add the appropriate volume of Collagenase solution to the tubules (1 mL per testis) and incubate the samples at room temperature for 3 min (*see* **Note 4**). Agitate by gently rocking the tube.
6. Allow the tubules to sediment by gravity. Remove and discard the supernatant.
7. Add HBSS to rinse the tubules and remove interstitial cells. Rock the tube several times and again allow the tubules to sediment by gravity. Remove and discard the supernatant.
8. Add the appropriate volume of DNase I solution to the tubules (0.5 mL per testis).
9. Add 2 mL 0.25% Trypsin/EDTA per testis to this solution. Pipette up and down with a 10 mL pipette, the solution will become cloudy as the cells begin to disperse.
10. Incubate at 37 °C for 5–10 min (e.g., in a bead bath). After 5 min, triturate (pipette up and down) vigorously using a 10 mL pipette, taking care not to aspirate the suspension into the pipette filter or pipettor or splash the suspension out of the tube.
11. Stop the trypsin digestion by adding FBS to 10% of the total volume in the tube containing the cell suspension (e.g., for 10 mL cell suspension, add 1.1 mL FBS).
12. Repeat the trituration in **step 10**.
13. Pipette the cell suspension into a 50 mL conical tube through a 70 μm basket strainer (prewet with 1 mL of MEMα +10% FBS medium).
14. Wash the digestion tube with 2 mL MEMα and pipette into the cell strainer/conical tube (*see* **Note 5**).
15. Mix well by inverting the capped tube several times and remove a 10 μL aliquot of cell suspension for counting.
16. Centrifuge the cells for 10 min at 600 g in the conical tube.
17. During the centrifugation, count the cells and determine cell viability using Trypan blue and a hemocytometer.
18. Remove the supernatant and resuspend the cells in MEMα +10% FBS to a final concentration of roughly 5×10^6 cells/mL and store on ice (*see* **Note 6**).
19. Count the cells and determine cell viability. For the most accurate cell count, the final diluted concentration (with Trypan blue) should be ~1×10^6/mL (with a fivefold dilution factor for this particular procedure).

3.1.3 Procedure for Testis Cell Isolation from P6 Mouse Testes

1. After obtaining the euthanized animals, dissect the testes using spring scissors and fine forceps and place the testes in HBSS in the well plate or culture dish. Users may find that dissecting the testes is easier using a dissecting microscope.
2. Determine and record the wet weight of each testis ($x \approx 3.3$ mg).
3. Remove the seminiferous cords from one testis by grasping the tunica albuginea at one pole with fine forceps, poking a small hole in the opposite pole and using the closed spring scissors to push the cords out of the hole with the side of the scissor blades. Users may find that removing the cords is easier using a dissecting microscope. Place the cords in HBSS in a second dish/well.
4. Move the dish/plate with seminiferous cords to the biosafety cabinet and transfer the cords to a 3 cm culture dish tube using forceps.
5. Add 0.5 mL DNase I solution to the cords (for a typical litter containing 3–6 males).
6. Add 4.5 mL 0.25% Trypsin/EDTA and pipette up and down with a 1 mL micropipette with the tip placed firmly against the intersection of the side and base of the dish, triturate 10× to disperse the cords.
7. Incubate at 37 °C in a culture incubator for 3 min (*see* **Note 7**).
8. After 3 min, add an additional 0.5 mL DNase I, triturate 10×, and incubate at 37 °C for an additional 3 min.
9. Again add 0.5 mL DNase I, mix by pipetting up and down and incubate at 37 °C for 1 min.
10. Stop the trypsin digestion by adding FBS to 10% of the volume of cell suspension (e.g., for 6 mL of cell suspension, add 660 μL FBS).
11. Strain the cell suspension into a 50 mL conical tube through a 40 μm basket (prewet using approximately 1 mL of MEMα +10% FBS).
12. Wash the digestion plate with 2 mL MEMα and strain this as well (*see* **Note 5**).
13. Mix the strained cells well by inverting the capped tube several times and remove a 10 μL aliquot of cell suspension for counting.
14. Centrifuge the cells for 10 min at 600 g in the conical tube.
15. During the centrifugation, count the cells and determine cell viability using Trypan blue and a hemocytometer.
16. Remove the supernatant and resuspend the cells in DPBS +10% FBS to a final concentration of roughly 5×10^6 cells/mL and store on ice (*see* **Note 6**).

17. Count the cells and determine cell viability. For the most accurate cell count, the final diluted concentration (with Trypan blue) should be ~1×10^6/mL (with a fivefold dilution factor for this particular procedure).

3.2 Pre-Enrichment of Spermatogonia (from Adult Testes Only)

3.2.1 Preparation

1. Prepare a DPBS +10% FBS solution, keep on ice.
2. Prepare a solution of 30% Percoll in DPBS +10% FBS (*see* **Note 8**). Keep solution on ice prior to use.

3.2.2 Procedure

1. Pipette the 30% Percoll solution into conical tubes (2.5 mL for 15 mL conical tube, 5 mL for 50 mL conical tube, *see* **Note 9**).
2. Carefully layer the adult testis cell suspension on top of the 30% Percoll solution using the lowest pipettor speed setting at a rate of ~0.1–0.2 mL/s to avoid mixing of the cell suspension with the Percoll. The phenol red dye in the MEMα medium will assist with visualizing the interface between the Percoll and cell suspension.
3. Centrifuge the cells for 8 min at 600 g without braking.
4. After centrifugation, the spermatogonia will be present in the cell pellet. Aspirate and discard the medium, interface, and Percoll layers without disrupting the cell pellet. Disperse the cell pellet by vigorously flicking the tube and resuspend in ice-cold DPBS +10% FBS (assuming a recovery of ~15% of the starting cell number) to a target cell concentration of 5×10^6/mL.
5. Count the cells. Dilute cells appropriately to 1×10^6/mL for counting.

3.3 Preparation for FACS Isolation of ID4-EGFP+ Undifferentiated Spermatogonia

Cells isolated from P6 mouse testes can be directly prepared for sorting, but cells from adult mouse testes must first be stained with antibodies that recognize the CD9 cell surface antigen to facilitate differential selection of spermatogonia. In both cases, aliquots of cells from wild-type (nontransgenic) C57BL/6 mice or ID4-EGFP- littermates are used as controls. For each fluorescence channel used on the FACS machine, a single-color compensation control tube must be included to establish positive fluorescence and compensate for spectral overlap. See Tables 1 and 2 below for an overview of control and sort tubes used for isolation of spermatogonia from adult or P6 mice, respectively. When working with fluorescently labeled cells, the best practice is to keep the reagents and cells in the dark (e.g., turn off the lights in the biosafety cabinet, transfer in foil-covered container) to minimize photobleaching.

Table 1
Control and sort tubes used for adult ID4-EGFP+ FACS

Tube	Cell origin	Staining	Concentration	Cell number
Unstained control	C57BL/6 or ID4-EGFP- littermate	–	–	0.5–1 × 10^6
EGFP compensation control	ID4-EGFP+	–	–	0.5–1 × 10^6
APC compensation control	C57BL/6 or ID4-EGFP- littermate	*Primary*: Biotin anti-CD9 antibody (Biolegend) *Secondary*: Streptavidin-APC (Biolegend)	1 μg/10^6 cells 0.5 μg/10^6 cells	0.5–1 × 10^6
PI compensation control	C57BL/6 or ID4-EGFP- littermate	*Viability*: Propidium iodide	5 μL/10^6 cells	0.5–1 × 10^6
Biotin isotype control	ID4-EGFP+	*Primary*: Biotin rat IgG2a, κ isotype control antibody (Biolegend) *Secondary*: Streptavidin-APC (Biolegend)	1 μg/10^6 cells 0.5 μg/10^6 cells	0.5–1 × 10^6
Sorting	ID4-EGFP+	*Primary*: Biotin anti-CD9 antibody (Biolegend) *Secondary*: Streptavidin-APC (Biolegend) *Viability*: Propidium iodide	1 μg/10^6 cells 0.5 μg/10^6 cells 5 μL/10^6 cells	All remaining cells

Table 2
Control and sort tubes used for P6 ID4-EGFP+ FACS

Tube	Cell origin	Staining	Concentration	Cell number
Unstained control	C57BL/6 or ID4-EGFP- littermate	–	–	0.5–1 × 10^6
EGFP compensation control	ID4-EGFP+	–	–	0.5–1 × 10^6
PI compensation control	C57BL/6 or ID4-EGFP- littermate	*Viability*: Propidium iodide	5 μL/10^6 cells	0.5–1 × 10^6
Sorting	ID4-EGFP+	*Viability*: Propidium iodide	5 μL/10^6 cells	All remaining cells

3.3.1 Preparation

1. Prepare a DPBS +10% FBS solution and keep on ice.
2. Label 1.5 mL microcentrifuge tubes and 5 mL FACS tubes (with and without 35 μm strainer cap) and keep on ice.
3. Chill microcentrifuge to 4 °C.

3.3.2 Procedure for Adult Cell Staining and Preparation for FACS

1. Calculate the volume of cell suspension to be added to each tube according to the required cell number and cell suspension concentration. For control tubes, total volume will be 300 μL, for the sort tube, volume will be dependent on the number of cells available (*see* Table 1). Calculate the amount of antibody/dye needed for each tube.
2. Add the appropriate volume of DPBS +10% FBS to the 1.5 mL control tubes (300 μL DPBS +10% FBS − X μL cell suspension).
3. Add the appropriate volume of cell suspension to each tube.
4. Add the appropriate primary antibodies, flick to mix (3×), and incubate for 20 min on ice.
5. Centrifuge the cells for 5 min at 600 g at 4 °C. Remove and discard the supernatant.
6. Flick tubes to disrupt the cell pellet.
7. Wash the cells by adding ice-cold DPBS +10% FBS (equal volume to staining).
8. Repeat **steps 5**–7 once.
9. Add the streptavidin-APC to appropriate control and sort tubes for detecting the primary antibodies (Table 1), flick to mix (3×), and incubate for 20 min on ice (*see* **Note 10**).
10. Centrifuge the cells for 5 min at 600 g at 4 °C. Remove and discard the supernatant.
11. Flick tubes to disrupt the cell pellet.
12. Wash the cells by adding ice-cold DPBS +10% FBS (equal volume to staining).
13. Repeat **steps 10–12** once.
14. Stain for viability by adding PI at a concentration of 5 μL per 10^6 cells to the appropriate tubes (*see* **Note 11**).
15. For control tubes, transfer the cells to the labeled, chilled 5 mL FACS tubes and store in the dark until FACS analysis.
16. For the sort tube, strain the labeled cell suspension through a 35 μm strainer cap 5 mL FACS tube to remove cell clumps and store in the dark and cold until FACS analysis (*see* **Note 12**).

3.3.3 Procedure for P6 Cell Preparation for FACS

1. Calculate the volume of cell suspension to be added to each tube. For control tubes, total volume will be 300 μL, for the sort tube, volume will be dependent on the number of cells available (*see* Table 2).

2. Add the appropriate volume of DPBS +10% FBS to the labeled, chilled 5 mL FACS control tubes (300 μL DPBS +10% FBS – X μL cell suspension).
3. Add the appropriate volume of cell suspension to each tube and flick to mix.
4. Stain for viability by adding PI at a concentration of 5 μL per 10^6 cells to the appropriate tubes (*see* **Note 11**).
5. For sort tube, strain the labeled cell suspension through a 35 μm strainer-cap 5 mL FACS tube to remove cell clumps (*see* **Note 12**).
6. Store tubes in the dark and cold until FACS analysis.

3.4 FACS Analysis

3.4.1 Preparation

1. Prepare a solution of 10% FBS in MEMα, keep on ice.
2. Label collection tubes for each sorted cell population that will be recovered.
3. Add MEMα +10% FBS solution to tubes as collection medium for recovery of sorted cells. The type/volume of tube will be dependent on the FACS machine configuration and number of anticipated sorted cells.

3.4.2 Procedure

1. Take the tubes to the FACS machine on ice and in the dark.
2. Use the negative and positive control tubes to perform compensation of the flow cytometer.
3. Set gates to exclude doublets and debris using the forward and side scatter properties as shown in Fig. 1a, b (for adult samples) and Fig. 2a, b (for P6 samples). Select for live cells by gating for cells lacking PI signal (Fig. 1c for adult and 2c for P6 samples). To select for undifferentiated spermatogonia in adult samples, set gates to include $CD9^{bright}$, ID4-EGFP+ cells (Fig. 1d). For P6 samples, undifferentiated spermatogonia can be isolated by gating for ID4-EGFP+ cells (Fig. 2d). For both adult and P6 cells, the ID4-EGFP+ fraction can be further subdivided into $ID4\text{-}EGFP^{bright}$ and $ID4\text{-}EGFP^{dim}$ subpopulations based on the intensity of EGFP signal, so that SSC-enriched ($ID4\text{-}EGFP^{bright}$) and progenitor-enriched (SSC-depleted, $ID4\text{-}EGFP^{dim}$) subpopulations can be obtained.
4. Run the experimental sample and collect the sorted cells into the collection tubes containing MEMα +10% FBS (*see* **Note 13**).
5. After extensive fluidics channel washing, perform a purity check by reanalyzing an aliquot of the sorted cells on the same FACS instrument (*see* Fig. 1e, f for adult samples and Fig. 2e, f for P6 samples).
6. Utilize sorted cells in downstream applications.

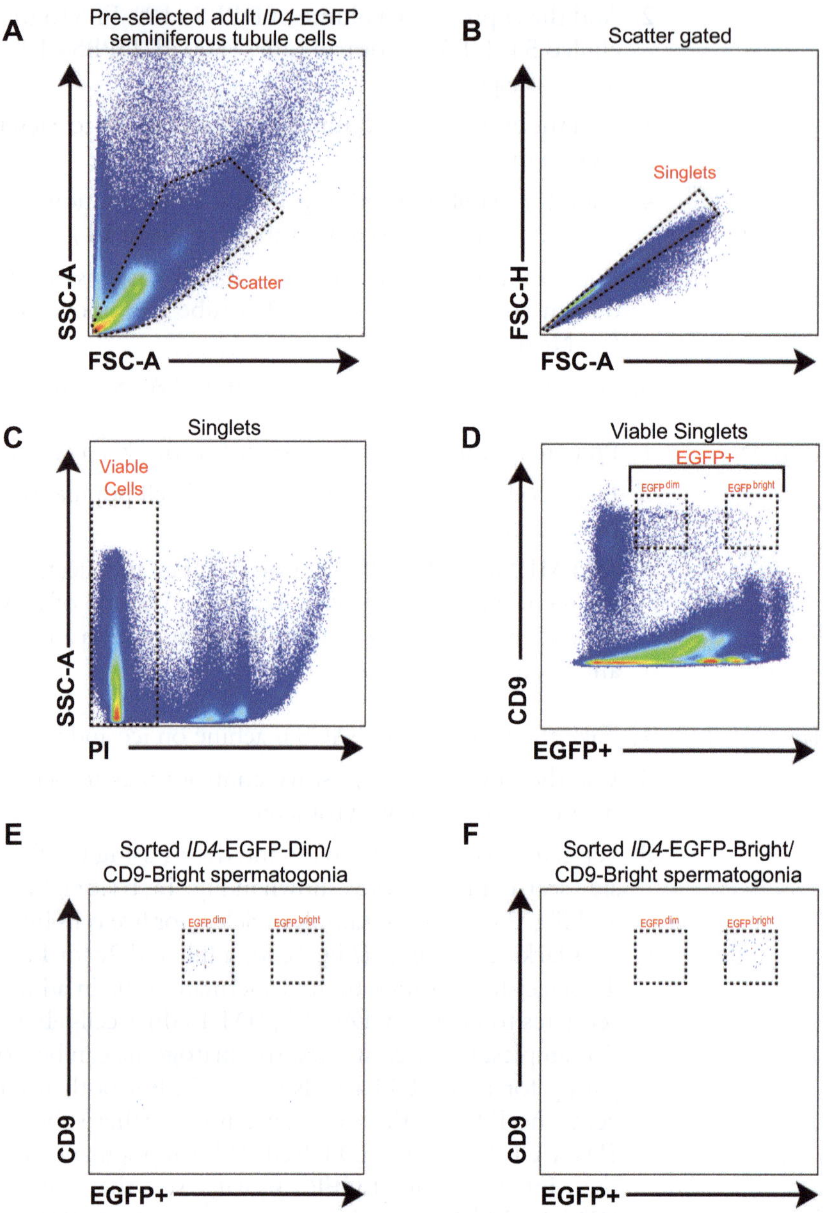

Fig. 1 Example FACS data for adult samples. (**a**) Forward and side scatter area properties are used to select cells of interest and exclude debris. (**b**) Forward scatter area and height are used to select for singlets. (**c**) Viable cells are selected for based on (absence of) PI staining. (**d**) Adult undifferentiated spermatogonia are isolated by gating for $CD9^{bright}$, ID4-GFP+ cells. The ID4+ cells can be further divided into a $CD9^{bright}$, ID4-GFP^{bright} population and a $CD9^{bright}$, ID4-GFP^{dim} population to obtain SSC-enriched and progenitor-enriched subpopulations, respectively. (**e**, **f**) To ensure proper sorting of cells, a postsort purity check can be performed on an aliquot of the $CD9^{bright}$, ID4-GFP^{bright} and $CD9^{bright}$, ID4-GFP^{dim} subpopulations

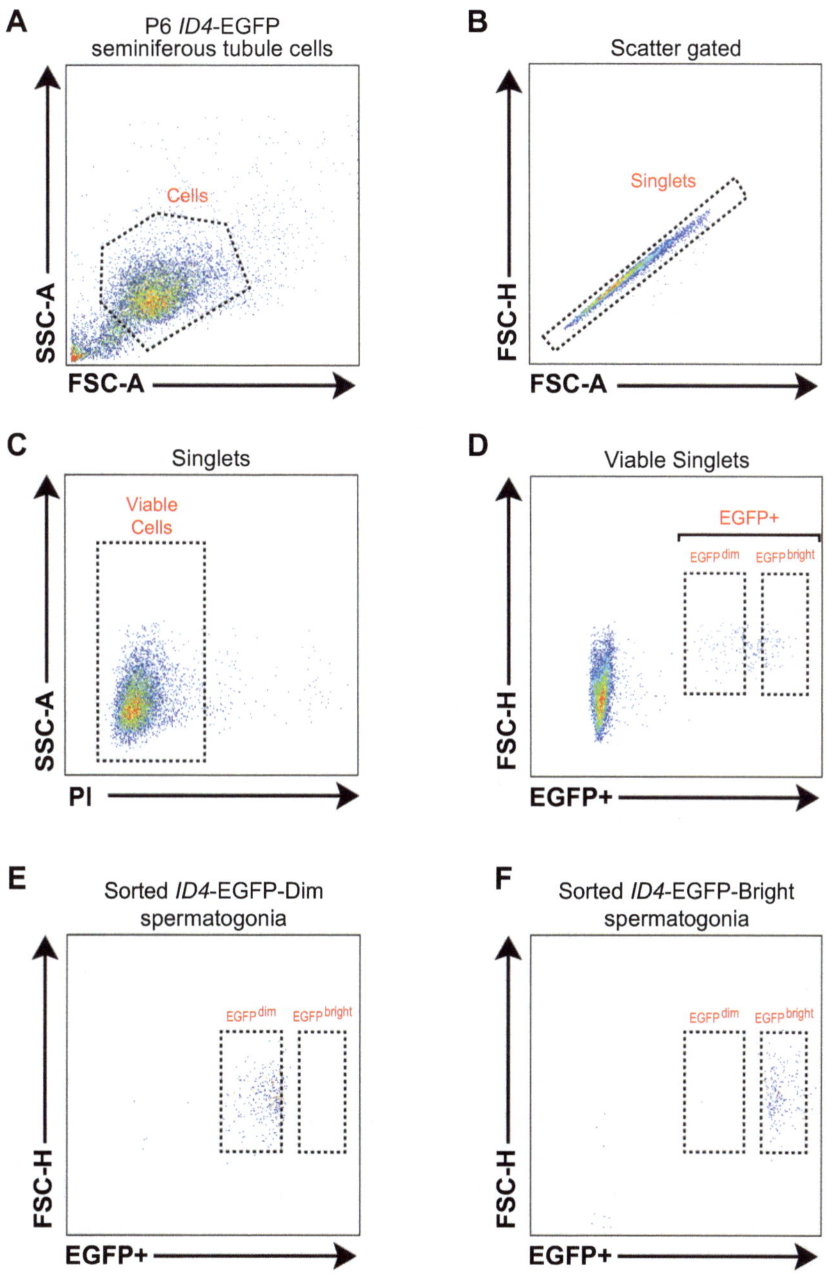

Fig. 2 Example FACS data for P6 samples. (**a**) Forward and side scatter area properties are used to select cells of interest and exclude debris. (**b**) Forward scatter area and height are used to select for singlets. (**c**) Viable cells are selected for based on (absence of) PI staining. (**d**) To obtain undifferentiated spermatogonia from P6 mice, ID4-GFP+ cells are selected for. The ID4+ cells can be further divided into an ID4-GFP^{bright} population and an ID4-GFP^{dim} population to obtain SSC-enriched and progenitor-enriched subpopulations, respectively. (**e**, **f**) To ensure proper sorting of cells, a postsort purity check can be performed on an aliquot of the CD9^{bright}, ID4-GFP^{bright} and CD9^{bright}, ID4-GFP^{dim} subpopulations

4 Notes

1. For experimental conditions, Id4-EGFP+ mice are needed. As negative controls, use Id4-EGFP- littermates or C57BL/6 mice. For adults from inbred strains, it is recommended to pool testes from three adult males. The approximate cell yield is 30–70 × 10^6 per adult animal.
2. For experimental conditions, Id4-EGFP+ mice are needed. As negative controls, use Id4-EGFP- littermates or C57BL/6 mice. The number of mice depends on the downstream application and availability. For isolation of P6 testis, it is recommended to only use a litter if there are ≥2 males, in which case testes are isolated from all male pups (excluding any runts, if present). The approximate cell yield is 4 × 10^6 per P6 animal.
3. The DNase I solution will turn yellow in phenol-red containing media and buffers, such as HBSS, since it is acidic.
4. Do not over-digest the testes with Collagenase IV. The desired result is free-floating tubules that are cleared of interstitial cells, but NOT a single-cell suspension at this step.
5. HBSS can also be used for the washes instead of MEMα.
6. For cryopreservation, around 40 × 10^6 cells/mL is targeted.
7. Monitor digestion under the microscope to check when a single-cell suspension has been achieved, in order to avoid over-digestion that will compromise viability.
8. Pre-enrichment of spermatogonia can be optimized by modifying the Percoll concentration between 24 and 34%.
9. The number of tubes depends on the cell input. For large input, 50 mL conical tubes can be used instead of 15 mL tubes.
10. Instead of streptavidin-APC, other streptavidin fluorochrome conjugates can also be used, depending on the available FACS instrument and the spectra of other fluorophores used in the experiment.
11. Instead of PI, other viability stains/reagents, such as 7-AAD or LIVE/DEAD Fixable stains (e.g., Violet), can also be used [10].
12. For some instruments, all tubes should be strained prior to loading the samples onto the FACS machine in order to prevent clumps in the fluidics.
13. Adult samples typically yield 20,000–30,000 cells per animal in the viable/singlet/EGFP+/CD9^{bright} fraction from Percoll pre-enriched cells.

Acknowledgments

This work was supported by NIH grants R01 HD90007 and U01 DA054179 to BPH. VvM was supported by an internship sponsored by Radboud University, the Netherland-America Foundation, and a Dekker-Padget Dutch2USA Internship Program Grant.

References

1. Phillips BT, Gassei K, Orwig KE (2010) Spermatogonial stem cell regulation and spermatogenesis. Philos Trans R Soc Lond Ser B Biol Sci 365(1546):1663–1678
2. Velte EK, Niedenberger BA, Serra ND, Singh A, Roa-DeLaCruz L, Hermann BP et al (2019) Differential RA responsiveness directs formation of functionally distinct spermatogonial populations at the initiation of spermatogenesis in the mouse. Development 146(12): dev173088
3. Nakata H, Wakayama T, Sonomura T, Honma S, Hatta T, Iseki S (2015) Three-dimensional structure of seminiferous tubules in the adult mouse. J Anat 227(5):686–694
4. Ozturk S, Guzeloglu-Kayisli O, Demir N, Sozen B, Ilbay O, Lalioti MD et al (2012) Epab and Pabpc1 are differentially expressed during male germ cell development. Reprod Sci 19(9):911–922
5. Chan F, Oatley MJ, Kaucher AV, Yang QE, Bieberich CJ, Shashikant CS et al (2014) Functional and molecular features of the Id4+ germline stem cell population in mouse testes. Genes Dev 28(12):1351–1362
6. Hermann BP, Cheng K, Singh A, Roa-De La Cruz L, Mutoji KN, Chen IC et al (2018) The mammalian spermatogenesis single-Cell transcriptome, from spermatogonial stem cells to spermatids. Cell Reports 25(6):1650–67.e8
7. Helsel AR, Yang QE, Oatley MJ, Lord T, Sablitzky F, Oatley JM (2017) ID4 levels dictate the stem cell state in mouse spermatogonia. Development 144(4):624–634
8. Oatley MJ, Kaucher AV, Racicot KE, Oatley JM (2011) Inhibitor of DNA binding 4 is expressed selectively by single spermatogonia in the male germline and regulates the self-renewal of spermatogonial stem cells in mice. Biol Reprod 85(2):347–356
9. Sun F, Xu Q, Zhao D, Degui Chen C (2015) Id4 marks spermatogonial stem cells in the mouse testis. Sci Rep 5:17594
10. Mutoji K, Singh A, Nguyen T, Gildersleeve H, Kaucher AV, Oatley MJ et al (2016) TSPAN8 expression distinguishes spermatogonial stem cells in the prepubertal mouse testis. Biol Reprod 95(6):117

Chapter 11

Spermatogonial Transplantation

Makoto Nagano and Xiangfan Zhang

Abstract

Spermatogonial transplantation is the unequivocal method to detect spermatogonial stem cells (SSCs) based strictly on the functional definition of stem cells – the cells' regenerative capacity. This method further allows for SSC quantification. A weakness of spermatogonial transplantation is its time-consuming nature; it takes 2 months to confirm the production of terminally differentiated cells in spermatogenesis, spermatozoa, in mice, which gives the assay endpoint. Using the mouse as the model system, we here describe the basic techniques of spermatogonial transplantation and provide practical guidance to successfully carry out this technique and to interpret data generated.

Key words Spermatogenesis, Stem cells, Microinjection, Regeneration, Retrospective functional assay, Stem cell quantification, Male fertility

1 Introduction

Spermatogenesis is a prototypical stem cell system where numerous sperm are produced and replenished everyday for a lifetime. Spermatogonial stem cells (SSCs) constitute the foundation of spermatogenesis and support this long-term high-throughput process of male gametogenesis [1].

Stem cells are defined by their function to regenerate and maintain an adult tissue or cell lineage [2]. SSCs continuously self-renew and produce committed progenitors that are destined to become sperm, thereby maintaining steady-state spermatogenesis or regenerating spermatogenesis after damage. Spermatogonial transplantation, developed by Ralph L. Brinster and his colleagues in 1994 [3, 4], is the method that specifically detects this cell functionality. Its procedural outline is as follows (Fig. 1). First, a test cell population is prepared in the form of a single-cell suspension (i.e., donor cells). Second, donor cells are microinjected into the lumen of seminiferous tubules in host testes where endogenous spermatogenesis is absent or abrogated. After a period of time, host testes are examined for re-establishment and maintenance of

Jon M. Oatley and Brian P. Hermann (eds.), *Spermatogonial Stem Cells: Methods and Protocols*, Methods in Molecular Biology, vol. 2656, https://doi.org/10.1007/978-1-0716-3139-3_11,

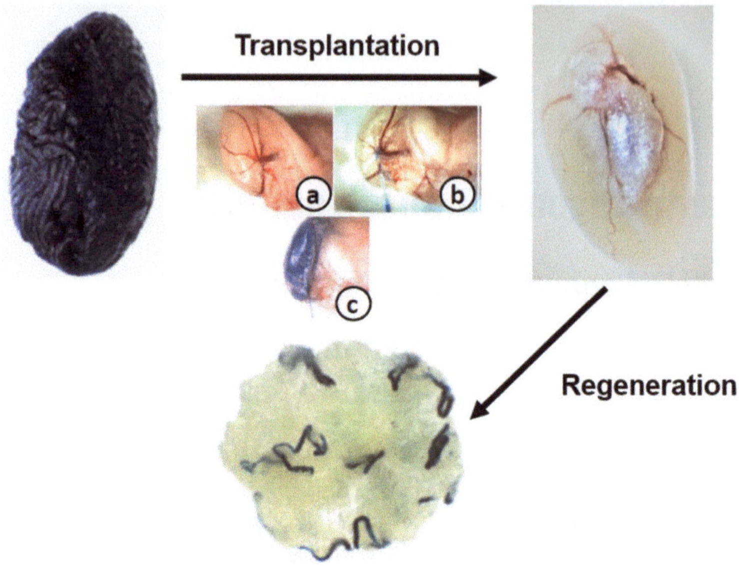

Fig. 1 Steps of spermatogonial transplantation. Testes derived from transgenic mice, which carry a genetic marker, provide an optimal donor cell source. Here shown is a testis of ROSA26 mice that express β-galactosidase gene in virtually all cells (*Left*). A single-cell suspension of donor testes is transplanted into testes of wild-type hosts that are made infertile chemically or genetically (*Right*). Three basic transplantation steps are shown. (**a**) Identify efferent duct; (**b**) Insert a microinjection needle into the duct; and (**c**) Inject donor cells; these steps are more detailed in Fig. 3. Colonies of regenerated, donor-derived spermatogenesis are found in the form of blue segments in host testes 2 months later (*Bottom*)

donor-derived spermatogenesis. Although nonstem male germ cells may engraft the host testis, they will be lost over time because they do not have the self-renewal capacity and the ability to continuously produce spermatozoa. Therefore, even if donor cells contain various germ cell types or include somatic cells, only SSCs selectively restore donor-derived spermatogenesis; nonstem germ cells will disappear though differentiation and somatic cells do not generate meiotic cells and spermatozoa.

Regeneration of donor-derived spermatogenesis occurs in the form of segments of host seminiferous tubules (Fig. 1). Under an optimal transplantation condition, each segment arises from one SSC [5] (*see* **Note 1**). Thus, SSCs that engrafted and regenerated spermatogenesis can be quantified by counting the number of the segments of donor-derived spermatogenesis in host testes. As such, spermatogonial transplantation generates both qualitative and quantitative parameters of SSC activity.

It should be noted that we detect SSCs using this transplantation technique by confirming a continuous production of spermatozoa, rather than directly observing SSCs; i.e., the readout of this functional assay is the terminally differentiated cell population. This is a common characteristic of functional assays of stem cells. For instance, hematopoietic stem cells are detected by confirming production of myeloid *and* lymphoid cells of donor origin in peripheral blood after transplantation [6], while embryonal carcinoma cells and embryonic stem cells are identified based on their ability to generate cell types derived from the three embryonic germ layers or an entire embryo [7–9]. This retrospective nature of stem cell assays is also applicable to spermatogonial transplantation. Thus, although SSCs can be detected unequivocally and quantitatively using the spermatogonial transplantation, when SSCs are detected and quantified after transplantation (Fig. 1), these cells are in host testes and do not remain in our hands. Prospective SSC analyses will require a set of convincing identification markers, but the fidelity and reliability of such marker molecules must be established functionally using spermatogonial transplantation prior to prospective analyses [10], exemplifying the critical importance of this technique in SSC research.

2 Materials

2.1 Host Preparation

1. Host mice: Adult or prepubertal mice. Wild-type strains or mutant strains. Immunocompatible to donor cells or immunodeficient (*see* **Notes 4** and **5**).
2. Busulfan: An alkylating agent that damages spermatogenesis [11]. To be dissolved first in dimethyl sulfoxide (DMSO) at 8 mg/mL (*see* **Note 6**).

2.2 Donor Cell Preparation

1. Donor cells: Adult, immature, or embryonic origin. Wild-type or mutant. From testes or tissue culture.
2. Dulbecco's Minimum Essential Medium (DMEM).
3. Fetal bovine serum (FBS).
4. Glutamine, 2 mM.
5. Penicillin, 100 units/mL.
6. Streptomycin, 100 μg/mL.
7. Hanks' Balanced Salt Solution (HBSS) or Dulbecco's Phosphate-Buffered Saline (PBS).
8. Collagenase IV, 1 mg/mL in HBSS or PBS, filter-sterilized.
9. Collagenase I, 1 mg/mL in HBSS or PBS, filter-sterilized.
10. 0.25% Trypsin–EDTA.
11. DNase I, 5 mg/mL in HBSS or PBS, filter-sterilized.
12. Cell strainer with a 70-μm pore size.
13. Trypan blue.

2.3 Transplantation

1. Standard surgical tools.
2. Anesthetics. Isoflurane or Ketamine cocktail (ketamine 100 mg/kg, xylazine 10 mg/kg, acepromazine 3 mg/kg). Or, as recommended by the institutional animal care committee.
3. Glass capillaries: 3-inch thin wall capillary with 1.0-mm outer diameter and 0.75-mm inner diameter. To be siliconized before preparation of microinjection needles using a standard puller (Fig. 2a).
4. 1-mL insulin syringe and 18G needle. The needle is cut to make a blunt end (Fig. 2b).
5. Silicon tubing with a 0.76-mm inner diameter (1.65 mm outer diameter). Connect it to the needle attached to the syringe (Fig. 2b).
6. Needle puller: Standard model to prepare microinjection needles.
7. Testis holding stages to stabilize host testes during cell transplantation (Fig. 2c).

3 Methods

3.1 Preparation of Busulfan

1. Shortly before i.p.-injecting busulfan into mice, slowly add sterile distilled water of the same volume as DMSO so that the aqueous phase overlays the DMSO layer, to make the final concentration of busulfan 4 mg/mL.
2. A moment prior to injection, mix the two layers to make a working concentration of 4 mg/mL. With this concentration, an i.p.-injection of 20 μL busulfan into a mouse of 20 g body weight (b.w.) makes the final concentration at 40 mg/kg b.w.
3. Periodically heat the busulfan solution to ~40 °C to avoid precipitation of busulfan. It may be helpful to have hot water in a beaker to heat a vial of busulfan solution during injection (*see* **Note 7**).

3.2 Preparation of Host Mice

1. Select a host mouse strain that is immunocompatible to donor cells or immunodeficient strains to avoid immunological rejection of donor cells.
2. I.p.-inject busulfan at 4–6 weeks of age.
3. Hosts can be used for spermatogonial transplantation from 4 weeks after busulfan injection. Animals up to 8 weeks after busulfan treatment are generally considered to be appropriate hosts, but those up to 10 weeks after the treatment can still

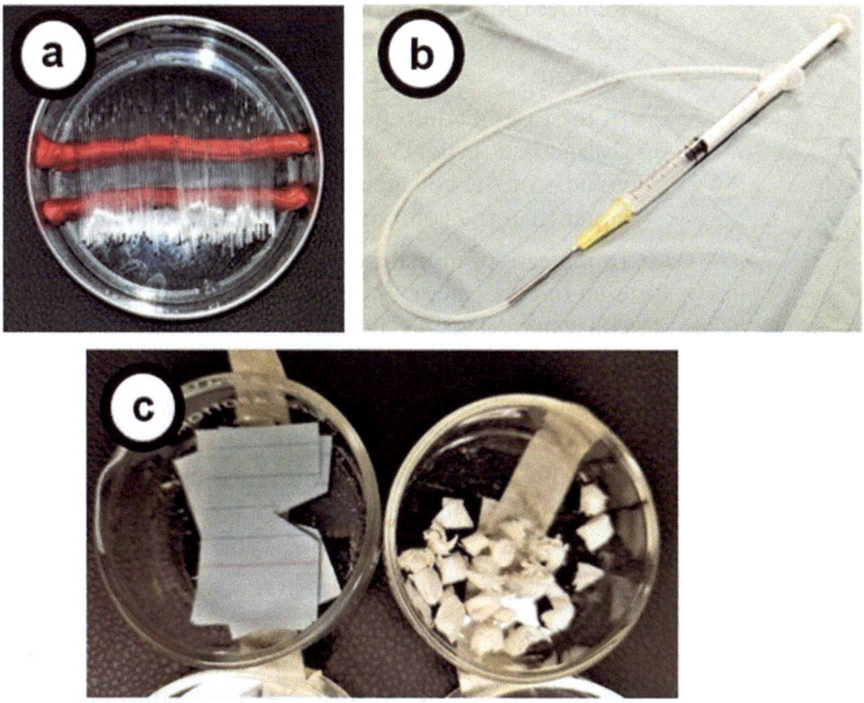

Fig. 2 Tools for transplantation. Shown are microinjection needles (**a**), an injection syringe with a tubing (**b**), and testis holding stages and swabs (**c**, left and right). The testis holding stages are prepared using mouse cards while swabs from paper fiber optic cleaning wipes, both of which are sterilized

serve as hosts. Such a restrictive time frame does not apply if genetically infertile mice are used as hosts. (*see* **Note 5**).

4. Similar to many stem cell types, SSCs represent a minute population of their specific cell lineage (germ cells) and the organ (testis). Thus, only a limited number of SSCs is expected in a donor cell sample, unless significantly enriched. Due to this scarcity of SSCs, the occurrence of SSC engraftment after transplantation follows the Poisson statistics, as seen in hematopoietic stem cells [6, 12]. In order to reliably quantify SSCs, therefore, it is ideal to use multiple host testes for one experimental group per experiment, with duplicate or triplicate experiments. In general, 5–6 host testes per group per experiment have been a standard for SSC quantification.

3.3 Preparation of Donor Cells

Donor cells in spermatogonial transplantation need to be single cells. A standard method for this cell preparation from whole testes is a two-step enzymatic digestion [13], first with collagenase, followed by trypsin, as follows. All procedures need to be carried out aseptically.

1. Remove tunica albuginea. Lightly loosen testes using forceps to make the subsequent steps of enzymatic digestion more efficient.
2. In the case of adult mouse testes, incubate them in 1 mg/mL collagenase Type IV dissolved in HBSS or PBS. The ideal volume of the collagenase solution is approximately 10 times larger than the volume of testes.
3. In the case of immature mouse testes, a mixture of 1 mg/mL each of collagenase Type IV and Type I is more effective than collagenase Type IV alone.
4. Let collagenase react at 37 °C for 10–15 min with periodic agitation or until the seminiferous tubules are sufficiently separated from the interstitium (*see* **Note 9**).
5. Let tubules sediment without centrifugation and then remove supernatant. This step removes interstitial cells.
6. To further remove interstitial cells, add HBSS or PBS and repeat **step 4**. Wash tubules in this way twice to three times in total.
7. Resuspend tubules with trypsin-EDTA supplemented with 1 mg/mL DNase I; this can be readily done by adding trypsin-EDTA and 5 mg/mL DNase I at the proportion of 4 to 1. Incubate at 37 °C for 5–10 min. Periodic agitation can hasten trypsinization. Gentle pipetting can also be effective, but care is needed not to overly damage cells. DNase I digests DNA derived from the cells that were damaged during trypsinization, avoiding cell clumping (*see* **Notes 10** and **11**).
8. Terminate trypsin reaction by adding FBS at 10% of the volume of trypsin. If cell clumps are noted at this step, transfer the clumps into the 5 mg/mL DNase I solution using a pipet. The clumps should be dissolved readily. Then, the entire solution can be combined with the original cell sample.
9. Centrifuge the cells at 500 × *g* for 5 min. Remove supernatant. Resuspend the cells in HBSS or PBS.
10. Filtrate the cells through a cell strainer with a 70-μm pore size. This step removes undigested tissue fragments or large cell clumps.
11. Repeat **step 8** to further remove trypsin.
12. Determine the number of cells harvested.
13. Resuspend donor cells in a medium (DMEM, glutamine, penicillin, streptomycin, and 10% FBS, *see* Subheading 2.2) with trypan blue at 10% of the volume. Trypan blue allows for visualization of a donor cell suspension during cell injection into the seminiferous tubules. The cell concentration is recommended to be up to 100×10^6 cells/mL (*see* **Notes 12–14**).

3.4 Microinjection of Donor Cells: Transplantation

Here presented is a basic protocol of manual microinjection without sophisticated microinjection apparatus. The preparation of such apparatus can be found in [4]. The procedure of microinjection is identical with or without microinjection apparatus.

1. Follow appropriate institutional standard operation protocols for animal anesthesia, analgesia, survival surgery, and health monitoring. For anesthesia, isoflurane can be used for adults (induction: 3–5% at 0.6–1.5 L/min; maintenance: 2–2.5% at 0.4–0.8 L/min) or pups (maintenance: 3–4%). If the Ketamine cocktail is used, it can be injected i.p. at 0.05–0.1 mL/10 g b.w. for adults or at 3.3 μL/g b.w. for pups of 6–14 days old.
2. Following the institutional animal care guidance, autoclave instruments, drapes, and other materials prior to surgical procedures.
3. Make a 1.5-cm incision on the midline of the host's abdomen. Exteriorize a testis and place it on a testis holding stage (Fig. 2c). The epididymis and fatty tissue are aligned in such a way that the opening to the rete testis becomes visible (Fig. 3a).
4. Carefully dissect the area between the testis and the epididymis to identify the efferent duct (Fig. 3b). No need to isolate and clean the efferent duct. When host mice are prepared as described in Subheading 3.2, they are young enough that fatty tissues have not overgrown; thus, disruption of the fat is often not necessary and can be even detrimental to the surgical operation. A simple alignment of fatty tissues should expose the duct for insertion of a microinjection needle.
5. Take a given volume of donor cell suspension in a microcentrifuge tube and aspirate it into the silicon tubing using a syringe plunger; then attach the microinjection needle to the silicon tubing (Fig. 3c).
6. Break the tip of the needle using forceps according to the size of the efferent duct. Then, gently apply a pressure to the plunger to transfer a donor cell solution from the tubing to the injection pipet (Fig. 3c).
7. Insert the microinjection pipet into the efferent duct around the mid-point between the testis and the epididymis (Fig. 3d).
8. Move the pipet along the duct until the tip of the pipet reaches the opening of the rete testis. When the pipet tip reaches the rete, a loss of friction is felt (due to the entry into a void space of the rete testis). The rim of the rete testis opening can be felt when the pipet tip is moved side ways.
9. Once the needle reaches the rete testis, you no longer need to hold the efferent with forceps. While holding the microinjection pipet with one hand, gently press the plunger of the

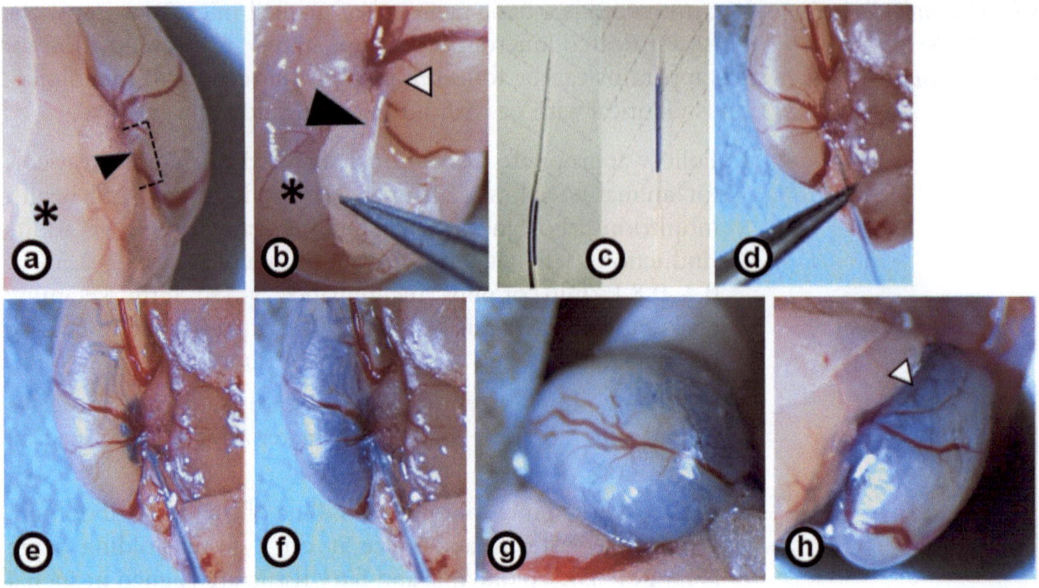

Fig. 3 Transplantation procedure. (**a**) A testis is placed on a testis holding stage (Fig. 2c) and aligned in such a way that rete testis and efferent duct become visible. The opening of the rete testis is indicated by a filled arrowhead and the region of the opening by hatched lines. Asterisk labels the epididymis. (**b**) The efferent duct is identified and gently pulled using forceps to give a tension to insert a microinjection pipet. The filled arrowhead indicates the efferent duct while the white arrowhead, opening of the rete testis. Asterisk labels the epididymis. (**c**) A donor cell suspension, which includes trypan blue, is loaded first into the tubing (left) and then transferred to the tip of the injection needle (right). The edge of the needle needs to be adjusted according to the size of the efferent duct beforehand. (**d**) Insert the microinjection pipet into the efferent duct. In this photograph, the pipet is seen over the forceps. (**e**) By gently applying pressure to the plunger, the donor cell suspension is introduced into the seminiferous tubule. Note that the rete testis is first filled by the cell suspension with trypan blue. (**f**) Donor cells then flow into the tubules. As cells are in the lumen of the tubules, the outline of the tubules is clearly identifiable. (**g**) Observe the extent of cell distribution. In general, a cell distribution over 60% of host testis surface is considered to be a success. (**h**) An example of failed injection. Here, a microinjection pipet punctured the rete testis, causing a leakage of a donor cell suspension into the interstitium. Note that the blue color is seen diffused into the interstitial space and tubule outlines are not clear (Compare to panels **e** and **f**)

syringe with the other. The donor cell suspension first fills the rete testis and then flows into the lumen of seminiferous tubules (Fig. 3e) (*see* **Note 16**). If the rete testis is punctured, a smear of blue dye spreads around the rete testis without confinement of the tubule lining (Fig. 3h) (*see* **Note 17**).

10. Remove the pipet from the duct. Up to ~10 μL of a cell suspension can be injected into the seminiferous tubules of a busulfan-treated adult mouse testis (*see* **Note 13**). In a successful injection, approximately 60 to 95% of the testis surfaces are observed to be occupied by blue tubules (Fig. 3g).
11. Gently return the testis to the abdomen and scrotum. Proceed to the second testis for microinjection. One injection pipet is

used only for one host testis even in the same host mouse; a new pipet is used for the testis. This is particularly important if you need to inject a different cell preparation/ group into the right and left testes. Changing an injection pipet for each host testis also allows for recording a volume of donor cell suspension for a given host testis.

12. Suture the muscle layer and then the skin at the incision site.
13. Monitor the recovery of host animals.
14. This procedure takes about 20 min per mouse from the time of anesthesia to incision closure.

3.5 Analysis of Host Testes

1. Donor-derived spermatogenesis occurs in the form of segments, which are called colonies (Figs. 1 and 4a). *See* **Note 18** for a definition of a colony, *and* **Note 14** about a range of colony numbers per host testis that is optimal for data analyses.
2. The production of spermatozoa, which is the consequence of normal SSC activity, is first observed 2 months after transplantation [14] (Fig. 4). Thus, the endpoint of spermatogonial transplantation is usually set at 2 months after transplantation (*see* **Notes 19** and **20**).
3. Colonies can be visualized distinctively if donor cells carry genetic markers, such as LacZ (as in Fig. 4a) or Green Fluorescent Protein. If donor cells do not carry such marker genes, they can be transplanted into the testis of mutant mice that lack endogenous spermatogenesis [4, 15–19]. Spermatogenesis established in these hosts after spermatogonial transplantation is derived only from donor SSCs, which can be detected by histological sections (*see* **Notes 4** and **5**).
4. If colonies of donor-derived spermatogenesis are visualized with LacZ staining or immunohistochemical staining, storage of host testes in neutral-buffered saline can permanently preserve the results of spermatogonial transplantation.

4 Notes

1. *Does a colony of donor-derived spermatogenesis arise from one SSC?:* Spermatogenesis is a one-way differentiation cascade and thus, SSCs produce only one type of terminally differentiated cells, spermatozoa. This unipotency distinguishes SSCs from multipotent or pluripotent stem cells that generate multiple types of terminally differentiated cells. Therefore, when donor cells are transplanted in such a way that only one colony is formed in a host testis, such a colony must have come from one SSC. It is known that under transplantation conditions

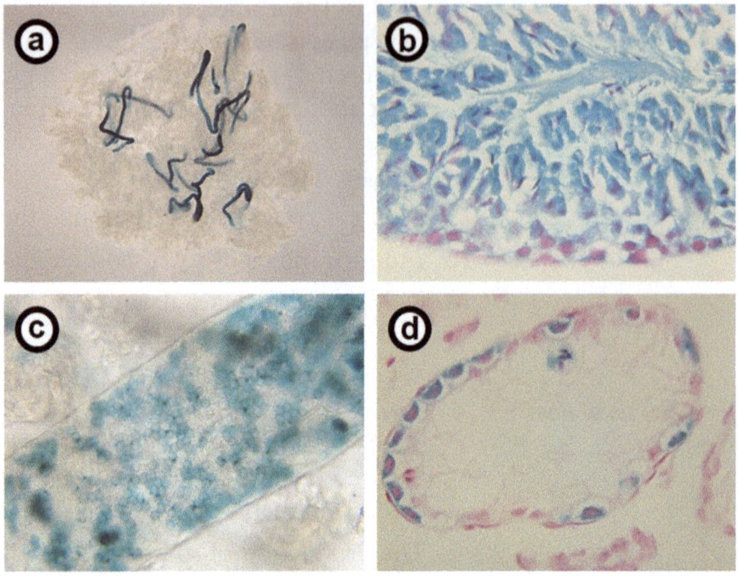

Fig. 4 Observation of transplantation results. (**a**) Colonies of donor-derived spermatogenesis are seen as distinct segments of host seminiferous tubules. Since donor cells carry the LacZ gene, donor spermatogenesis is stained blue on the background of unstained wild-type host tubules. Here shown is the host testis 2 months after transplantation. (**b**) Histology of a colony 2 months after transplantation. The heads and tails of elongated spermatids and spermatozoa are visible at this time, confirming that active SSCs existed in a sample at the time of transplantation. (**c**) Donor-derived spermatogenesis 1 month after transplantation. At this time, donor cells generate a monolayer sheet of spermatogonia on the basal membrane in distinct segments of seminiferous tubules. (**d**) Histology of a 1-month colony. Donor-derived spermatogonia are spread around the circumference of the tubules. (Images from ref. 14)

that result in up to four colonies per host testis, 100% of colonies are derived from one SSC (i.e., clonal colonies) [5]; a genetic study confirmed this "4 colonies per host testis" principle [20]. Likewise, 90% or more of colonies are clonal if up to 20 colonies emerge in a host testis after transplantation [5]. The clonality of colonies of donor-derived spermatogenesis justifies the quantitative measurement of SSCs using spermatogonial transplantation – the number of SSCs that successfully regenerated spermatogenesis is measured by simply counting the number of colonies. The clonality declines with higher donor cell concentrations and more colonies per host testis [5]. A decline of clonal colonies can also be observed with a longer time of incubation after transplantation, presumably as colonies become longer and fuse each other [14].

2. *Do all SSCs transplanted colonize and regenerate spermatogenesis?:* It has been reported that when adult busulfan-treated

testes are used as hosts, one out of eight SSCs (12%) in a donor cell preparation successfully survives, colonizes, and regenerates donor-derived spermatogenesis upon transplantation [21, 22]. This value of SSC homing is similar when adult and immature donor cells are used, as long as adult mice serve as hosts [21, 22]. Based on this SSC home efficiency, the absolute number of SSCs can be calculated by dividing colony numbers by 0.12.

3. *Why and how do SSCs migrate from the lumen of seminiferous tubules to their niches on the basal membrane?:* It is remarkable that SSCs introduced into the lumen of seminiferous tubules migrate to the basal membrane. How it occurs has not been well documented. It is known, however, the greatest roadblock of SSC migration, or "homing", to their niches after transplantation is the Sertoli cell junctions [23]. Sertoli cells bind to each other through tight junctions, forming the blood-testis barrier (BTB). Thus, more SSCs are known to engraft when immature mice (up to 14 days of age) are used as hosts, because BTBs have not been established [23]. Approximately tenfold more SSCs colonize pup host testes without BTBs, compared to adult host testes with established BTBs [23]. Interestingly, since the SSC homing efficiency is estimated to be 12% [21, 22], it can be suggested that pup testes without the BTB may allow for 100% of SSCs present in a donor cell preparation to home and regenerate spermatogenesis after transplantation.
4. Choice of host mice: It is ideal that host mice are immunocompatible to donor cells to avoid potential rejection after transplantation. Allografts in mice using Nanos-knockout hosts or those in farm animals have been successful without clearly detrimental effects suggestive of immunorejection ([6], Chapters 17 and 18). Yet, with the lack of systematic investigations into this issue, no convincing conclusion has been reached at this moment whether or not immunological rejection occurs after spermatogonial transplantation. In this regards, immunodeficient mice serve as universal hosts.
5. Genetically infertile mice devoid of germ cells are also excellent hosts. Since they do not have endogenous spermatogenesis, any gametes produced in these mutant hosts after transplantation are derived from donor SSCs. The use of these mutants makes it unnecessary to treat host mice with busulfan. Such mutant mouse strains include compound mutants involving the W locus (e.g., W/Wv, Wv/W54) [4, 15, 17, 19], Nanos2 knockout [16], and Jsd [18] mice. W/Wv mice have been widely used, while a recent report [16] shows that Nanos2 knockout animals provide robust and effective hosts.

6. Busulfan dose and metabolism: Busulfan is an alkylating reagent and damages not only spermatogenesis but also the bone marrow. Care should be taken to handle this compound by following institutional guidelines for human health. The bone marrow sensitivity to busulfan differs depending on a mouse strain, making it necessary to empirically determine an appropriate dose of busulfan, a dose that causes sufficient damage to endogenous spermatogenesis but allows for bone marrow recovery and avoids the death of treated animals. Some guidelines are 50 mg/kg for C57BL/6, 129, and C57BL/6 × 129 F1 mice as hosts while 40–44 mg/kg for Ncr nude mice. A study using rats reports that busulfan is metabolically inactivated in vivo and excreted in the urine. Approximately 95–96% of injected busulfan become inactivated within 32 h [24].
7. It is important to avoid the diluted solution of busulfan from precipitating. A dry bath with a 50-mL conical block set at 37 ° C gives another good option for this purpose.
8. Kinetics of SSC depletion by busulfan: Over 99% of SSCs disappear within 3 days after injection of busulfan at the dose of 45 mg/kg [25, 26]. Similarly, 95% of SSCs are lost within 3 days after busulfan at 15 and 30 mg/kg [25, 26]. Regardless of busulfan doses examined, males become infertile in 4 weeks [25]. The ability to sire offspring through natural mating is restored sooner with a smaller busulfan dose (~12, 26, and 30 weeks after busulfan at 15, 30, and 45 mg/kg, respectively) [25].
9. Collagenase digestion is the most important step in preparing single cells from testes. If this step is properly done, the trypsin digestion step can be readily and quickly completed, avoiding over-digestion of samples and clogging of microinjection needles (*see* **Notes 10** and **16**).
10. Over-reaction with enzymes, particularly with trypsin, makes it difficult to prepare a clean single-cell suspension, because DNA released from damaged cells causes the formation of cell clumps. During trypsin digestion, it is helpful to periodically monitor the digestion process by examining under an inverted microscope. Over-digested donor cells may cause clogging of microinjection needled (*see* **Note 15**) even after filtration using a cell strainer. It is an option to add 0.1–0.5 mg/mL of DNase I, using the leftover enzyme solution from Subheading 3.3, **steps 7** and **8**, in the transplantation preparation of donor cells to reduce cell aggregation.
11. Cultured cells can be digested using trypsin to prepare a single-cell preparation for spermatogonial transplantation.
12. An appropriate donor cell concentration needs to be determined empirically in a preliminary test (if possible). A

suspension of adult intact testis cells of 129 × C57BL/6 mice (ROSA26) at 100 × 10^6/mL usually gives around 20 colonies per host testis [14]. This value changes depending on what type of donor cells are used. For instance, cells derived from SSC amplification culture will be highly enriched for SSCs and thus need to be diluted significantly. Various parameters can affect SSC concentrations, such as mouse strains, age, cell sorting, and experimental manipulation (e.g., chemical administration, experimental cryptorchidism).

13. It is important to consider the volume of host testes when spermatogonial transplantation is planned. In general, host testes treated well with busulfan can accommodate 6–8 μL of donor cell suspension, with a maximum volume of 10 μL/testis. Note that the testis size differs depending on the strain or the age. For instance, a pup testis (up to 14 days old) can carry 3–5 μL of donor cell suspension, but a testis of younger mouse may accept only 2 μL.
14. When adult hosts are used in spermatogonial transplantation, up to 20 colonies/host testis is a good range of colony numbers that give confident quantitative analyses [5]. Too many colonies make it impossible to quantify SSCs. The result of 0 colonies/ testis is problematic for statistics. In such cases, commonly used data transformation methods, such as the logarithmic (N + 1) transformation, need to be applied.
15. SSC cryopreservation: SSCs can be cryopreserved in the same manner as with somatic cells using 10% FBS and 10% DMSO in DMEM [27]. After thawing, SSCs successfully colonize host mice and regenerate spermatogenesis.
16. Clogged injection needle: The microinjection needle sometimes becomes clogged, when, for example, a donor cell concentration is high. Needle clogging can also be caused by insufficient removal of tissue debris or by dead cells in a donor cell preparation. If it happens, retrieve the needle from the efferent duct and break the tip of a needle; thereafter, the needle can be re-inserted through the efferent duct to the rete testis to continue microinjection.
17. Surgical damage to the rete testis: If the rete testis is punctured by a microinjection needle, such a testis needs to be eliminated from an experiment. This surgical error cannot be overcome.
18. Definition of the colony: In the original study of colony formation after spermatogonial transplantation [14], a colony was defined as a tubule segment where at least 50% of the basal surface of the tubule is occupied by the spermatogonial network with at least 0.1 mm in length.
19. Patterns of colony formation: SSCs first produce chains of spermatogonia on the basal membrane of the host

seminiferous epithelia. As spermatogonia proliferate, they constitute a network of spermatogonia and form a monolayer of spermatogonia (Fig. 4c, d). This structure expands circumferentially as well as along the tubule. Then, meiotic differentiation starts in the middle portion of the colony, which spreads toward both ends of a colony [14]. This sequence of events is similar to the process of spermatogenic regeneration after irradiation [28, 29].

20. Kinetics of spermatogenic regeneration: It is known that colony numbers remain similar from 1 month to 4 months after transplantation, while the colony length increases with time [14]. It therefore appears that SSCs do not remain dormant after transplantation. This observation also suggests that colony counting can technically be done 1 month after transplantation, although most colonies at this point are formed only by a single layer of spermatogonia.
21. Fertility restoration after transplantation: Fertility can spontaneously be restored after transplantation, and host males can regain fertility to sire offspring [3, 15, 19]. It is possible that spermatogonial transplantation as descried here could damage the efferent duct, making fertility restoration difficult, although such a possibility has not been well examined. Spermatogonial transplantation in mice can also be done through peripheral seminiferous tubules [3, 4, 30], which should reduce this possibility.
22. Regenerative capacity of SSCs derived from aging animals or embryos: Aged and atrophied testes of up to 14-month-old B6 x 129 donors contain a cell population that regenerates complete spermatogenesis in busulfan-treated host testes [31]. However, regenerative cells are not found in atrophied testes of 23-month-old or older donors [31]. As host testes, atrophied testes at one year of age can support spermatogenic regeneration, but those at two years old cannot [31]. Hence, aging of spermatogenesis occurs both on SSCs and the testis environment. In the case of embryonic donor cells, transplantation of genital ridge cells at 14.5 embryonic day into adult host testes consistently results in donor-derived spermatogenesis (unpublished). Notably, it has been reported that spermatogenic regeneration can be observed when primordial germ cells at 8.5 embryonic days are transplanted in adult host mouse testes [32].

Acknowledgments

Our research is supported by the Canadian Institutes of Health Research (PJT-156305 and 153233). The authors wish to express their appreciation to the Animal Resources Center of the RI-MUHC for technical supports.

References

1. Valli H, Phillips BT, Gassei K, Nagano MC, Orwig KE (2015) Spermatogonial stem cells and spermatogogenic lineage development. In: Plant TM, Zeleznik AJ (eds) Knobil and Neill's physiology of reproduction, 4th edn. Elsevier, San Diego, pp 595–635
2. Weissman IL (2000) Translating stem and progenitor cell biology to the clinic: barriers and opportunities. Science 287:1442–1446
3. Brinster RL, Avarbock MR (1994) Germline transmission of donor haplotype following spermatogonial transplantation. Proc Natl Acad Sci U S A 91:11303–11307
4. Brinster RL, Zimmermann JW (1994) Spermatogenesis following male germ-cell transplantation. Proc Natl Acad Sci U S A 91: 11298–11302
5. Zhang X, Ebata KT, Nagano MC (2003) Genetic analysis of the clonal origin of regenerating mouse spermatogenesis following transplantation. Biol Reprod 69:1872–1878
6. Szilvassy SJ, Nicolini FE, Eaves CJ, Miller CL (2002) Quantitation of murine and human hematopoietic stem cells by limiting-dilution analysis in competitively repopulated hosts. In: Klug CA, Jordan CT (eds) Methods in molecular medicine, hematopoietic stem cell protocols. Humana Press, Totowa, pp 167–187
7. Stevens LC (1960) Embryonic potency of embryoid bodies derived from a transplantable testicular teratoma of the mouse. Dev Biol 2: 285–297
8. Kelinsmith LJ, Pierce GB (1964) Multipotentiality of single embryonal carcinoma cells. Cancer Res 24:1544–1551
9. Nagy A, Rossant J, Nagy R, Abramow-Newerly W, Roder JC (1993) Derivation of completely cell culture-derived mice from early-passage embryonic stem cells. Proc Natl Acad Sci U S A 90:8424–8428
10. Nagano MC, Yeh JR (2013) The identity and fate decision control of spermatogonial stem cells: where is the point of no return? Curr Top Dev Biol 102:61–95
11. Bucci LR, Meistrich ML (1987) Effects of busulfan on murine spermatogenesis: cytotoxicity, sterility, sperm abnormalities, and dominant lethal mutations. Mutat Res 176:259–268
12. Taswell C (1981) Limiting dilution assays for determination of immunocompetent cell frequencies. I. Data analysis. J Immunol 126: 1614–1619
13. Bellve AR (1993) Purification, culture, and fractionation of spermatogenic cells. In: Wassarman PM, DePamphilis ML (eds) Method in enzymology, vol 225. Academic Press Inc., New York, pp 84–113
14. Nagano M, Avarbock MR, Brinster RL (1999) Pattern and kinetics of mouse donor spermatogonial stem cell colonization in recipient testes. Biol Reprod 60:1429–1436
15. Nagano M, Brinster CJ, Orwig KE, Ryu BY, Avarbock MR, Brinster RL (2001) Transgenic mice produced by retroviral transduction of male germ-line stem cells. Proc Natl Acad Sci U S A 98:13090–13095
16. Ciccarelli M, Giassetti MI, Miao D, Oatley MJ, Robbins C, Lopez-Biladeau B, Waqas MS, Tibary A, Whitelaw B, Lillico S, Park CH, Park KE, Telugu B, Fan Z, Liu Y, Regouski M, Polejaeva IA, Oatley JM (2020) Donor-derived spermatogenesis following stem cell transplantation in sterile NANOS2 knockout males. Proc Natl Acad Sci U S A 117:24195–24204
17. Ogawa T, Dobrinski I, Avarbock MR, Brinster RL (2000) Transplantation of male germ line stem cells restores fertility in infertile mice. Nat Med 6:29–34
18. Boettger-Tong HL, Johnston DS, Russell LD, Griswold MD, Bishop CE (2000) Juvenile spermatogonial depletion (jsd) mutant seminiferous tubules are capable of supporting transplanted spermatogenesis. Biol Reprod 63: 1185–1191
19. Kanatsu-Shinohara M, Morimoto H, Shinohara T (2016) Fertility of male germline stem cells following spermatogonial transplantation in infertile mouse models. Biol Reprod 94:112
20. Kanatsu-Shinohara M, Inoue K, Miki H, Ogonuki N, Takehashi M, Morimoto T, Ogura A, Shinohara T (2006) Clonal origin of germ cell colonies after spermatogonial transplantation in mice. Biol Reprod 75:68–74
21. Nagano MC (2003) Homing efficiency and proliferation kinetics of male germ line stem cells following transplantation in mice. Biol Reprod 69:701–707
22. Ebata KT, Zhang X, Nagano MC (2007) Male germ line stem cells have an altered potential to proliferate and differentiate during postnatal

development in mice. Biol Reprod 76:841–847

23. Shinohara T, Orwig KE, Avarbock MR, Brinster RL (2001) Remodeling of the postnatal mouse testis is accompanied by dramatic changes in stem cell number and niche accessibility. Proc Natl Acad Sci U S A 98:6186–6191
24. Peng CT (1957) Distribution and metabolic fate of S35-labeled myleran (busulfan) in normal and tumor-bearing rats. J Pharmacol Exp Ther 120:229–238
25. Zohni K, Zhang X, Tan SL, Chan P, Nagano MC (2012) The efficiency of male fertility restoration is dependent on the recovery kinetics of spermatogonial stem cells after cytotoxic treatment with busulfan in mice. Hum Reprod 27:44–53
26. Kanatsu-Shinohara M, Ogonuki N, Inoue K, Ogura A, Toyokuni S, Shinohara T (2003) Restoration of fertility in infertile mice by transplantation of cryopreserved male germline stem cells. Hum Reprod 18:2660–2667
27. Avarbock MR, Brinster CJ, Brinster RL (1996) Reconstitution of spermatogenesis from frozen spermatogonial stem cells. Nat Med 2:693–696
28. van den Aardweg GJ, de Ruiter-Bootsma AL, Kramer MF, Davids JA (1983) Growth and differentiation of spermatogenetic colonies in the mouse testis after irradiation with fission neutrons. Radiat Res 94:447–463
29. van den Aardweg GJ, de Ruiter-Bootsma AL, Kramer MF, Davids JA (1982) Growth of spermatogenetic colonies in the mouse testis after irradiation with fission neutrons. Radiat Res 89:150–165
30. Ogawa T, Aréchaga JM, Avarbock MR, Brinster RL (1997) Transplantation of testis germinal cells into mouse seminiferous tubules. Int J Dev Biol 41:111–122
31. Zhang X, Ebata KT, Robaire B, Nagano MC (2006) Aging of male germ line stem cells in mice. Biol Reprod 74:119–124
32. Chuma S, Kanatsu-Shinohara M, Inoue K, Ogonuki N, Miki H, Toyokuni S, Hosokawa M, Nakatsuji N, Ogura A, Shinohara T (2005) Spermatogenesis from epiblast and primordial germ cells following transplantation into postnatal mouse testis. Development 132:117–122

Chapter 12

Viral Transduction of Mammalian Spermatogonial Stem Cells

Makiko Yamada, Yanyun Luo, and Marco Seandel

Abstract

Lentiviral vectors have been major tools for genetic manipulation of spermatogonial stem cells (SSCs) in vitro. Adeno-associated viral vectors are promising emerging tools for in vivo SSC transduction that are less invasive, compared to lentivirus, since AAV DNA is not integrated into the host genome and the host genome remains intact. In this chapter, we describe protocols using lentiviral and adeno-associated viral vectors to transduce SSCs in vitro and vivo, respectively.

Key words Transgenesis, Lentivirus, Adeno-associated virus, Mouse, Testis, Seminiferous tubule

1 Introduction

Viral transduction refers to a process by which genetic material (DNA) is introduced into cells by viral vectors (Fig. 1). The main purposes of viral transduction in the field of spermatogonial stem cells (SSCs) are to (a) analyze gene functions, (b) generate transgenic research tools, and (c) deliver gene therapy to correct spermatogenetic defects or genetic abnormalities in germ cells. Each of which will be discussed in the following paragraphs.

Viral transduction has been used most frequently for functional analysis of genes in rodent SSCs. This is due to the easy accessibility of rodent SSCs in vitro for genetic manipulations, such as knockdown, knockout, overexpression, and gene editing. Rodent SSCs can be cultured routinely in vitro for either short or long periods (Chapters 6 and 7) and transferred back to recipient testes for reconstitution of spermatogenesis (Chapter 5). To this end, in vitro genetic manipulation by viral transduction can be followed by the transplantation assay in order to assess the outcome. For example, if one knocks down a gene encoding a protein that promotes self-renewal of SSCs by shRNA via lentiviral transduction in vitro and then transfers these SSCs back into recipient testes, the

Jon M. Oatley and Brian P. Hermann (eds.), *Spermatogonial Stem Cells: Methods and Protocols*, Methods in Molecular Biology, vol. 2656, https://doi.org/10.1007/978-1-0716-3139-3_12,

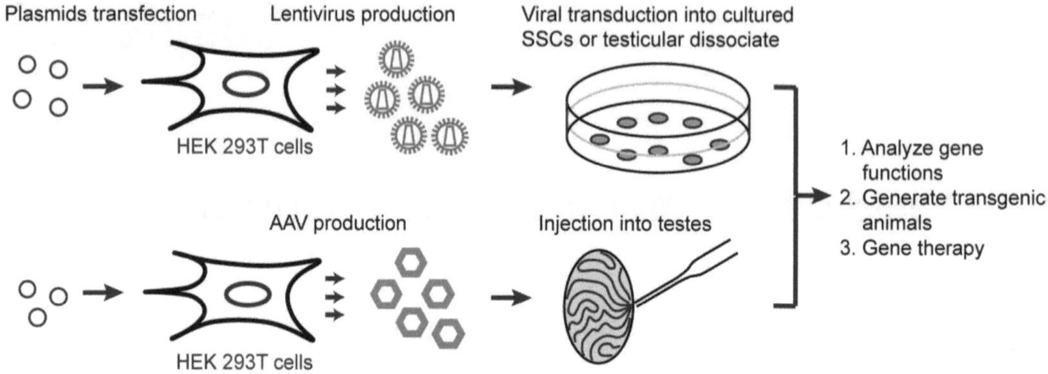

Fig. 1 Diagram of viral transduction procedures and their purposes. In this chapter, methods of lentivirus and adeno-associated virus (AAV) production and in vitro/vivo transduction, respectively, are discussed

expected outcome would be reduced colonization of these SSCs in the testis. On the other hand, SSCs in vivo are somewhat shielded from exposure to injected viruses, due to the location of SSCs on the basal side of the blood testis-barrier, between the seminiferous tubule basement membrane and Sertoli cell tight junctions. Also, Sertoli cells may act like "sponges" for reagents that are introduced into seminiferous tubules. Thus, viral transduction of SSCs in vivo may be inefficient, possibly due to (a) suboptimal viral dose effectively available in this compartment and (b) strong immune response to virus, as observed in adenoviral transduction [1, 2]. There are two routes to deliver viral particles to SSCs in vivo: adluminal side (via seminiferous tubule injection) and basal side (via interstitial injection) (Fig. 2). In either technique, the in vivo transduction efficiency of SSCs is considerably lower compared to that of Sertoli cells or other cell types in the testis [1, 3].

In order to generate transgenic animals, manipulation by viral transduction of parental male germline stem cells can be performed [4–7]. Although parental male germline manipulation is not a frequently employed protocol for mouse engineering, it has served well in generation of transgenic rats and domestic animals, since pronuclear DNA injection was not as efficient as in mice and embryonic stem cell lines were not available in these mammals [5–8]. Lastly, the ultimate application of viral transduction is gene therapy. Clinical genetic manipulation of SSCs is still at the conceptual stage but potentially holds great promise in reproductive medicine. Transgenesis by viral vectors could be used to treat genetic disorders, such as infertility caused by spermatogenetic failure or for correction of genetic abnormalities that are transmitted to offspring. However, a major concern for germline gene therapy would be random integration of viral DNA into the genome of a cell that will ultimately contribute genetic information to offspring.

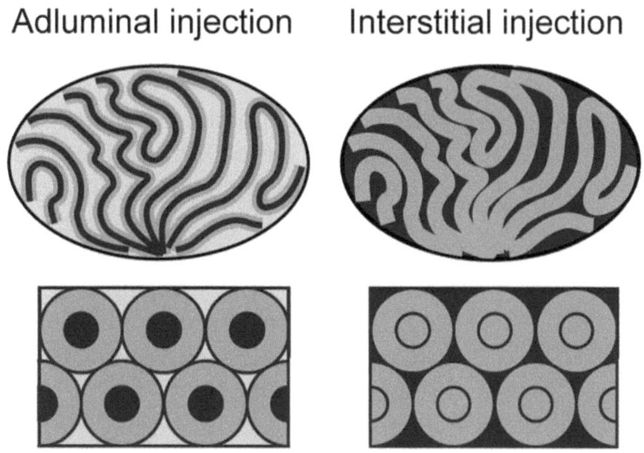

Fig. 2 Two approaches for in vivo virus delivery. For adluminal injection, viral solution is injected through efferent ducts. The adluminal side will be filled with Trypan blue (black) after injection. Lower panels are cross sections of the seminiferous tubules. For interstitial injection, Trypan blue solution diffuse through the interstitial space, i.e., between the seminiferous tubules

This risk would need to be mitigated by more sophisticated approaches to make germline gene therapy a reality.

Different viral vectors have been tested for their suitability for driving transgenesis of rodent SSCs in vitro and vivo. The first viral vector applied to mouse SSCs was a Molony murine leukemia virus-based vector with low transduction efficiency [9, 4, 10]. The development of replication-deficient, self-inactivating, pseudotyped lentiviruses created a path to substantial progress in genetic manipulation of mammalian SSCs [11, 12]. In general, compared to conventional retroviral vectors, lentivirus yields higher transduction efficiency in nondividing cells or slow-cycling cells and more stable expression of transgenes [11, 13]. Lentivirus has since become a major tool for in vitro SSC transgenesis, e.g., gene knockdown, knockout, and overexpression [14–17]. Thus, in this chapter, we describe a protocol of the lentivirus transduction into SSCs, which is routinely performed in our laboratory. Although not as frequently used as in vitro, in vivo lentiviral transduction seems to be efficient enough to produce offspring carrying transduced genes [18].

A promising viral vector emerging in the field of SSCs, especially in vivo, is recombinant adeno-associated viruses (AAVs), for which this chapter also includes a protocol to transduce SSCs. The AAV system is currently the leading vector for in vivo gene therapy for monogenic disorders due to its low potential for insertional mutagenesis, since the AAV genome predominantly persists as episomes and does not integrate into host genomes (reviewed in [19]). However, AAV integration into host genomes has been observed in germ cells, as well as in other organ systems. One group reported

AAV2 integrations into germ cells and subsequent vertical transmission to offspring in the setting of in vitro exposure [7, 8]. The other studies using in vivo systems using AAV1 and AAV9 did not observe genome integration into SSCs or offspring [3, 20]. Since the application of AAV to the SSC system is relatively recent and there is limited literature, more systematic evaluations may be required to address efficiency of transduction (robustness and stability of transgene expression), tropism of different serotypes, and extent of genomic integration in SSCs.

In this chapter, we describe protocols using lentiviral and adeno-associated viral vectors to transduce SSCs in vitro and vivo, respectively (Fig. 1).

2 Materials

2.1 Lentivirus Production

Lentiviral handling requires the biosafety level 2 (BSL-2) precautions agreed by the Centers for Disease Control and Prevention, National Institute of Health, and U.S. Public Health Service. The BSL-2 precautions entail personnel training in lentiviral handling and designated biosafety level cabinets, in addition to the BSL-1 requirements. All handling, storage, and disposal of biohazard materials must be in accordance with regulations of institutional biosafety committee (IBC), local, and state governments, and those of OSHA and EPA.

1. Transfer plasmid with your gene of interest (GOI) or shRNA: e.g., pCCL-PGK-EGFP (*see* **Note 1**).
2. Envelope plasmid: pMD2.G (VSV-G) (Addgene #12259).
3. Packaging plasmid 1: pMDLg/pRRE (GAG/POL) (Addgene #12251).
4. Packaging plasmid 2: pRSV-Rev (REV) (Addgene #12253).
5. Dulbecco's modified Eagle medium/10% Fetal Bovine Serum (FBS) (DMEM-10): DMEM 500 mL, FBS 50 mL, 200 mM L (+)-Glutamine 5 mL, and Penicillin/Streptomycin 5 mL (*see* **Note 2**). Mix all the ingredients and filter-sterilize with 0.22 μm PES filter.
6. 0.05% trypsin, 0.53 mM EDTA.
7. Dulbecco's Phosphate-buffered saline (PBS).
8. 1× Opti-MEM I Reduced serum medium.
9. HEK293T cell lines.
10. Sterile10 or 15-cm tissue culture dishes, polystyrene.
11. Sterile 50 mL polypropylene conical tubes.
12. Sterile 15 mL polypropylene conical tubes.
13. Autoclaved 1.5 mL microcentrifuge tubes.

14. 0.22 μm polyethersulfone (PES) sterilizing filter.
15. Sterile syringe filter with 0.45 μm PES membrane.
16. Sterile syringe.
17. Polyethylenimine (PEI) (linear, MW 25,000). Make up 1 g/L solution in distilled water, neutralize the pH to 7.0, and filter sterilize. It is quite stable at 4 °C up to 9 months. Store the aliquots at −80 °C. Do not repeat freeze-thaw cycles.
18. Lenti X™ concentrator (Takara).
19. Centrifuge with swing-bucket rotors, fixed-angle rotors, and centrifugation speed of up to 20,000 × g.
20. CO_2 incubator.
21. Biosafety cabinet equipped with high efficiency particulate air filters, laminar airflow, and directional airflow.
22. Water bath (37 °C).

2.2 Lentiviral Transduction of SSCs In Vitro

1. CSP (complete StemPRO-34 medium with additives for SSC culture): StemPRO-34,500 mL, StemPRO-34 Nutrient Supplement 13 mL, 1% FBS, 0.5% Bovine serum albumin (BSA) powder, 1× Nonessential amino acids, 1× MEM vitamin solution, 1× L-glutamine, 1× Antibiotics-antimycotics cocktail (penicillin, streptomycin, and amphotericin B), 6 mg/mL D (+)glucose [6 mg/mL], 30 ng/mL β-estradiol, 60 ng/mL Progesterone, 100 μg/mL Bovine holo-transferrin, 25 μg/mL Insulin, 20 ng/mL Human GDNF, 10 ng/mL Human bFGF, additional supplements (60 μM putrescine, 30 nM sodium selenite, 30 μg/mL pyruvic acid, 1 μg/mL DL-lactic acid, 50 μM β-mercaptoethanol, 100 μM ascorbic acid, 10 μg/mL d-biotin).
2. JK-1 cell line [21].
3. Primary cultures of spermatogonial stem cells (*see* **Note 3**).
4. 0.4% gelatin solution: Dissolve 4 g of gelatin from porcine skin in 1 L of distilled water and autoclave.
5. Polybrene 1000×: 10 mg/mL.
6. Puromycin.
7. Sterile 12 or 24-well plates.
8. Hemocytometer.

2.3 AAV Virus Production

AAV virus handling does not usually require the BSL-2 precautions but is generally performed under BSL-1 conditions. However, if AAV constructs are produced in the presence of a helper virus, such as adenoviruses or herpes viruses, and/or contain a potentially harmful transgene such as genes encoding tumorigenic or toxic molecules, BSL-2 would be appropriate [22].

Table 1
Iodixanol gradient solutions

% Iodixanol	60% (mL)	2 M NaCl (mL)	10× PBS-MK (mL)	H_2O (mL)	Phenol red (μL)	Total volume (mL)
15	50	100	20	30	300	200
25	62.5		15	72.5	225	150
40	80		12	28	0	120
60	120		0	0	0	120

1. Transfer plasmid* containing the GOI/shRNA flanked by inverted terminal repeats.
2. pCapRep*: plasmid encoding Cap and Rep, which defines specific serotype.
3. pHelper*: Adeno helper plasmid.

   *These three plasmids are available at Penn Vector Core, Cell Biolabs, Addgene, etc.
4. Polyethylenimine (PEI) (linear, MW 25,000): Prepare stock solution in the same way as **item 17** in Subheading 2.1.
5. 1.5 M NaCl.
6. 2 M NaCl.
7. 2.5 M NaCl.
8. 5 M NaCl.
9. Lysis Buffer: 50 mM Tris–Cl, 150 mM NaCl, 2 mM $MgCl_2$. Mix 10 mL of 1 M Tris–Cl, 20 mL of 1.5 M NaCl, 0.4 mL of 1 M $MgCl_2$, adjust pH to 8.0, bring to 200 mL, filter through 0.22 μm, and store at RT.
10. 40% PEG/2.5 N NaCl: Dissolve 160 g of PEG 8000 (Polyethylene glycol; MW 8000) and 58.44 g of NaCl in distilled water, bring to 400 mL, filter through 0.45 μm filter, and store at 4 ° C.
11. 10% sodium deoxycholate.
12. Benzonase endonuclease (250 units/μL).
13. Iodixanol gradient solutions (Table 1): 60% iodixanol OptiPrep TM Density Gradient Medium, 10× PBS-MK (10× PBS; 10 mM $MgCl_2$; 25 mM KCl), and Phenol Red.
14. Ultracentrifuge tubes Quick-Seal, Polypropylene, 39 mL, 25×89 mm and tube rack: Beckman Coulter.

    500 mL sterile conical tubes
15. 5/10/20 mL syringe.
16. Regular bevel sterile18 gauge needles.

17. Tube sealer ("Quick Charge Soldering Iron Kit"; Iso-Tip, Altoona, WI, USA).
18. Ultracentrifuge Rotor (Beckmann 70 Ti or equivalent).
19. Ultracentrifuge (Beckmann 70 Ti or equivalent).
20. Amicon Ultra-15 centrifugal filter units, molecular weight cutoff (MWC) 100 kDa.
21. Bio-Rad QX 200 Droplet Digital PCR system.
22. SYBR™ Select Master Mix.
23. DNase I.
24. Proteinase K.
25. Primers (forward primer 5′-GGA ACC CCT AGT GAT GGA GTT-3′ and reverse primer 5′ -CGG CCT CAG TGA GCG A-3′).

2.4 AAV Transduction of SSC In Vivo

For detailed materials required for in vivo injection of virus, please *see* Chapter 5, since in vivo viral injection can be accomplished using the same technique as SSC injection. The exception is that recipient mice do not need to be busulfan-treated. However, larger volumes of reagents (up to 10 μL) can be easily introduced in the busulfan-treated testis compared to an untreated testis, without excessively increasing interstitial pressure.

3 Methods

3.1 Lentivirus Production

1. Prepare 293 T cells for transfection. On the day of transfection, cells should be 65–80% confluent. For example, on the day before transfection, plate 293 T cells in a dish at approximately 40% confluency (i.e., ~eight millions cells for 10 cm dish).
2. One to two hours before transfection, feed cells with fresh DMEM-10.
3. Wash the cells with warm Opti-MEM and add 6 mL (for 10 cm dish) of DMEM-10 without antibiotics and antimycotics (*see* **Note 2**). Put the cells in the incubator.
4. Prepare DNA-PEI mixture (*see* **Note 4**). Dilute 50 μL of PEI into a total volume of 1 mL of Opti-MEM.
5. Dilute DNA (15 μg of Transfer plasmid, 3 μg of VSV-G, 5 μg of GAG/POL, and 2.5 μg of REV) into a total volume of 1 mL of Opti-MEM.
6. Add the diluted PEI mix to the diluted DNA, mix gently by pipetting 3–5 times, and incubate the mixture at room temperature for 30 min.

7. Add 2 mL of the DNA-PEI mixture to the dish dropwise. Final volume will be 8 mL. Take care not to disrupt the adherent cells. Return the dish to the incubator.
8. After incubation for 5–6 h, replace the media with 10 mL of fresh DMEM-10 with antibiotics.
9. 48 h after transfection, collect the supernatant in 15 mL conical tubes. Collect again after another 24–48 h if the cells are still healthy. The virus concentration may vary between the collections.
10. Centrifuge at 2000 × g for 5 min. Collect the supernatant and filter through 0.45 μm filter (*see* **Note 5**).
11. Concentrate virus (optional). Combine 1 volume of Lenti-X Concentrator with 3 volumes of supernatant. Mix gently by inversion.
12. Incubate mixture at 4 °C for 30 min.
13. Centrifuge at 1500 × g, at 4 °C for 45 min. Carefully remove the supernatant. Take care not to disturb the white pellet.
14. Resuspend the pellet in 1/10 to 1/100 of the original volume using DMEM-10, PBS, or specific medium for SSC culture.
15. Immediately titrate virus (*see* **Note 6**), aliquot in single-use volume (e.g., ~100 μL), and store at −80 °C.

3.2 Lentiviral Transduction of SSCs In Vitro

1. Wash SSCs with PBS and trypsinize them together with feeder cells (e.g., JK-1) at 37 °C for 5 min.
2. Centrifuge at 300 × g at RT for 5 min.
3. In the meantime, coat a 10 cm dish with 0.4% gelatin solution at RT for 5 min and remove gelatin.
4. Remove the supernatant, resuspend the pellet in either DMEM-10 or CSP, and plate the cell suspension on the gelatin-coated dish.
5. Incubate at 37 °C for 1–2 h to allow the feeder cells to adhere the plate whereas SSCs remain floating in the medium (*see* **Note 7**).
6. After 1–2 h, collect floating SSCs, spin down at 300 × g at RT for 5 min, and resuspend in small volume of CSP.
7. Count cells.
8. Thaw virus quickly at 37 °C in water bath.
9. Infect (e.g., For 24-well plate, each well requires 400 μL of final volume of medium) (*see* **Note 8**). Add 400,000 SSCs (e.g., 270 μL of 1,500,000/mL of SSCs) into a tube.
10. Add 110 μL of CSP.
11. Add 0.4 μL of Polybrene.

12. Add 50% virus: 20 μL of 10× concentrated virus (*see* **Note 9**).
13. Mix gently by pipetting and plate.
14. Incubate overnight at 37 °C in incubator with 5% CO_2 in humidified condition.
15. The next day, gently triturate off SSCs from bottom of the well by gentle pipetting, spin down at 300 × g for 5 min, resuspend in fresh CSP, and plate on the feeders. Normally, we split the cells at 1:2 or greater, since the cell concentration used for infection is higher than that for maintenance.
16. Drug resistance selection can be initiated after ~48 h of viral infection (*see* **Note 10**).
17. Proceed to phenotype analyses (*see* **Note 11**).

3.3 AAV Virus Production

Subheading 3.3 is adapted from ref. 23.

3.3.1 Transfection

1. Prepare HEK293T cells (follow the same protocol as lentiviral production Subheading 3.1).
2. One to two hours before transfection, feed cells with 20 mL of fresh DMEM-10 (15-cm dish).
3. Prepare DNA-PEI mixture (*see* **Note 12**). Calculate required volume of the three plasmids in order to obtain a total of 21 μg of DNA with a 1:1:1 molar ratio. This should equal ~1 μg of DNA per mL of medium (21 μg/22 mL).
4. Mix DNA with 200 μL of 1.5 M NaCl, 63 μL of PEI, and 1.737 mL of sterile H_2O in a sterile tube.
5. Vortex and incubate the mixture at RT for 30 min.
6. A total of 2 mL of the mixture is added to each plate dropwise. Final volume will be 22 mL. Incubate at 37 °C, 5% CO_2 for 72 h (*see* **Note 13**).

3.3.2 Virus Harvesting

1. At 72 h after transfection, collect cell culture medium and harvest transfected cells by scraping, respectively.
2. Add 40% PEG/2.5 N NaCl to the supernatant to a final concentration of 8%, and incubate on ice for 2 h. Resuspend the collected transfected cell pellet in 14 mL of lysis buffer and incubate at 4 °C for 2 h.
3. After 2 h of incubation, centrifuge the supernatant at 4 °C, 2500 × g, for 30 min, and collect the PEG-precipitated pellet.
4. Combine the cell lysate with supernatant precipitate, add 750 μL of 10% sodium deoxycholate to final concentration 0.5%, and add benzonase to a final concentration of 50 units/mL. Incubate at 37 °C for 30 min.

5. After 30 min, add 3 mL of 5 M NaCl to the sample, incubate at 50 °C for 30 min, and treat with three cycles of freezing and thawing between −80 °C and 37 °C. After the last thaw, freeze the sample at −20 °C overnight (*see* **Note 14**).

3.3.3 Iodixanol Gradient Centrifugation Purification

1. Thaw the virus lysates from Subheading 3.3.2, **step 5** by centrifuging at 12,000 × g for 30 min at 4 °C, and extract the lysate from the pellet debris by means of an 18 g needle attached to a 20 mL syringe (*see* **Note 15**).
2. Pipette the lysate into a Beckman 25 × 89 quick seal tube and immediately load with 9 mL of 15%, 6 mL of 25%, 5 mL of 40%, and 5 mL of 60% iodixanol using an 18 g needle attached to a 10 mL syringe, in this order. Add 1× PBS to top off the tube completely.
3. The sealed tube containing the prepared iodixanol gradient and lysate is centrifuged at 500,000 × g using rotor 70Ti for 1.5 h at 18 °C.
4. After ultracentrifugation, insert an 18 g needle into the top of the tube to allow air to enter. Another 18 g needle attached a 10 mL syringe should be inserted at the interface of the 40% and 60% iodixanol layers with the bevel of the needle facing upwards.
5. Collect approximately 2.5–4 mL of volume from the 40% iodixanol layer which should contain the extracted AAV particles (*see* **Note 16**).

3.3.4 Virus Dialysis and Concentration

1. Add the extracted virus to 10 mL of PBS previously dispensed into the top portion of an Amicon 15 100,000 MWCO concentration unit and centrifuge at 3000 × g for 10 min.
2. After the solution passes through the filter, add 10 mL of PBS, and repeat washing and spinning 4 times.
3. After the final wash, spin the remaining solution until 100–150 μL remains.
4. Aliquot the virus and store at −80 °C.

3.3.5 Determination of Viral Titers

1. Extract the viral DNA from 1 μL of purified virus by treating with 0.5 U DNase I to digest any contaminating unpackaged DNA at 25 °C for 15 min.
2. Add 10 μg Proteinase K and incubate at 65 °C for 1 h.
3. Inactivate Proteinase K at 95 °C for 20 min.
4. Determine the vector genome copy number by qPCR using SYBR™ Select Master Mix with primers for the ITRs common to the AAV transfer vector plasmids. The cycling conditions are: 95 °C 10 min, then cycled 40 times at 95 °C for 15 s, 60 °C for 30 s, and 72 °C for 30 s.

5. To generate a standard curve, an AAV transfer vector plasmid is used in serial dilutions from 1×10^7 to 1×10^3 genome copies, performed in triplicate. A no-template negative control and positive control with known genome copy number are also performed in triplicate.

3.4 In Vivo Microinjection of AAV

For detailed descriptions, please *see* Chapter 5. Recipient mice should be 8 to 15 weeks of age and not busulfan-treated. For the tubular injections, virus particles are introduced into the seminiferous tubules via the efferent ducts. Each injection fills 75–85% of all seminiferous tubules with approximately 4–10 μL virus particles with the 0.4% Trypan Blue solution. The same amount of virus is recommended using for interstitial injection [3, 20]. For the interstitial injection, the injection needle can be inserted randomly into the testis to release the solution into the interstitium. The other method is to insert the needle to the rete testis and advanced further for a couple of millimeters by passing through the rete testis. Slowly the solution should be released until the interstitium is filled with Trypan Blue as shown in Fig. 2. In either approach, it is extremely rare that the needle is inserted into the seminiferous tubules. Both tubular and interstitial injection can transduce the germ and Sertoli cells. AAV9 is highly recommended to use for the SSCs in vivo transfection to improve the AAV-mediated gene transduction efficiency [3].

4 Notes

1. In our laboratory, we most often use the human phosphoglycerate kinase 1 (PGK) promoter to express genes in the lentivirus system, particularly for functional analysis of genes [16, 17, 24]. Other promoters used successfully are the human elongation factor 1α (EF1α) promoter and the CAG promoter (consisting of the cytomegalovirus immediate-early enhancer, the chicken β-actin promoter, and rabbit β-globin polyadenylation signal) [25, 26]. Also, the puromycin or G418 resistance gene is often inserted into the transfer vector to enrich the transduced SSCs [26].
2. Antibiotics are routinely added to media for cell maintenance. However, during transfection, it is anecdotally recommended to avoid antibiotics, because one may worry about altered membrane permeability due to PEI, which may allow more antibiotics access to the cell interior. Polyene antimycotics, such as amphotericin B, which is included in antibiotic-antimycotic cocktail and often used for SSC culture should be definitely avoided.

3. To establish primary cultures of mouse SSCs, *see* Chapter 7 and refs. 27, 28.
4. PEI works more efficiently with HEK293 cell lines compared to $CaPO_4$ precipitation, another common method for transient transfection [29]. In HEK293 cell lines, ratios of PEI to DNA (w/w) are generally 2:1 to 5:1 [30, 31]. This ratio has to be optimized if the virus titer is lower than expected. For example, if the most of virus-treated SSCs die after puromycin selection, one can speculate that the virus titer is low. If 1/3 of cells are gone after drug selection, one can retrospectively speculate that the multiplicity of infection (MOI: number of transducing viral particle per cell) is about 1, i.e., the survived cells received 1 or 2 viral particles, given that viral particle allocation to cells follows the Poisson distribution. Since the total amount of DNA and ratios of four plasmids vary between protocols, these should be optimized. The final DNA concentration can be 1–5 μg per 1 mL of medium. The molar ratio of plasmids could be equimolar or the ratio of transfer plasmid could be higher sometimes than that of the rest of three plasmids.
5. Recommended are cellulose acetate or polyethersulfone (PES) filters. Avoid nitrocellulose filters because viral surface proteins can be bound to the membrane.
6. The viral titers can be quantified by two methods: functional and physical titer. Functional titer reflects actual number of infectious viral particles and expressed as transduction units per mL (TU/mL) and is determined by counting infected cell numbers by FACS (in case a fluorescence marker gene is transduced) or colony forming assay under antibiotics selection (Puro or G418). The functional titer is calculated as:

$$\text{Functional titer (TU/mL)} = \text{Positive cell fraction} \times \frac{\text{Total cell numbers}}{\text{Volume of virus (mL)}}$$

   Physical titer represents physical number of viral particles, expressed as viral particles per mL (VP/mL), and is measured by either quantification of viral RNA, genome (qPCR) or surface protein (ELISA), for which fast kits are commercially available. For fast determination, a direct quantification of the viral genomes targeting a conserved region of virus by qPCR is most prevalent.
7. We recommend that the virus transduction should be performed without feeders, since large numbers of feeder cells can reduce transduction efficiency and a huge amount of virus will be wasted if feeders are co-cultured.

8. The cell numbers and viral volume should be adjusted according to your conditions, including viral titers, viral cargo size, quality of virus, viral toxicity to the cells, and optimal MOI to obtain a desired phenotype. If the virus volume to infect SSCs is too high relative to CSP, viral suspension should be made with CSP but not DMEM-10 or PBS, in advance (Subheading 3.1, **step 14**).
9. Empirically, we start from this volume for which the MOI is ≥4, since no cell death is observed after drug selection (i.e., almost all the cells receive at least one viral particle). This is an example using pLKO (shRNA) and up to 50–80% of knockdown of target gene expression can be obtained with this volume of virus. However, in general, overexpression viral vectors with huge inserts usually require more viral volume in order to effectively overexpress target genes.
10. For example, for SSCs carrying the puromycin resistance gene, we use 400 ng/mL of puromycin for at least 4 days. With this dose, control nontransduced SSCs start to die from day 3 and are cleared within a week, whereas the transduced cells rarely die. It is important to treat the control nontransduced cells with puromycin in parallel for comparison. If the experimental group shows massive cell death, viral titers should be revisited and the viral amount for transduction should be optimized. With this concentration of puromycin, the feeder cells are viable.
11. One can proceed to transplantation assays after viral transduction. Normally, virus is cleared within a week after transduction. However, if the SSCs will be transplanted into a recipient testis after overnight transduction, extensive washing is necessary in order to eliminate residual virus before transplantation. This has been an issue in the past with residual virus making its way into the injection solution and transducing residual host SSCs that were not eliminated by busulfan or irradiation of the recipient.
12. In this protocol, the PEI/DNA ratio is 3. However, this ratio should be adjusted if the final virus titer is low.
13. Incubation periods may vary depending on AAV serotype. AAV2 is often harvested at 48–72 h but viral production seems to remain high at 96 h after transfection [32].
14. This section of the whole protocol until this point takes approximately 1 day. The following steps take one more day.
15. Carefully transfer the supernatant to the iodixanol gradient tube without disturbing the pellet, otherwise the purity of the final AAV particles is affected.

16. The total volume of the AAV sample should be around 2.5–4 mL, depending on the final outcome of the centrifugation. Avoid collecting more than 4 mL of volume from the 25% to 40% interface, otherwise contamination of the debris or undesired proteins in the final AAV particles may occur.

References

1. Ikawa M, Tergaonkar V, Ogura A, Ogonuki N, Inoue K, Verma IM (2002) Restoration of spermatogenesis by lentiviral gene transfer: offspring from infertile mice. Proc Natl Acad Sci U S A 99(11):7524–7529. https://doi.org/10.1073/pnas.072207299
2. Kojima Y, Sasaki S, Umemoto Y, Hashimoto Y, Hayashi Y, Kohri K (2003) Effects of adenovirus mediated gene transfer to mouse testis in vivo on spermatogenesis and next generation. J Urol 170(5):2109–2114. https://doi.org/10.1097/01.ju.0000092898.91658.08
3. Watanabe S, Kanatsu-Shinohara M, Ogonuki N, Matoba S, Ogura A, Shinohara T (2018) In vivo genetic manipulation of Spermatogonial stem cells and their microenvironment by adeno-associated viruses. Stem Cell Reports 10(5):1551–1564. https://doi.org/10.1016/j.stemcr.2018.03.005
4. Nagano M, Brinster CJ, Orwig KE, Ryu BY, Avarbock MR, Brinster RL (2001) Transgenic mice produced by retroviral transduction of male germ-line stem cells. Proc Natl Acad Sci U S A 98(23):13090–13095. https://doi.org/10.1073/pnas.231473498
5. Hamra FK, Gatlin J, Chapman KM, Grellhesl DM, Garcia JV, Hammer RE, Garbers DL (2002) Production of transgenic rats by lentiviral transduction of male germ-line stem cells. Proc Natl Acad Sci U S A 99(23):14931–14936. https://doi.org/10.1073/pnas.222561399
6. Ryu BY, Orwig KE, Oatley JM, Lin CC, Chang LJ, Avarbock MR, Brinster RL (2007) Efficient generation of transgenic rats through the male germline using lentiviral transduction and transplantation of spermatogonial stem cells. J Androl 28(2):353–360. https://doi.org/10.2164/jandrol.106.001511
7. Honaramooz A, Megee S, Zeng W, Destrempes MM, Overton SA, Luo J, Galantino-Homer H, Modelski M, Chen F, Blash S, Melican DT, Gavin WG, Ayres S, Yang F, Wang PJ, Echelard Y, Dobrinski I (2008) Adeno-associated virus (AAV)-mediated transduction of male germ line stem cells results in transgene transmission after germ cell transplantation. FASEB J 22(2):374–382. https://doi.org/10.1096/fj.07-8935com
8. Zeng W, Tang L, Bondareva A, Honaramooz A, Tanco V, Dores C, Megee S, Modelski M, Rodriguez-Sosa JR, Paczkowski M, Silva E, Wheeler M, Krisher RL, Dobrinski I (2013) Viral transduction of male germline stem cells results in transgene transmission after germ cell transplantation in pigs. Biol Reprod 88(1):27. https://doi.org/10.1095/biolreprod.112.104422
9. Nagano M, Shinohara T, Avarbock MR, Brinster RL (2000) Retrovirus-mediated gene delivery into male germ line stem cells. FEBS Lett 475(1):7–10. https://doi.org/10.1016/s0014-5793(00)01606-9
10. Kanatsu-Shinohara M, Toyokuni S, Shinohara T (2004) Transgenic mice produced by retroviral transduction of male germ line stem cells in vivo. Biol Reprod 71(4):1202–1207. https://doi.org/10.1095/biolreprod.104.031294
11. Naldini L, Blomer U, Gallay P, Ory D, Mulligan R, Gage FH, Verma IM, Trono D (1996) In vivo gene delivery and stable transduction of nondividing cells by a lentiviral vector. Science 272(5259):263–267. https://doi.org/10.1126/science.272.5259.263
12. Tiscornia G, Singer O, Verma IM (2006) Production and purification of lentiviral vectors. Nat Protoc 1(1):241–245. https://doi.org/10.1038/nprot.2006.37
13. Ellis J (2005) Silencing and variegation of gammaretrovirus and lentivirus vectors. Hum Gene Ther 16(11):1241–1246. https://doi.org/10.1089/hum.2005.16.1241
14. Dann CT, Alvarado AL, Molyneux LA, Denard BS, Garbers DL, Porteus MH (2008) Spermatogonial stem cell self-renewal requires OCT4, a factor downregulated during retinoic acid-induced differentiation. Stem Cells 26(11):2928–2937. https://doi.org/10.1634/stemcells.2008-0134
15. Ishii K, Kanatsu-Shinohara M, Toyokuni S, Shinohara T (2012) FGF2 mediates mouse spermatogonial stem cell self-renewal via upregulation of Etv5 and Bcl6b through MAP2K1

activation. Development 139(10):1734–1743. https://doi.org/10.1242/dev.076539

16. Martin LA, Assif N, Gilbert M, Wijewarnasuriya D, Seandel M (2014) Enhanced fitness of adult spermatogonial stem cells bearing a paternal age-associated FGFR2 mutation. Stem Cell Reports 3(2): 219–226. https://doi.org/10.1016/j.stemcr.2014.06.007
17. N'Tumba-Byn T, Yamada M, Seandel M (2020) Loss of tyrosine kinase receptor Ephb2 impairs proliferation and stem cell activity of spermatogonia in culturedagger. Biol Reprod 102(4):950–962. https://doi.org/10.1093/biolre/ioz222
18. Sehgal L, Thorat R, Khapare N, Mukhopadhaya A, Sawant M, Dalal SN (2011) Lentiviral-mediated transgenesis by in vivo manipulation of spermatogonial stem cells. PLoS One 6(7):e21975. https://doi.org/10.1371/journal.pone.0021975
19. Wang D, Tai PWL, Gao G (2019) Adeno-associated virus vector as a platform for gene therapy delivery. Nat Rev Drug Discov 18(5): 358–378. https://doi.org/10.1038/s41573-019-0012-9
20. Watanabe S, Kanatsu-Shinohara M, Ogonuki N, Matoba S, Ogura A, Shinohara T (2017) Adeno-associated virus-mediated delivery of genes to mouse spermatogonial stem cells. Biol Reprod 96(1):221–231. https://doi.org/10.1095/biolreprod.116.143495
21. Kim J, Seandel M, Falciatori I, Wen D, Rafii S (2008) CD34+ testicular stromal cells support long-term expansion of embryonic and adult stem and progenitor cells. Stem Cells 26(10): 2516–2522. https://doi.org/10.1634/stemcells.2008-0379
22. Collins DE, Reuter JD, Rush HG, Villano JS (2017) Viral vector biosafety in laboratory animal research. Comp Med 67(3):215–221
23. Huang X, Hartley AV, Yin Y, Herskowitz JH, Lah JJ, Ressler KJ (2013) AAV2 production with optimized N/P ratio and PEI-mediated transfection results in low toxicity and high titer for in vitro and in vivo applications. J Virol Methods 193(2):270–277. https://doi.org/10.1016/j.jviromet.2013.06.008
24. Kanatsu-Shinohara M, Naoki H, Shinohara T (2016) Nonrandom germline transmission of mouse spermatogonial stem cells. Dev Cell 38(3):248–261. https://doi.org/10.1016/j.devcel.2016.07.011
25. Hasegawa K, Saga Y (2012) Retinoic acid signaling in Sertoli cells regulates organization of the blood-testis barrier through cyclical changes in gene expression. Development 139(23):4347–4355. https://doi.org/10.1242/dev.080119
26. Kanatsu-Shinohara M, Morimoto H, Shinohara T (2012) Enrichment of mouse spermatogonial stem cells by melanoma cell adhesion molecule expression. Biol Reprod 87(6):139. https://doi.org/10.1095/biolreprod.112.103861
27. Kanatsu-Shinohara M, Ogonuki N, Inoue K, Miki H, Ogura A, Toyokuni S, Shinohara T (2003) Long-term proliferation in culture and germline transmission of mouse male germline stem cells. Biol Reprod 69(2):612–616. https://doi.org/10.1095/biolreprod.103.017012
28. Martin LA, Seandel M (2013) Serial enrichment of spermatogonial stem and progenitor cells (SSCs) in culture for derivation of long-term adult mouse SSC lines. J Vis Exp 72: e50017. https://doi.org/10.3791/50017
29. Tang Y, Garson K, Li L, Vanderhyden BC (2015) Optimization of lentiviral vector production using polyethylenimine-mediated transfection. Oncol Lett 9(1):55–62. https://doi.org/10.3892/ol.2014.2684
30. Ehrhardt C, Schmolke M, Matzke A, Knoblauch A, Will C, Wixler V, Ludwig S (2006) Polyethylenimine, a cost-effective transfection reagent. Signal Transduct 6(3): 179–184. https://doi.org/10.1002/sita.200500073
31. Longo PA, Kavran JM, Kim MS, Leahy DJ (2013) Transient mammalian cell transfection with polyethylenimine (PEI). Methods Enzymol 529:227–240. https://doi.org/10.1016/B978-0-12-418687-3.00018-5
32. Kimura T, Ferran B, Tsukahara Y, Shang Q, Desai S, Fedoce A, Pimentel DR, Luptak I, Adachi T, Ido Y, Matsui R, Bachschmid MM (2019) Production of adeno-associated virus vectors for in vitro and in vivo applications. Sci Rep 9(1):13601. https://doi.org/10.1038/s41598-019-49624-w

Chapter 13

Pharmacological Treatment of Neonatal and Juvenile Mice to Study Spermatogenesis

Oleksandr Kirsanov and Christopher B. Geyer

Abstract

The delivery, to newborn and juvenile mice, of drugs and other compounds that manipulate the physiology or cellular/molecular state -e.g., by activating or inhibiting signaling pathways) is a powerful, yet under-utilized approach to studying spermatogenesis. Here, we provide detailed protocols we have optimized in our laboratory for safely and effectively feeding and injecting mice and discuss troubleshooting approaches.

Key words Feeding, Injection, Newborn/neonatal, Spermatogenesis

1 Introduction

Mammalian spermatogenesis is a complex developmental process that occurs within the testis in the germinal, or seminiferous epithelium. This complex specialized environment includes a number of known and uncharacterized autocrine, paracrine, and endocrine signaling interactions between somatic cells and the germline that have not been faithfully recapitulated in vitro. Due to this limitation, our laboratory is one of many who rely heavily on mice as a model to study germ cell development. The gene products and signaling pathways that regulate spermatogenesis can be manipulated in vivo in two distinct, but complementary approaches. In the first approach, transgenic mice can be generated by inserting or deleting segments of DNA—and this reverse genetics approach has yielded invaluable insights into germ cell behavior, but suffers from significant disadvantages. These include high costs, extended timelines, and risks of unforeseen complications (e.g., position effect variegation, off-target effects, copy number variation, functional compensation, etc.). In the field of mammalian spermatogenesis, researchers are sorely limited in their ability to induce recombination of floxed alleles due to the paucity of effective Cre recombinase-expressing lines, most of which have limitations

Jon M. Oatley and Brian P. Hermann (eds.), *Spermatogonial Stem Cells: Methods and Protocols*, Methods in Molecular Biology, vol. 2656, https://doi.org/10.1007/978-1-0716-3139-3_13,

(e.g., incomplete penetrance, "leaky" expression outside the germline) that limit their usefulness to study postnatal spermatogenesis. In the second, or pharmacologic approach, molecular compounds that inhibit or activate specific gene products or cellular pathways can be administered in vivo. Key advantages of the pharmacologic approach include the ability to assess: (1) acute responses (limiting the potential for functional compensation); (2) differential dose responses; (3) similarity to human toxicant exposures or drug treatments; (4) strain-dependent response variation. There are also disadvantages, which include cross-reactivity (off-target effects) and whole animal toxicity. However, many compounds exhibit high specificity and safety profiles and are thus excellent tools to define the molecular mechanisms regulating spermatogenesis. The Geyer laboratory studies the program of differentiation that all mitotically dividing spermatogonia must complete prior to entering meiosis as spermatocytes. In the course of these studies, we have spent considerable time optimizing the delivery of drugs and other compounds to neonatal and juvenile mice. Here, we present our detailed protocols, and have arbitrarily designated mice aged P0-13 as "neonatal" and P14-35 as "juvenile" (P14-P35) based on how these animals are handled for the described treatments.

2 Materials

Feeding needles are cleaned with small volumes of distilled water, followed by small volumes of 70% ethanol. Needles are then air-dried using a high-pressure air nozzle. Single-use syringes and needles are sterilized by the manufacturer and discarded after use on each individual mouse.

2.1 Feeding of Mice, Aged Postnatal Day (P) 1-P13

1. Slide warmer.
2. Balance.
3. 2 × P200 empty pipette tip boxes.
4. Delicate task wipers (our lab uses Kimwipes).
5. Plastic weigh boats—3 1/2 × 3 1/2 × 1" D.
6. 24-gauge straight reusable oral gavage needle.
7. P200 unfiltered pipette tips.
8. P20 pipette.
9. Vehicle solution—our lab uses Dimethyl Sulfoxide (DMSO).

2.2 Subcutaneous Injections of Mice, Aged P7-P13

1. Slide warmer.
2. P20 and P100 pipettes.
3. 2 × P200 empty pipette tip boxes.

4. Kimwipes.
5. 1 cc syringe.
6. 30-gauge needles, 1/2 inch long.
7. P200 unfiltered pipette tip.
8. Vehicle solution—our lab uses Dimethyl Sulfoxide (DMSO).

2.3 Intraperitoneal (IP) Injections of Juvenile Mice (>P14)

1. P100 pipette.
2. P200 unfiltered pipette tip.
3. P1000 empty pipette tip box.
4. Kimwipes.
5. 1 cc syringe.
6. 30-gauge needles, 1/2 inch long.
7. Two clean cages, one for pups preinjection, the other for pups postinjection.
8. Vehicle solution—our lab uses Dimethyl Sulfoxide (DMSO).

3 Methods

3.1 Preparing the Feeding Needle and the Treatment Station

1. Approximately 15 min prior to beginning procedures, place Kimwipe-lined pipette tip boxes on the slide warmer (set to ~40 °C).
2. Remove compounds to be administered along with vehicle controls from storage (e.g., 4 °C, −20 °C, −80 °C) and equilibrate to 37 °C in a warm water bath.
3. Line two empty pipette tip boxes with Kimwipes (to absorb urine) and place boxes on a slide warmer preheated to 40 °C. Place an additional Kimwipe directly on the slide warmer, where each pup will be placed temporarily for observation immediately after compound administration (Fig. 1).
4. Make a clean and straight cut to remove ~2 cm from the narrow end of an unfiltered P200 pipette tip (Fig. 2) and discard the tip.
5. Insert the cut side of the pipette tip into a 24-gauge reusable feeding needle and place the pipette tip onto the P20 pipette (Fig. 2).

3.2 Oral Feeding of Neonatal (P1-P13)

Feeding is the preferred method (over injections) to administer compounds to neonatal mice. This is because mouse pups aged P0-P13 cannot see (eyes are closed), and are thus entirely receptive to feeding. We have given injections to neonatal mice, but have found that before ~P7 they have difficulty healing the area of injection.

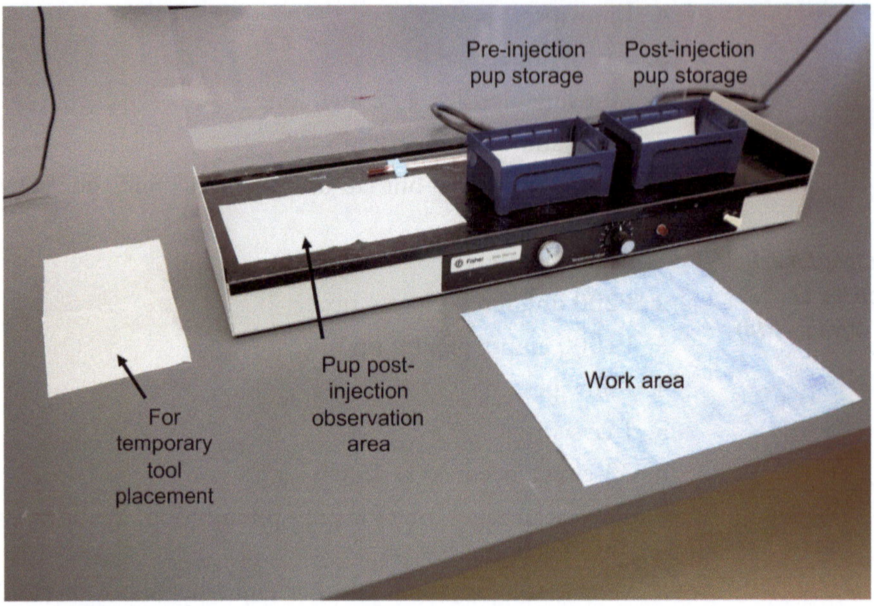

Fig. 1 Feedings and injection station setup. Two pipette boxes (preinjection and postinjection) are lined up with Kimwipes and placed on the right side of the preheated slide warmer. One additional Kimwipe is placed on the left side of the slide warmer for postinjection/feeding observation and another on the bench for temporary tool placement. An absorbent pad is placed in the working area in front of the slide warmer

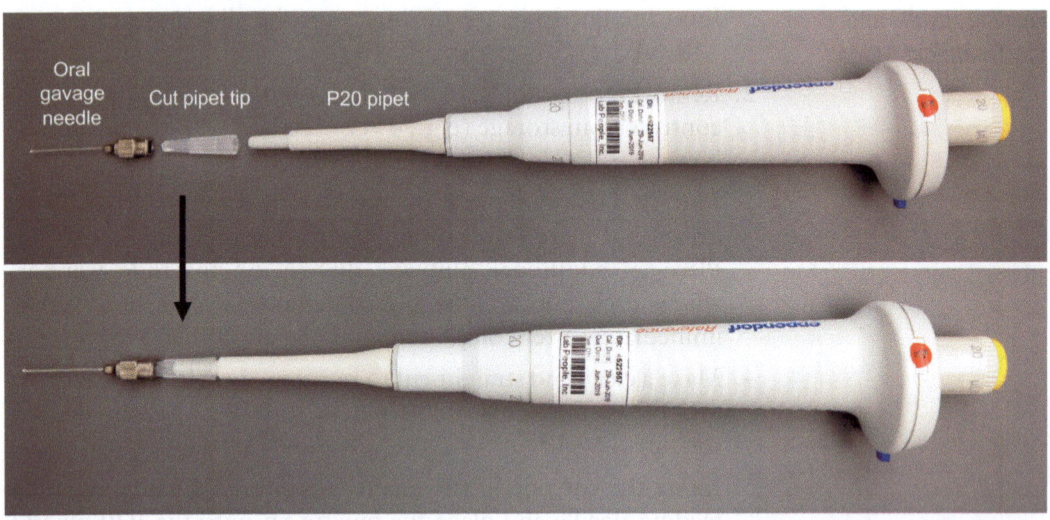

Fig. 2 Oral feeding needle setup. The cut side (narrow end) of the unfiltered P200 pipette tip is first inserted into a 24-gauge reusable feeding needle. Then, the wide end of the pipette tip is placed onto P20 pipet

1. Place all pups from litter to be treated in first (pretreatment) prewarmed pipette tip box on the slide warmer. For mouse weights, strain-specific numbers of pups per litter, and tips for housing and handling, *see* **Notes 1–4**.

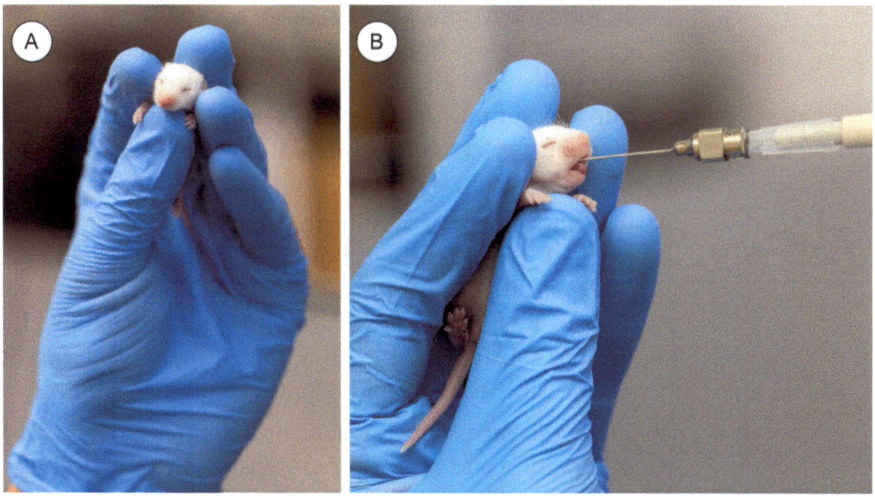

Fig. 3 Neonatal mouse handling technique for feeding. A neonatal pup is held at a 45-degree angle, with the head pointing up. The handler's thumb is on the ventral side, the middle finger on the dorsal side, and the ring finger and index fingers support the lateral sides of the animal (**a**). The tip of the oral feeding needle is inserted into the pup's mouth and the solution is slowly ejected (**b**)

2. Select first pup and quickly place it in a plastic weigh boat on a laboratory balance to record body weight. Use value to determine proper amount of solution to administer (*see* **Note 9**).
3. Using feeding pipette prepared in Subheading 3.1, pipette solution up and down and ensure no air bubbles enter the feeding needle. This is critical, as large trapped air bubbles in the needle may cause pups to choke.
4. Once pipette with solution is ready, gently grasp pup and hold it at a 45-degree angle, with the head pointing up, with thumb on the ventral side, middle finger on the dorsal side, ring finger and index finger supporting left and right lateral sides (Fig. 3a). Carefully note signs of discomfort while holding neonatal mice such as urination and excessive movement.
5. Prior to feeding, brace your arms to eliminate potential shaking—rest a flat portion of each elbow on a horizontal surface (e.g., table or bench), and you should rest the hand with pipette against your face.
6. Carefully insert the feeding needle into pup's mouth and begin to slowly depress the pipette to eject the solution (Fig. 3b). This must be done in small increments, approximately 2–3 μL at a time. Ideally, pups will gently lick and then swallow the solution from the tip of the needle.
7. Periodically remove the feeding needle from the mouth to allow pups to swallow. DO NOT draw up any air into the needle until done administering the solution. If pup is struggling to swallow, gently massage its stomach with your thumb.

If too much solution has been administered, pup will have excess solution outside its mouth as a small drop. The pup will still swallow it, but this will require more time. Observe pup coloration—if it turns a shade of gray, this is a sign of choking—refer to *see* **Note 10** for tips on how to assist pup during choking events or health complications.

8. Continue solution administration until the first pipette stop is reached. Once finished, hold pup on palm for 30 s to ensure most of the solution has been swallowed. Approximately 1–2 μL loss of solution is common during feeding. If the loss is more substantial (e.g., 3–5 μL), euthanize the mouse as it is not possible to determine the exact amount administered.
9. Place this first pup on a Kimwipe on the slide warmer for observation and repeat these steps for the second pup.
10. When done feeding the next pup, place it on a fresh Kimwipe and move previously fed pup into the second (post-treatment) prewarmed pipette tip box.
11. Repeat above steps until remaining pups have been fed.
12. Return all pups back to their housing cage. If more litters are to be fed, *see* **Note 5**.

3.3 Subcutaneous Injections of Neonatal Mice (P7-P13)

1. Prime the 1 cc syringe and 30-gauge needle by aspirating compound suspended in vehicle. Draw the solution halfway up the needle hub, but do not allow the solution to enter the syringe. Repeat until there are no air bubbles coming out of the needle. Empty the solution back into the tube.
2. Draw up exactly desired volume (we routinely use 10 μL) using a P20 pipette. Using a syringe with loaded needle (as shown in Fig. 4), aspirate the entire volume directly from the loaded pipette tip, ensuring large air bubbles do not enter the needle hub. While small air bubbles are not ideal, they are acceptable because they typically escape as needle is removed from the mouse. Large air bubbles typically create a separation of solution in the needle hub, and must be avoided at all times, as they have potential to cause pups to choke and may cause significant solution loss as the needle is withdrawn.
3. Hold pup firmly (but gently) between index finger and thumb, with ring finger on the ventral side (*see* Fig. 4).
4. Inject solution by inserting the needle about halfway into the small skin-fold over the interscapular area. Bevel of the needle must be oriented dorsally (*see* **Note 6**).
5. Inject the solution by depressing plunger firmly, and hold for 5–10 s to ensure all of the solution exits the needle.

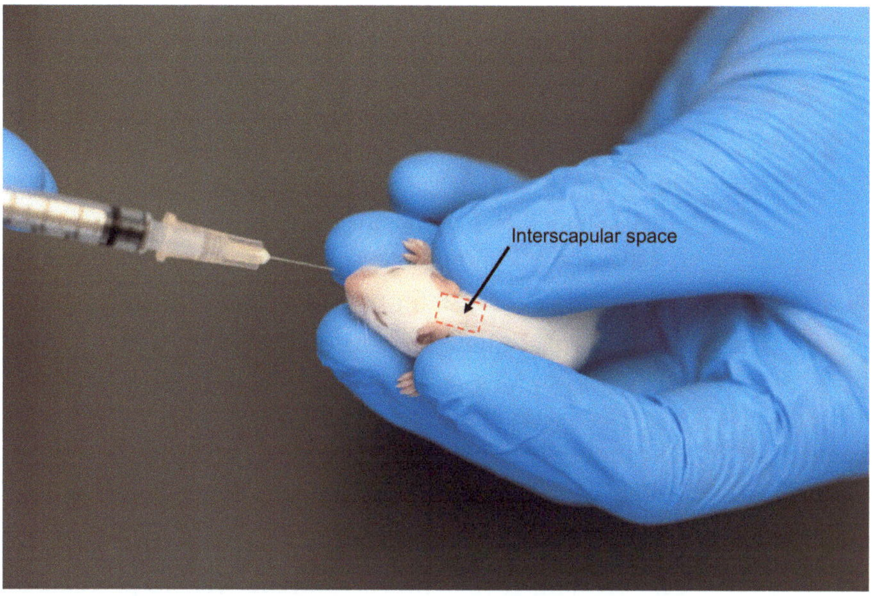

Fig. 4 Subcutaneous injection technique for neonatal mouse. The pup is held firmly between the index finger and the thumb, with the middle finger on the ventral side and the ring finger supporting the thumb. The dashed red square represents the interscapular space where injections should be given for best results

6. Slowly and carefully remove the needle. If excessive solution leaks out (this happens rarely), the pup should be euthanized, as it is impossible to determine the amount lost.

3.4 Intraperitoneal (IP) Injections of Juvenile Mice (>P14)

1. Weigh each mouse in a P1000 pipette tip box lined with a Kimwipe and calculate necessary compound dose resuspended in vehicle. Transfer mouse into first clean cage (preinjection) prior to solution administration. For the IP injections station setup, *see* **Note** 7.
2. The solution will be drawn into the same syringe and needle combination described in Subheading 3.3. Note, however, the P200 tip may need to be slightly shortened to aid drawing up the solution, as required volume will increase. Set aside the prepared syringe until ready to inject. For the upper/lower limits of compounds to be injected and examples from our laboratory, *see* **Notes 8** and **9**.
3. Gently remove mouse from the cage and restrain at a 45-degree angle with the head facing downwards. This will shift abdominal viscera anteriorly and thus reduce risk of accidentally puncturing organs during injection.
4. Inject solution by inserting the 30-gauge needle one-third to halfway into one of the four lower abdominal quadrants, with bevel oriented dorsally. Be careful to not insert the needle all the way in, as it may damage abdominal organs. Retract the

plunger to ensure negative pressure and proceed with the injection by depressing the plunger. The speed of injection depends on the volume and viscosity of the resuspended compound. Keep the needle inserted for ~7 s to ensure the solution is fully expelled.

5. Remove the needle and discard into an appropriate sharps container. Return the mouse into its cage. Observe for potential complications, as described *see* **Note 3** in Subheading 4.
6. Repeat **steps 1–5** for remaining mice.
7. Return all animals back to the housing cage.
8. Replace pre- and postinjection cages with clean ones.

3.5 Validation of Feeding and Injections Results

The mouse-to-mouse variation should be rather low due to the care taken to ensure each animal receives equivalent dosages (*see* **Note 11**). However, results should be verified by (e.g., qRT-PCR, histological analysis, immunohistochemistry [1, 2], or immunofluorescent staining [3]). When performing feeding procedures for the first time, it is especially important to use a positive control, or a compound that gives a well-known effect. This will ensure that you are administering a solution correctly. Our lab uses the mammalian target of rapamycin complex 1 (mTORC1) inhibitor rapamycin, as its efficacy can be readily assessed [4, 5]. This is especially important when training new laboratory members in these techniques.

4 Notes

There are multiple critical considerations to follow while feeding and injecting mice:

1. The best outcomes come from using outbred strains (e.g., CD-1), as pups are hardier and dams much less likely to cannibalize pups after handling and treatment. When possible, initiate treatments at P1, as P0 pups are more likely to die or be cannibalized after treatment.
2. Whenever possible, standardize litter sizes to ~10—this is especially important with outbred strains that routinely have large litters. This ensures nearly equivalent maternal care, which reduces developmental variation between pups. In our laboratory, which studies spermatogenesis, litter sizes are reduced by euthanizing female pups. By P1, pups can be reliably sexed based on anogenital distance differences.
3. The body weights of vehicle- and compound-treated mice should be recorded daily. If mice fail to gain weight, then this will be evident, and veterinary staff should be consulted. Euthanasia may be an option to ensure animal welfare. As a

reference, in our laboratory using outbred CD-1 mice: P1 ≈ 1.5–2 grams, they gain ≈ 0.5–1 grams per day, P10 ≈ 7–9 grams.

4. *It is essential during this entire procedure to keep pups warm.* To ensure this, pups are removed from the cage one at a time and placed into a prewarmed pipette tip box lined with Kimwipes on the slide warmer. Warm pups are more relaxed during feeding and injections.
5. Change gloves between litters. Clean pipette tip boxes with ethanol and change Kimwipes between litters. Clean reusable feeding needle with distilled water and 70% ethanol between litters. Dry needles using a laboratory high-pressure air-nozzle.
6. For subcutaneous injections, needle bevel must be oriented toward the dorsum of the mouse to be injected. Injections are done into the largest potential space in developing pups, which is the interscapular space on the upper back. Both hands must be rested on the table to prevent shaking. Pups cannot be reunited with mom for 20 min, as solution must have time to absorb before dam inevitably "scruffs" each pup at the injection site with her mouth to move it to the cage nest. Our laboratory uses DMSO as vehicle for compound dissolution.
7. Prepare two clean cages—one for preinjection and one for postinjection mice. Mice >P14 must be kept in a closed preinjection cage prior to handling, as they can readily jump out and escape or injure themselves. After injection, place each mouse into the same clean postinjection cage until the entire litter has been treated (no warming devices are necessary) and ready to be returned to the housing cage with mom. Our laboratory uses DMSO as vehicle for compound dissolution.
8. WIN 18,446 (BDAD). Prepare in a laminar flow hood. Check molecular weight (MW: 370.61) of this RA synthesis inhibitor prior to mixing, as this can vary slightly by manufacturer. Empty the contents of the BDAD vial (50 mg) into a 15 mL conical tube and add 1.35 mL sterile DMSO to obtain 100 mM solution. Add DMSO in increments to aid in dissolving. Make aliquots of 150–200 μL and store in −20 °C freezer. For oral administration (100 μg/g body weight), multiply mouse body weight × 2.7 = total amount to administer.
9. All-trans Retinoic Acid (ATRA). Prepare in a laminar flow hood. 10 mg of ATRA comes in an ampule. Add 1 mL DMSO to ampule in 100 μL increments to make 100 μg/μL stock. For complete dissolution, ampule may need to be incubated at 37 °C. For the RA working solution, add 6.32 μL stock to 120 μL DMSO to get a concentration of 5 μg/μL, or

add 13.4 μL of stock to 120 μL to get a concentration of 10.05 μg/μL. Store aliquots in black 1.5 mL microcentrifuge tubes in a box at −20 °C (reminder: RA is light-sensitive).

10. For feeding and injecting, it works best to administer volumes ≥2 μL. The upper limit, in our experience for feeding at any time is 5× body weight (g). If larger volumes must be administered, then consult with your institutional veterinarian.

11. While complications arise rarely, they are possible when treating multiple large litters on the same day. To minimize mistakes and avoid potential complications with mouse health, dedicate sufficient time to perform treatments and follow the protocols carefully. As with all procedures involving mice, consult your institutional veterinarian with any questions, concerns, or complications that arise. The following are potential issues, and how they might be addressed. (A) Bleeding at the site of injection—apply pressure until bleeding stops and clean with gauze and ethanol. (B) Inflammation of the peritoneal cavity—this may occur if gut was punctured or nonsterile substances were injected. Observe the animal for a 24 h period. If inflammation does not decrease, euthanasia should be considered. (C) Choking during feeding—typical signs of choking include lack of movement and discoloration (turning a shade of gray). This can often be resolved by laying the animal on its side and gently massaging the stomach area with a finger until pup regains normal pink coloration and movement. Place the pup on the slide warmer and observe for ~15 min before returning it to the housing cage. If the animal remains a shade of gray, euthanasia should be considered.

Acknowledgments

All procedures using animals adhere to guidelines outlined in the National Research Council Guide for the Care and Use of Laboratory Animals and were approved by the Animal Care and Use Committee at East Carolina University (approval A3469-01). The authors thank previous Geyer laboratory graduate students (Drs. Jonathan Busada, Ellen Harrington, and Nicholas Serra) who began the treatment procedures we describe here. The authors also thank Rhett Butler from East Carolina University News Services for photography. This work was supported by the NIH/-NICHD (R01 HD090083 to C.B.G.) as well as the Male Contraceptive Initiative (fellowship supporting O.K.).

References

1. Alturkistani HA, Tashkandi FM, Mohammedsaleh ZM (2015) Histological stains: a literature review and case study. Glob J Health Sci 8:72–79
2. Ehmcke J, Schlatt S (2008) In: Hou SX, Singh SR (eds) Germline stem cells. Humana Press, Totowa, NJ. pp. 109–118
3. Niedenberger BA, Geyer CB (2018) Advanced immunostaining approaches to study early male germ cell development. Stem Cell Res 27:162–168
4. Kirsanov O et al (2020) The rapamycin analog Everolimus reversibly impairs male germ cell differentiation and fertility in the mouse†. Biol Reprod 103:1132–1143
5. Busada JT, Niedenberger BA, Velte EK, Keiper BD, Geyer CB (2015) Mammalian target of rapamycin complex 1 (mTORC1) is required for mouse spermatogonial differentiation in vivo. Dev Biol 407:90–102

Chapter 14

Models and Methods for Evaluating Regeneration of Spermatogenesis After Cytotoxic Treatments

Gunapala Shetty

Abstract

Cytotoxic exposure, predominantly during radiation and/or chemotherapy treatment for cancer, interferes with fertility in men. While moderate doses cause temporary azoospermia allowing eventual recovery of spermatogenesis, higher doses of sterilizing agents can cause permanent sterility by killing the spermatogonial stem cells (SSCs). In this chapter, the methods involved in the following aspects of cytotoxic regeneration are described: (i) designing rodent and non-human primate models for regeneration of spermatogenesis after cytotoxic treatment by radiation and chemotherapy; (ii) analysis of SSCs with respect to the impact of the cytotoxic treatment, including analysis of spermatogonial clones, scoring the testicular section to analyze the extent of spermatogenic recovery, preparation of testicular and epididymal sperm, and collection of semen in non-human primates for sperm analysis; and (iii) preparation and delivery of a GnRH antagonist and steroids for enhancement or induction of spermatogonial differentiation, leading to the regeneration of spermatogenesis, largely applicable in the rat model.

Key words Cytotoxic treatment, Spermatogonial stem cell, Spermatogonia, Spermatogenesis

1 Introduction

In the clinical scenario, the gonads become susceptible to the sterilizing effects of radiation either when the localized or whole body irradiation is given, depending on whether the disease is localized or systemic [1]. When the localized treatment is given to body parts other than the gonads, although the gonads are usually shielded, they are expected to receive partial scatter doses of radiation. In animal models, either the whole body or only testicular region are exposed to precise doses of radiation to model its effects on spermatogenesis. A large number of chemotherapy agents, in addition to ionizing radiation, are prescribed to cancer patients [2] (Table 1), which are cytotoxic and potentially gonadotoxic. Since they target the actively dividing cells in general, as the malignant cells, they are expected to be gonadotoxic primarily killing the actively proliferating spermatogonia.

Jon M. Oatley and Brian P. Hermann (eds.), *Spermatogonial Stem Cells: Methods and Protocols*, Methods in Molecular Biology, vol. 2656, https://doi.org/10.1007/978-1-0716-3139-3_14,

Table 1
Classes of cytotoxic and potentially genotoxic agents used in cancer therapy [2]

Class of agent	Action on genome	Examples
Alkylating (and other cross-linking) agents	Form adducts and cross-links on DNA	Cyclophosphamide, procarbazine, cisplatin, carboplatin, melphalan, busulfan, ifosfamide, nitrogen mustard, chlorambucil, dacarbazine, mitomycin C, thiotepa, BCNU, CCNU
Topoisomerase inhibitors	Block the re-ligation of DNA during unwinding or strand unknotting	Doxorubicin, etoposide, mitoxantrone, teniposide, merbarone, idarubicin, amsacrine
Microtubule inhibitors or stabilizers	Inhibit formation or depolymerization of microtubules at mitosis or meiosis	Vincristine, vinblastine, paclitaxel
DNA strand breakage	Produce single- and double-strand breaks in DNA	Ionizing radiation, bleomycin
Nucleoside analogs and antimetabolites	Misincorporation, block biosynthesis of nucleotides, epigenetic alterations	5-fluorouracil, methotrexate, cytosine arabinoside, 6-mercaptopurine, 5-aza-deoxycytidine

Abbreviations: *BCNU* 1,3-bis(2-chloroethyl)-1-nitrosourea, *CCNU* 1-(2-chloroethyl)-3-cyclohexyl-1-nitrosourea. Reprinted from Meistrich, 2019 https://doi.org/10.1111/andr.12740 with permission

Gonadotropin and testosterone suppression in rats before or after cytotoxic therapy has been shown to enhance the recovery of endogenous spermatogenesis [3, 4]. The mechanism involves testosterone suppression-induced reversal, of a block in differentiation of surviving spermatogonia caused by damage to the somatic environment [5, 6]. Studies involved treatments with GnRH analogs [5], androgens [7], estradiol [8], or ethane dimethane sulfonate (EDS) [9] to manipulate testosterone levels. In addition to the enhancement of endogenous spermatogenic recovery, hormone suppression treatment has also been shown to augment recovery from the transplanted spermatogonial stem cells after cytotoxic treatment in mice [10] and monkeys [11].

In this chapter, the following basic methods that can be followed after testicular cytotoxic exposure have been provided: designing rodent and non-human primate models for regeneration of spermatogenesis, analysis of spermatogonial differentiation leading to spermatogenic recovery, and preparation and delivery of a model GnRH antagonist and steroids, for enhancement or induction of spermatogonial differentiation.

2 Materials

2.1 Radiation Treatments

1. Superflabs of different sizes.
2. Pressed wood, the measurement of each can be normally 1.5 × 1.5 ft. with about 0.5 cm thickness, but can be variable.
3. Irradiator (radiation sources such as ^{137}Cs, ^{60}Co or X-rays).
4. Anesthesia (inhalants such as isoflurane and injectables such as ketamine or telazol).
5. 1 mL syringes.
6. Syringe needles, 25 G 5/8″ are good for mouse ip or sc injections.
7. Thermoluminescent detectors, if used.

2.2 Chemotherapy Treatments

1. Busulfan, cyclophosphamide; for non-human primates, pharmacy grade drugs are used.
2. In rodents, the powder can be weighed and diluted in diluents described, for injections.
3. 22G gavage needle; 1 mL syringes; 25G 3/8″ needle.
4. Saline.
5. Sterile water.
6. Custom 5-7Fr silicone indwelling central venous catheters.
7. Spectra or Spectra Optia apheresis machine.
8. Granulocyte colony-stimulating factor (G-CSF)/Neulasta.

2.3 Spermatogenic Recovery Analysis

1. Supplies for processing of tissues for histology.
2. Light microscope with phase contrast optics, fluorescent microscope, with attached camera in all is useful.
3. Anesthesia (such as ketamine–xylazine cocktail).
4. Surgical instruments with forceps and scissors of suitable sizes.
5. Sterile water.
6. D-PBS.
7. DMEM/F12.
8. Polytron homogenizer.
9. Sonication instrument.
10. Petridishes 50 mm, 100 mm.
11. Microcentrifuge tubes 1 mL, 2 mL.
12. 70 μm filters.
13. Hemacytometer; Cell-Vu chamber.

14. Cryovials.
15. Electroejaculator.
16. Water bath.

2.4 Gonadotropin and Testosterone Suppression

1. GnRH antagonist. Acyline is available from the Contraceptive Development Program of the NICHD, Rockville, MD, USA.
2. Sterile water.
3. 1 mL syringes.
4. Syringe needles 25G 3/8″, 18G 1″.
5. Silastic capsule.
6. Pasteur pipette.
7. Silicone rubber (sealant).
8. 70% ethanol, betadine.
9. Sutures.
10. PBS containing 1% BSA and 0.1% sodium azide.

3 Methods

3.1 Mammalian Models for Radiation Therapy

3.1.1 Mouse Model for Radiation Therapy

1. A ^{137}Cs gamma-ray unit with a slot for inserting a metal shield module for animals or a ^{60}Co gamma-ray unit is preferred.
2. When using a ^{137}Cs gamma-ray unit for radiation exposure only to the testis and surrounding area, restrain the unanesthetized mouse in plastic chambers (Fig. 1a) and then place into the metal shield module (brass or lead) with about 3-cm diameter hole (Fig. 1b), so that only the lower abdominal and scrotal area of the animal can be irradiated.
3. Note that a metal shield is required only while using a cesium irradiator. The cobalt unit normally has a collimator to define field size.
4. Fractionated radiation regimens have been shown to be more effective than a single dose in depleting germ cells and produce less unwanted adverse effects [12].
5. Doses can be chosen based on the recoveries of spermatogenesis and fertility after different single doses of irradiation [13].

3.1.2 Rat Model for Radiation Therapy

1. The lower part of the body of rats including the testis can be irradiated, until maximum doses of 10 Gy using an irradiation unit such as a ^{60}Co gamma-ray unit without any adverse effects on health conditions [14].
2. Place the anesthetized rats on their backs (Fig. 2) on a minimum of 5″ press wood to provide backscatter for radiation uniformity.

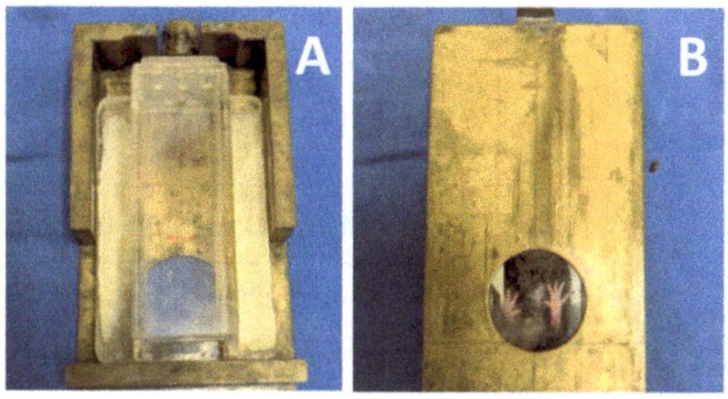

Fig. 1 Restraining of unanaesthetized mouse for ^{137}Cs irradiation with a plastic chamber (**a**) that can be fitted to a metal shield module, made of brass, with a hole for radiation exposure. The back side of the module is shown (**b**) with the testis portion of the mouse being exposed (**b**). Note that the brass may work for cesium source, which has a lower energy but may not work so well for Cobalt

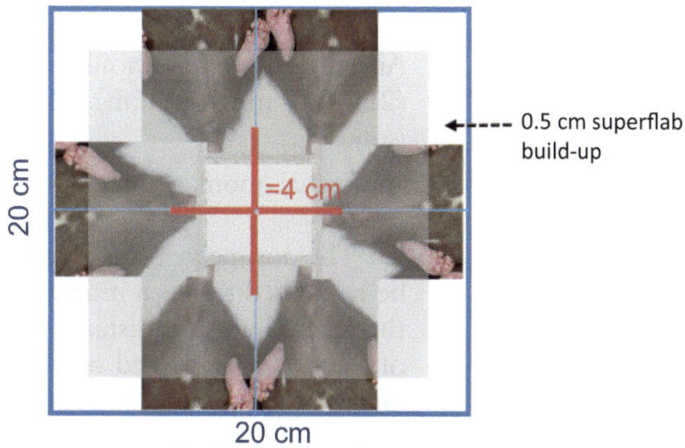

Fig. 2 Positioning of rats for irradiation. A 0.5 mm bolus needs to be placed in the center covering the testes of all rats. In this case, the field size is chosen as 20 × 20 cm

3. Place 5 mm of tissue-equivalent bolus material over the scrotum to provide a build-up layer.
4. The approximate field can be selected distally from a line about 5 cm below the diaphragm or about 6.5 cm above the base of the scrotum.
5. Four to six rats can be irradiated at a time, ideally with a field size of 20 × 20 cm.
6. When calculating the dosimetry (duration of radiation exposure), the distance from the radiation source to the top of the superflab on the testis, the field size, and the determined dose rate of the source (Gy/min) at the time of irradiation are considered.

7. Example: To irradiate six rats underneath 5 cm superflab on 5″ press wood with 80 cm distance from the source to the top of the superflab.

Depth of calculation (testis)	1 cm
Field size dependent factor 20 × 20 cm	1.045
Back scatter factor for 20 × 20 cm field	1.078
Percent depth dose @depth of 1 cm	98.3%
Off-axis factor	0.995
Total factor	**1.102**

Irradiation time (min): Given dose (cGy)/(standard output in cGy/min) (Total factor)

3.1.3 Non-human Primate Model for Radiation Therapy

1. Anesthetize the monkeys with ketamine or telazol (2.2–4.4 mg/kg; i.m). Maintain anesthesia with 1–3% isoflurane.
2. Individual monkeys need to be irradiated separately to the testes.

   The dosimetry to give a particular dose of radiation should be calculated for each monkey testes based on the thickness of the testis, field size, distance of the radiation source to the top of the superflab placed on testes, and the standard output of radiation (cGy/min) at the time of radiation.
3. The precise dosimetry can be performed in a simulated phantom made from 1.5% agarose (Fig. 3) [11]. In this case, a radiochromic film can be laid in midsagittal plane of the agarose scrotum, which can be processed to estimate the radiation exposure.
4. Place a tissue-equivalent bolus material (e.g., superflab) (5 mm thick) over the scrotum to provide a build-up layer.
5. For dose uniformity, both postero-anterior (PA) and antero-posterior (AP) positions can be used in adult monkeys, with half of the dose given in each position.

3.2 Mammalian Models for Chemo Therapy

3.2.1 Mouse Model for Busulfan

1. First dissolve busulfan in dimethyl sulfoxide (DMSO) and then add an equal volume of sterile water to obtain a final busulfan concentration of 3 mg/mL.
2. Maintain the aqueous dilution at a temperature slightly greater than 37 °C until injection to prevent the busulfan from crystallizing and to keep it in solution.

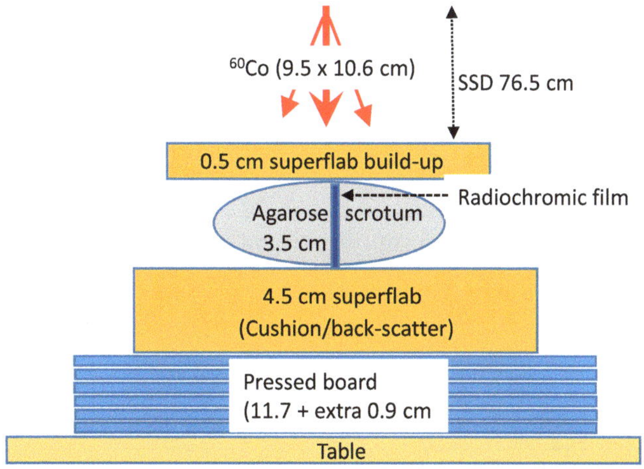

Fig. 3 Schematics and setup to determine the dosimetry while irradiating monkey testis. A radiochromic film (such as Gafchromic® EBT2 model), which is widely used for dosimetry, is laid in midsagittal plane of an agarose scrotum. . Underneath the scrotum, 4.5 cm superflab plus about 20 cm-thick pressed board is placed. A 0.5 cm superflab is also placed on the top. This arrangement reproduces the desired geometry established in the dry run for a monkey with a source to surface distance (SSD; 76.5 cm in this case) to top of the superflab. Field size need to be set ($x = 9.5$ cm, and $y = 10.6$ cm in this case). Set timings in minutes for AP and PA irradiation, respectively, to deliver an estimated total dose in cGy. The actual radiation dose received into the mid region of the testis can be ascertained by analyzing the exposed radiochromic film

3. Using a 25 G 5/8″ needle, inject intraperitoneally in a maximum volume of 0.5 mL to produce depletion of endogenous spermatogenesis.
4. A variable dose of 20–60 mg/kg have been used in various studies in mice [15–17]. A dose of 60 mg/kg is known to produce about 18% of the tubules to have differentiating germ cells in nude mice [16].

3.2.2 Mouse Model for Cyclophosphamide

1. Cyclophosphamide containing 75 mg mannitol per 100 mg cyclophosphamide can be dissolved in sterile water at an appropriate concentration.
2. Previous studies involved treatment with cyclophosphamide (either through oral (p.o., 22G gavage needle) or intraperitoneal (i.p.) administration.
3. For single injections, doses at 150 mg/kg [18] or 200 mg/kg [19], either p.o. or i.p. are reasonable considering its effects on spermatogonia and animal survival.
4. To evaluate the effects of multidose treatments, mice can be given up to 7 doses of cyclophosphamide, at a dose of 150 mg/kg each, at 4 or 5 day intervals, either through p.o. or i.p. routes.

Table 2
Assessment intervals for sperm counts chosen based on predicted recovery of sperm production from different cell types potentially affected by cyclophosphamide in mice

Days after last injection	Latest stage of cell capable of producing testicular spermatozoa at indicated time[a]
34	A_2 spermatogonia stage IX
37	$A_{aligned}$spermatogonia, stage VI
40	A_{paired}spermatogonia, stage I
45	A_{single}spermatogonia, stage VI
56	A_{single}spermatogonia, earlier cycle

[a]Based on different stage of the seminiferous epithelial cycle in mice

5. The time points to harvest the testis after cytotoxic treatment to directly analyze the short-term cytotoxic effects on spermatogonia should be based on the different stages of seminiferous epithelial cycle.
6. Particular time points for analysis of testicular sperm count should be chosen based on the time intervals for the different cell types to produce testicular sperm [19, 20] (Table 2) and the agent/dosing chosen based on the sensitivities of cell types of interest (Table 6).

3.2.3 Non-human Primate Model for Chemotherapy (Busulfan) [21–26]

1. In non-human primates, either single doses of busulfan at 4–12 mg/kg [21, 22] or multiple doses of (4–9.6 mg/kg each time) [25–27] have been tried to test the effects on spermatogenesis.
2. Dilute Busulfex in physiological saline and administer intravenously at 0.6 mg/mL over 10–20 min.
3. Although a single dose of 4 mg/kg is tolerable and myelosuppressive effects are minimal, doses beyond 8 mg/kg need restoration of hematopoietic system by peripheral blood stem cell (PBSC) transplant, to support animal vitality and survival. The procedure for the PBSC transplant is as below [22].
4. First, autologous blood needs to be collected (~5 mL/kg/week) for few weeks and pooled to prime the apheresis tubing set.
5. For this purpose, custom 5-7Fr silicone indwelling central venous catheters (Access Technologies, Skokie, IL) can be used, which needs to be placed in the right internal jugular vein.
6. Simultaneously, starting few days before busulfan injection, hematopoietic stem cells from the bone marrow can be mobilized into the general circulation, by giving daily subcutaneous

injections of cytokines (10 μg/kg/day G-CSF plus 200 ug/kg/day SCF or 20 μg/kg/day G-CSF alone) [28].

7. PBSC collections ideally have to be performed a day before busulfan injection by apheresis using either a Spectra or Spectra Optia apheresis machine (CaridianBCT), utilizing the pooled blood to prime the apheresis tubing set.
8. Typically at least 250×10^6 total white blood cells per kg body weight are needed for transplant according to published recommendations [28].
9. Twenty-four hours after completing apheresis, animals can be treated with busulfan, followed by transfusion with the collected PBSCs, a day later.
10. To stimulate rapid expansion of engrafted stem cells and hematopoietic recovery, animals can be given one or more subcutaneous injections of long-acting G-CSF (300 ug/kg; Neulasta, Amgen).
11. It is important that animals are monitored twice weekly for hematopoietic deficits by complete blood count and to provide appropriate supplemental veterinary care.

3.3 Analysis of Spermatogenic Recovery After Cytotoxic Treatment

3.3.1 Determination of Sizes and Densities of a Spermatogonial Clones

1. The clonal size of the A spermatogonia can be determined in tubule whole mounts after the cytotoxic exposure as demonstrated in transgenic mice expressing EGFP in the spermatogonia [29] or in irradiated rats [30, 31].
2. Dissect the tubules from the testis of the experimental animal (please see procedure for preparation of tubules for whole mounts in other chapter).
   (a) Use tubules from mice with GFP transgene driven by a spermatogonial gene or fluorescently stain with spermatogonia marker (such as GFRa1, CDH1) and mount on slides.
   (b) Alternatively, stain the isolated tubules with hematoxylin and mount on slides.
3. Per animal, randomly choose 1–3 cm in length of tubule fragments and count the number of clones along the length of the tubule.
4. Use photomicrographs to count the numbers of clones per unit length to determine the clone density.
5. Undifferentiated A spermatogonia lying within 20 μm of each other and showing the same morphology are generally considered to be belonging to the same clone, but rigorous proof has shown that it has to be done by visualizing intercellular bridges, such as with Tex14 staining.
6. The A spermatogonia that belong to the same clone are connected by intercellular bridges, and therefore, the cells

comprising a clone will go through the cell cycle in a synchronous fashion.

7. As mitosis is of short duration, neighboring cells simultaneously in mitosis will most likely belong to the same clone.

3.3.2 Counts of Repopulated Tubules in the Testis

1. These counts must be done at a time greater that the duration of spermatogenesis (from spermatogonial stem cell to sperm).
2. In each testicular cross section, score the repopulated and total tubular cross sections that are round or nearly round.
3. When more detail is desired, one can assign them a category based on the latest germ cell maturation stage present (A spermatogonia to condensed spermatids; Table 4).
4. Choose the 40× objective on a bright field microscope and start scoring at the left top corner of the section.
5. Using the Y-axis stage controller, move toward the left lower side of the section and start counting the number of round or oval tubular sections, and for the meantime, assign them a category based on the latest germ cell maturation stage (Sertoli cells only, or from A spermatogonia to condensed spermatids; Table 4).
6. A good criterion to avoid misclassifying tubules is only count a tubule as repopulating if there are at least three clearly differentiated germ cells (B spermatogonia or beyond) present, or if assigning the tubule to a specific differentiation stage, that there are three cells of that stage of maturity.
7. Once the lowest edge of the section is reached, using X-axis controller move slightly to the right side of the section (about 100 μm), making sure that the tubules already scored are not repeated, and then move upward scoring the sections.
8. Repeat until a total of about 100–200 tubules are scored in mice/rats. In monkeys, due to the bigger size of the testis, it is advised to score at least 1000 tubules choosing 4–5 randomly chosen tissue pieces from different regions of the testis.
9. The ratio of the total number of the differentiated tubules (tubules with B spermatogonia or beyond) vs total number of tubules counted is designated as tubule differentiation index (TDI), previously designated as RI (repopulation index) [13].
10. Provided there is not a block in spermatogonial differentiation, as is prominent in the rat, the relative number of surviving stem cells per tubular cross section, referred to as the stem cell survival index, SSI, can be calculated from the RI by applying the Poisson distribution rules [32] $SSI = -\log_e (1 - RI)$.

3.3.3 Histological Quantification of Spermatogonial Germ Cells Relative to Sertoli Cells After Cytotoxic Treatment

1. For best results, use tissues fixed in Bouin's fixative and embedded in methacrylate for this purpose.
2. Quantitative germ cell comparison between control and treated testes can be done by determining the ratio of each generation of premeiotic germ cells to Sertoli cells (with nucleoli visible) in seminiferous tubules.
3. Selecting the round or oval tubule cross sections, count the number of each type of premeiotic germ cells, being sure that enough tubules are counted to ensure a representative sampling of all stages of the cycle of the seminiferous epithelium.
4. Calculate the averages and SE (between animals) of these ratios.

3.3.4 Sample Preparation for Testicular Sperm Head Counts

1. Immediately after euthanasia, place the testes with the intact tunica albuginea in mice and without tunica in rats, into distilled water (for mice, approximately 1 testis/mL).
2. Homogenize each sample with a Polytron homogenizer for 30 s and sonicate using a sonicator (example: Branson sonifier fitted with a microprobe), for 2–3 minutes.
3. Remove an aliquot for sperm head counts.
4. Count the sperm heads, i.e., sonication-resistant spermatid nuclei in **steps 12–16** of development, in a hemacytometer using a microscope with phase contrast optics.
5. Count at least 200 sperm heads. If the counts are very low, count a volume of 9×10^{-4} (one whole side of hemacytometer).

3.3.5 Mouse Epididymal Sperm Count

1. Euthanize the mice and dissect out the epididymis first, before removing the testis.
2. Cut the cauda epididymis, place in a small Petri dish.
3. Add 100 μL of DMEM/F12 and mince thoroughly using small scissors.
4. Add another 900 μL of DMEM/F12 and repeatedly pipette the preparation ~10 times, using 1 mL pipette.
5. Allow at room temperature for 30 min.
6. Filter the preparation using 70 μm filter.
7. First, observe the motility and then count the sperm.
8. If the sperm are moving fast, dilute with water to arrest motility so that they can be counted.
9. For analyzing motility, record the percentage of sperm that are showing [1] progressive forward motility, [2] just wiggling, or [3] immotile.

3.3.6 Rhesus "Swim-Up" Sperm Preparation from the Epididymis

1. Pre-warm D-PBS, Test Yolk Buffer freezing medium, and a 250 mL beaker containing water to 37 °C.
2. Isolate a 0.5 g portion of the epididymal tail (cauda). Note from which testis the epididymis is from and keep Left and Right separate, if possible.
3. Mince epididymis in 0.5 mL pre-warmed D-PBS (37 °C) with small scissors in 1.5 mL microcentrifuge tube.
4. Incubate sample at 37 °C for 1 h., allowing sperm to "swim-up" into the buffer (invert tube once at 30 min. Large pieces of tissue will sediment by gravity over the remaining 30 min, do not centrifuge!)
5. Pre-wet a 70 μm cell strainer basket (BD 352350) with 500ul D-PBS over a 50 mL conical tube.
6. Gently remove supernatant buffer containing sperm with a 1 mL pipette and filter through cell strainer into 50 mL conical tube.
7. Add 0.5 mL pre-warmed D-PBS to the epididymal tissue, invert 3× to mix, allow tissue to settle, remove supernatant, and filter through cell strainer.
8. Discard tissue and cell strainer.
9. Measure the volume of D-PBS + sperm in 50 mL conical tube. Bring total volume to 2 mL with pre-warmed D-PBS. (*Ideally, a sperm analysis should be performed at this step (motility, morphology, contamination with blood cells,* etc.)

3.3.7 Freezing Monkey Sperm for Cryopreservation

1. Drop-wise, add an equal volume of pre-warmed Test Yolk Buffer freezing medium (i.e., 2 mL) over a 30 s period. Mix thoroughly after each drop of freezing medium is added to avoid osmotic shock to the sperm.
2. Allow mixture to equilibrate for approximately 10 min at room temperature.
3. Transfer final mixture into cryovials or cryostraws.
4. Chill samples for 1 h in the refrigerator (2–5 °C).
5. Expose to liquid nitrogen vapor for 30–60 min either by suspending them in the liquid nitrogen storage tank above the liquid level or by placing them in the vapor phase in a small temporary liquid nitrogen dewar.
6. Transfer to the liquid nitrogen tank for storage at −196 °C.

3.3.8 Semen Collection and Analysis in Monkeys Using a Rectal Probe

1. The electroejaculator that we used is from Standard Precision Electronics, Inc. It has a manual control to regulate the intensity and duration of the electrical stimuli. The rectal probe, composed of a resin rod, is 5/8″ in diameter and X 5″ long.

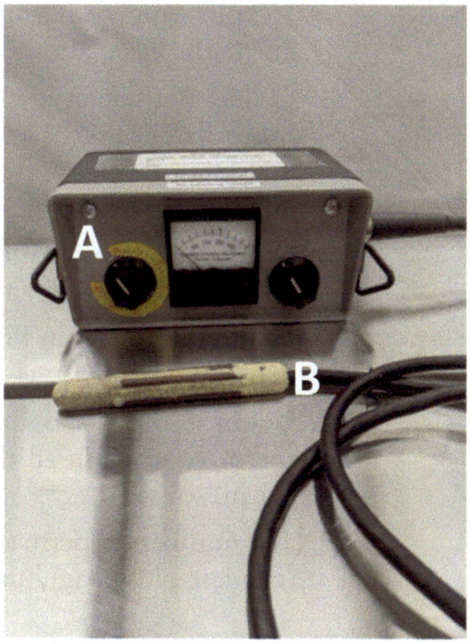

Fig. 4 Electroejaculator (from standard precision electronics) showing the control unit (**a**) and the rectal probe (**b**) with three electrodes extending lengthwise

Three metal sheets are placed parallel to the probe in a linear position (Fig. 4).

2. *BATT* position without the use of *System Operational Checker (SOC)* is used for normal operation. (First time use: set in the EXT position and connect the system operational checker (SOC) to the power control, which is then connected to the probe. However, use of SOC is not enough to elicit response as we found earlier.
3. The battery has to be checked the day before electroejaculation is scheduled. TEST position of the unit (right side knob) has to be used to check the battery voltage. The meter should indicate 350, if it is less than 300, it has to be charged. The charging has to be done in the OFF position.
4. The electroejaculation procedure is explained below [11].
5. Place the anesthetized monkey in ventral recumbency, with hindlegs and penis off the table, supported by a person.
6. Lubricate the rectal probe with a lubricant inserted gently in to the rectum by facing the side with the electrodes to the abdominal side so that the electrodes are adjacent to the prostate.
7. In a rhythmic motion, stimulate the monkey by turning the power knob of the ejaculator clock wise, held for a second, returned to zero, and again held for a second. The stimulation needs to be applied every 3–5 s initially at 10 volts, and if the ejaculation does not occur, increase the voltage gradually to

15 volts. Apply a maximum of about 20 stimulations at a time unless the ejaculation occurs before that.

8. Collect the semen in a 50 mL graduated polypropylene tube. Immediately after ejaculation, evaluate the physical characteristics of the semen and estimate the volume of the liquefied and unliquified coagulum.
9. During semen evaluation, keep the ejaculate in a water bath at 37 °C for an hour for the liquefaction of the coagulum.
10. Dilute the sample in PBS or HEPES buffer to count if it is too dense.
11. Count using disposable Cell-Vu chamber. The number of sperm counted in all the 100 grids × 5 × 10^4 is the number per mL of semen.
12. The number of sperm is expressed as number per total ejaculate (exudate plus coagulum) and per mL of ejaculate.

3.4 Gonadotropin and Testosterone Suppression Treatments

3.4.1 Treatment with a GnRH Antagonist Acyline (Rodents)

1. Prepare the GnRH antagonist (Acyline) in sterile water at a concentration of 2 mg/mL.
2. Deliver the drug by subcutaneous injections. In rats and mice, a 25G 5/8″ needle can be used for injections.
3. In mice the standardized effective dose is an initial dose of 20 mg/kg body weight followed by maintenance doses of 10 mg/kg body weight given every other week [33].
4. In rats, a dose of 1.5 mg/kg given every week is effective for a significant suppression of testicular testosterone levels [34].

3.4.2 Silastic Capsule Implants for Steroid Delivery

1. Decide which Silastic Medical Grade Tubing to use (can be purchased from Thermo Fisher; Dow Corning Corporation brand).

   0.078″ I.D. × 0.125″ O.D. is ideal for rats and mice

   Decide on length of capsule: 2 cm has surface area of 199.80 mm^2
2. Cut tubing to desired length. Allow 3–5 mm on each side. Wash the cut tubing pieces in hot soapy water solution (use non-oily, mild soap). Rinse copiously in hot running water for approximately 3 min. Rinse again with cold distilled or Milli-Q water.
3. Seal the ends with silicone (Clear 100% Silicone Rubber #732 Multi-Purpose Sealant).
4. Fill a disposable 1 mL syringe with silicone from 3 above.
5. Plug one end of tubing (push adhesive from syringe into tubing ~3–5 mm). Let dry overnight.
6. Take a glass Pasteur pipette and cut constricted end just below cone-shaped area and cut tip off as large as possible (tubing

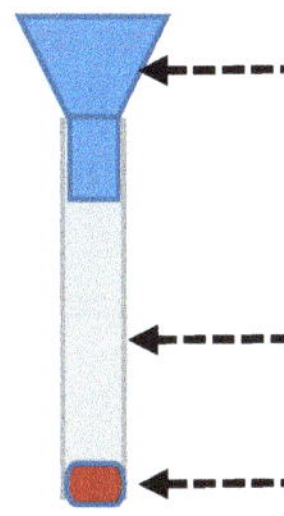

Fig. 5 Diagram showing the filling of the silastic tubing with substances such as steroids used for subcutaneous delivery of drugs

must slip over it). A 1 mL pipette tip can also be used suitably trimmed at the tip.

7. Flame both ends of pipette to remove rough edges. Let cool before attaching tubing.
8. Put tubing on pipette. Fill larger portion of Pasteur pipette with substance (Fig. 5).
9. Use a plugged capillary tube to push and pack the substance to the base of the tubing to the Silicone plug.
10. Take an 18-G needle and file approximately 5 mm from the bevel tip and sand to remove any sharp edges.
11. Place on syringe filled with silicone sealant and insert the needle into the tube to the base of the substance and fill as the needle is slowly being removed. Make sure that adhesive is in contact with tubing walls and that there is no air bubbles.
12. Dry overnight.
13. Rinse in methanol.
14. Use the following solution to store the implants: PBS containing 1% BSA and 0.1% sodium azide.
15. The silastic capsules can be implanted to animals as follows.
16. Shave 5–10 mm a section on the animals' back with a double-edged razor blade.
17. Rinse thoroughly in 70% ethanol or betadine.
18. Cleanse area thoroughly. Grasp a portion of the skin on the back with flat-tipped forceps. Make a small cut with the razor blade. Use a trocar or a pair of narrow forceps to make a subcutaneous opening at least 2.5× the length of the capsule being inserted. Do not push downward. Stay near the surface as possible.
19. Suture or staple the opening. Make sure that the implanted capsule is not near the open incision.
20. Check animals several times before going home and check the next morning to make sure the implants have not come out.

Also, make sure the incision is closed and the implants are not exposed. Re-cage the next morning, if there are signs that the animals were fighting.

4 Notes

1. With fractionated doses, the mouse survival would be better with similar degree of stem cell killing [13], but there may be strain differences in response of mice to fractionated doses of irradiation [35]. The longer fractionation (about 4 days) will result in less skin and gut damage with good stem cell killing (Subheading 3.1.1).
2. To minimize damage to other organs, irradiation units with more precise deliveries to smaller organs may also be considered (Subheading 3.1.1).
3. No toxicities in single doses up to 10.0 Gy were observed in rats [14] (Subheading 3.1.2).
4. Lewis and Brown Norway (BN) are good models for studying effects of radiation on spermatogonial differentiation in rats, since these strains are most sensitive to radiation-induced block in the differentiation of spermatogonia [36] (Subheading 3.1.2).
5. To study the effect of radiation on the block in the spermatogonial differentiation in vivo, the testes can be harvested at the earliest 10 weeks after irradiation [37]. During this period, they are to be treated with or without other agents to be tested (Subheading 3.1.2).
6. Radiation source such as cobalt-60 gamma [11], cesium-137 [10], photon beams [38], or an X-ray unit [39] can be used (Subheading 3.1.3).
7. The actual doses of radiation, received by the testis, can also be determined by placing two to four thermoluminescent detectors on the anterior and posterior aspects of the scrotum [40] (Subheading 3.1.3).
8. Radiation sensitivities of testicular stem cells in animal models are provided in Table 5 (Subheading 3.1.3).
9. Busulfan is an antineoplastic agent that produces lethality through hematopoietic toxicity. Previously, 5% lethality was observed in C3H mice with a dose of 40 mg/kg [15] (Subheading 3.2.1).
10. The expected extent of loss of testicular germ cells and their recovery with time after busulfan treatment in mice has been provided in Table 6 (Subheading 3.2.1).

Table 3
Comparison of mouse testicular stem cell sensitivities to alkylating and other cytotoxic agents (modified from [15])

Agent	LD_{50} stem cell/LD_{50} differentiated spermatogonia[a]	LD_{50} stem cell/LD_{50} animal mortality
Busulfan (in 50% DMSO)	0.8	<0.20[b]
Adriamycin	3	0.19
BCNU	4	0.70
Mitomycin C	6	0.63
Gamma radiation	6	0.43
Chlorambucil	7	0.87
Procarbazine	8	0.57
Thio-TEPA	17	0.43
Mechlorethamine	>4[c]	>0.91[c]
Cyclophosphamide	>7[c]	>0.61[c]
Cis-platinum	>9[c]	0.98[c]
Cytosine arabinoside	>440[c]	>1.1[c]

[a]LD_{50} of stem cells and differentiated spermatogonia found from testicular sperm head counts at 56 and 29 days posttreatment, respectively
[b]Upper limit of ratio, as LD_{50} for animal mortality was not reached at 40 mg/kg
[c]Lower limit of ratio, as LD_{50} of stem cells was not reached at doses lethal to mice

Table 4
Plan for scoring the testicular seminiferous tubules in sections to estimate the recovery of spermatogenesis, by determining the Tubule Differentiation Indices (TDI), after cytotoxic exposure (use 40× objective)

1	2	3	4	5	6	7	8	9	10	11
Spermatogenesis advanced to the stage								Total	TDI (%)	Rows & Directions scored
Sertoli Cell Only	A spermatogonia	Mitotic (1)	B spermatogonia (3)	Spermatocyte (3)	Round Spermatid (3)	Elongating Spermatid (3)	Condensed Spermatid (10)			
								a	b	

Numbers in the parenthesis represent the minimum number to be present in a tubule section, if to be assigned in that latest stage
[a]Sum columns 1–8
[b]Sum columns 4–8/column 9

Table 5
Interspecies differences in the survival and functional ability of stem spermatogonia after irradiation in adults

Species	Radiation dose (time after)	Repopulating tubules (%)	Non-repopulating tubules with A spermatogonia
Rat	6 Gy (10 weeks)	<1	13%
Mouse	10 Gy (6 wks)	18	1%
Monkey	6.7 Gy (13 weeks)	1	0%
Human	6 Gy (single dose) (13 weeks)	0	0.7%*

[a]Number of spermatogonia counted per Sertoli cell and given as percent of control

Table 6
The expected loss and recovery of different germ cell types after a moderate dose of 20 mg/kg of busulfan treatment in mice based on the literature [17, 47–49]

days	As/pr	Aal	A1-A4	Int-B	Pl-Z	EP	LP	RS	ES	LS
4	?	constant	constant	absent	normal	normal				
8	7%	17%	17%	5%	5-7%	39%	normal			
16	20%	Slight increase*	Low*	↑	<10%	↓toP7	normal			
24	25%	↓	↑	Low	Low	Low	absent	absent	normal	normal
36	30%	30%	30%	30%	30%	Low	absent	absent	absent	absent
56	40%	40%	40%	40%	40%	~30%	~20%	low	low	1%

Abbreviations: A single/A paired (As/pr); A aligned (Aal); Intermediate-B spermatogonia (Int-B); Preleptotene (PL); Leptotene (L); Zygotene (Z); Early Pachytene (EP); Late Pachytene (LP); Round Spermatid (RS); Elongating Spermatid (ES); Elongated Spermatid (LS)

11. Comparison of testicular stem cell sensitivities to alkylating agents and other cytotoxic agents in mice is provided in Table 3 (Subheading 3.2.2).
12. In contrast to busulfan and radiation, cyclophosphamide produces very little spermatogonial stem cell killing effects in rodents [19] (Subheading 3.2.2).
13. The counts of repopulated tubules in the testis after cytotoxic exposure provide a measure of the number of surviving, fully functional stem cells (Subheading 3.3.2).
14. A stereology software (Stereo Investigator, MBF Bioscience) can be used for the systematic random choosing of the tubules

for scoring. In this case, the scored tubules can be marked avoiding multiple scoring of tubules (Subheading 3.3.2).

15. Alternatively, using the stereology software (Stereo Investigator, MBF Bioscience), the absolute numbers of specific germ cell types per testis can be determined by optical dissector method [41, 42]. This requires thick sections embedded in methacrylate (25 μm) from different regions of the testis selected in systematic random method (Subheading 3.3.3).
16. Testicular sperm count is more reproducible than epididymal sperm counts and not affected by hormonally related changes in epididymal sperm transport.
17. To ensure the best quality sperm, it is important that the epididymides are not chilled before sperm isolation (Subheading 3.3.6).
18. Hormonal suppression treatments can be initiated either few weeks before, both before and after, or after cytotoxic treatment to enhance endogenous recovery after cytotoxic treatment in rodents [5, 10, 43] (Subheading 3.4.1).
19. Similarly, to possibly enhance the spermatogenic recovery from the transplanted spermatogonial stem cells, hormone suppression treatment has been attempted either before or both before and after transplantation in the cytotoxic models of rodents and non-human primates [6, 10, 11, 44–46] (Subheading 3.4.1).
20. The duration of treatment in mice and rats can be 4–13 weeks. In non-human primates, the optimal treatment time useful for enhancing the recovery from the transplanted cells has been 8 weeks, given just before transplantation [11] (Subheading 3.4.1).
21. Implants can be stored in the solution above for approximately 8–9 months or longer (Subheading 3.4.2).
22. Do not store implant capsules containing chemicals together. You will run a risk of cross-contamination (Subheading 3.4.2).
23. The reason for the storage of the tubing in a solution is to prevent a surge in release of a chemical coming to the surface during the time between implant preparation and implantation (Subheading 3.4.2).
24. The capsules can be re-used within a period of 8–9 months of preparation, if the chemicals inside (steroids) are still fully packed.

Acknowledgement

The author is thankful to Dr. Marvin L. Meistrich for editorial assistance and valuable suggestions on the technical aspects of this manuscript.

References

1. Meistrich ML (2013) Effects of chemotherapy and radiotherapy on spermatogenesis in humans. Fertil Steril 100(5):1180–1186. https://doi.org/10.1016/j.fertnstert.2013.08.010
2. Meistrich ML (2020) Risks of genetic damage in offspring conceived using spermatozoa produced during chemotherapy or radiotherapy. Andrology 8(3):545–558. https://doi.org/10.1111/andr.12740
3. Meistrich ML, Shetty G (2003) Suppression of testosterone stimulates recovery of spermatogenesis after cancer treatment. Int J Androl 26:141–146
4. Shetty G, Meistrich ML (2005) Hormonal approaches to preservation and restoration of male fertility after cancer treatment. J Natl Cancer Inst Monogr 34:36–39
5. Shetty G, Wilson G, Huhtaniemi I, Shuttlesworth GA, Reissmann T, Meistrich ML (2000) Gonadotropin-releasing hormone analogs stimulate and testosterone inhibits the recovery of spermatogenesis in irradiated rats. Endocrinology 141(5):1735–1745
6. Zhang Z, Shao S, Meistrich ML (2007) The radiation-induced block in spermatogonial differentiation is due to damage to the somatic environment, not the germ cells. J Cell Physiol 211:149–158
7. Shetty G, Wilson G, Hardy MP, Niu E, Huhtaniemi I, Meistrich ML (2002) Inhibition of recovery of spermatogenesis in irradiated rats by different androgens. Endocrinology 143:3385–3396
8. Porter KL, Shetty G, Shuttlesworth G, Weng CY, Huhtaniemi I, Pakarinen P, Meistrich ML (2009) Estrogen enhances recovery from radiation-induced spermatogonial arrest in rat testes. J Androl 30(4):440–451
9. Shetty G, Zhou W, Weng CC, Shao SH, Meistrich ML (2016) Leydig cells contribute to the inhibition of spermatogonial differentiation after irradiation of the rat. Andrology 4(3):412–424. https://doi.org/10.1111/andr.12168
10. Wang G, Shao SH, Weng CC, Wei C, Meistrich ML (2010) Hormonal suppression restores fertility in irradiated mice from both endogenous and donor-derived stem spermatogonia. Toxicol Sci 117(1):225–237
11. Shetty G, Uthamanthil RK, Zhou W, Shao SH, Weng CC, Tailor RC, Hermann BP, Orwig KE, Meistrich ML (2013) Hormone suppression with GnRH antagonist promotes spermatogenic recovery from transplanted spermatogonial stem cells in irradiated cynomolgus monkeys. Andrology 1(6):886–898. https://doi.org/10.1111/j.2047-2927.2013.00126.x
12. Creemers LB, Meng X, den Ouden K, van Pelt AM, Izadyar F, Santoro M, Sariola H, de Rooij DG (2002) Transplantation of germ cells from glial cell line-derived neurotrophic factor-overexpressing mice to host testes depleted of endogenous spermatogenesis by fractionated irradiation. Biol Reprod 66(6):1579–1584
13. Meistrich ML, Hunter N, Suzuki N, Trostle PK, Withers HR (1978) Gradual regeneration of mouse testicular stem cells after ionizing radiation. Radiat Res 74:349–362
14. Abuelhija M, Weng CC, Shetty G, Meistrich ML (2013) Rat models of post-irradiation recovery of spermatogenesis: interstrain differences. Andrology 1(2):206–215. https://doi.org/10.1111/j.2047-2927.2012.00034.x
15. Bucci LR, Meistrich ML (1987) Effects of busulfan on murine spermatogenesis: cytotoxicity, sterility, sperm abnormalities, and dominant lethal mutations. Mutat Res 176:259–268
16. Zhang Z, Shao S, Meistrich ML (2006) Irradiated mouse testes efficiently support spermatogenesis derived from donor germ cells of mice and rats. J Androl 27(3):365–375
17. Parekh PA, Garcia TX, Waheeb R, Jain V, Gandhi P, Meistrich ML, Shetty G, Hofmann MC (2019) Undifferentiated spermatogonia regulate Cyp26b1 expression through NOTCH signaling and drive germ cell differentiation. FASEB J 33(7):8423–8435. https://doi.org/10.1096/fj.201802361R
18. Drumond AL, Weng CC, Wang G, Chiarini-Garcia H, Eras-Garcia L, Meistrich ML (2011) Effects of multiple doses of cyclophosphamide on mouse testes: accessing the germ cells lost, and the functional damage of stem cells. Reprod Toxicol 32(4):395–406. https://doi.org/10.1016/j.reprotox.2011.09.010. S0890-6238(11)00381-9 [pii]
19. Lu CC, Meistrich ML (1979) Cytotoxic effects of chemotherapeutic drugs on mouse testis cells. Cancer Res 39:3575–3582
20. de Rooij DG (1998) Stem cells in the testis. Int J Exp Pathol 79(2):67–80
21. Hermann BP, Sukhwani M, Lin CC, Sheng Y, Tomko J, Rodriguez M, Shuttleworth JJ, McFarland D, Hobbs RM, Pandolfi PP, Schatten GP, Orwig KE (2007) Characterization, cryopreservation, and ablation of spermatogonial stem cells in adult rhesus macaques. Stem Cells 25(9):2330–2338

22. Hermann BP, Sukhwani M, Winkler F, Pascarella JN, Peters KA, Sheng Y, Valli H, Rodriguez M, Ezzelarab M, Dargo G, Peterson K, Masterson K, Ramsey C, Ward T, Lienesch M, Volk A, Cooper DK, Thomson AW, Kiss JE, Penedo MC, Schatten GP, Mitalipov S, Orwig KE (2012) Spermatogonial stem cell transplantation into rhesus testes regenerates spermatogenesis producing functional sperm. Cell Stem Cell 11(5):715–726. https://doi.org/10.1016/j.stem.2012.07.017. S1934-5909(12)00475-4 [pii]
23. Kuramoto K, Follman D, Hematti P, Sellers S, Laukkanen MO, Seggewiss R, Metzger ME, Krouse A, Donahue RE, von Kalle C, Dunbar CE (2004) The impact of low-dose busulfan on clonal dynamics in nonhuman primates. Blood 104(5):1273–1280. https://doi.org/10.1182/blood-2003-08-2935
24. Kang EM, Hsieh MM, Metzger M, Krouse A, Donahue RE, Sadelain M, Tisdale JF (2006) Busulfan pharmacokinetics, toxicity, and low-dose conditioning for autologous transplantation of genetically modified hematopoietic stem cells in the rhesus macaque model. Exp Hematol 34(2):132–139. https://doi.org/10.1016/j.exphem.2005.10.010
25. Xie J, Larochelle A, Maric I, Faulhaber M, Donahue RE, Dunbar CE (2010) Repetitive busulfan administration after hematopoietic stem cell gene therapy associated with a dominant HDAC7 clone in a nonhuman primate. Hum Gene Ther 21(6):695–703. https://doi.org/10.1089/hum.2009.191
26. Uchida N, Nassehi T, Drysdale CM, Gamer J, Yapundich M, Bonifacino AC, Krouse AE, Linde N, Hsieh MM, Donahue RE, Dunbar CE, Kean LS, Tisdale JF (2019) Busulfan combined with immunosuppression allows efficient engraftment of gene-modified cells in a rhesus macaque model. Mol Ther 27(9):1586–1596. https://doi.org/10.1016/j.ymthe.2019.05.022
27. Mahmud N, Khanal A, Taioli S, Koca E, Gaitonde S, Petro B, Sweiss K, Halliday L, Wang X, Patel P, Rondelli D (2018) Preclinical IV busulfan dose-finding study to induce reversible myeloablation in a non-human primate model. PLoS One 13(11):e0206980. https://doi.org/10.1371/journal.pone.0206980
28. Donahue RE, Kuramoto K, Dunbar CE (2005) Sources and methods to obtain stem and progenitor cells. In: Current protocols in immunology, vol Unit 22A.1. Wiley, Hoboken
29. Nakagawa T, Sharma M, Nabeshima Y, Braun RE, Yoshida S (2010) Functional hierarchy and reversibility within the murine spermatogenic stem cell compartment. Science 328(5974): 62–67. https://doi.org/10.1126/science.1182868
30. Shuttlesworth GA, de Rooij DG, Huhtaniemi I, Reissmann T, Russell LD, Shetty G, Wilson G, Meistrich ML (2000) Enhancement of a spermatogonial proliferation and differentiation in irradiated rats by GnRH antagonist administration. Endocrinology 141(1):37–49
31. Kitadate Y, Jorg DJ, Tokue M, Maruyama A, Ichikawa R, Tsuchiya S, Segi-Nishida E, Nakagawa T, Uchida A, Kimura-Yoshida C, Mizuno S, Sugiyama F, Azami T, Ema M, Noda C, Kobayashi S, Matsuo I, Kanai Y, Nagasawa T, Sugimoto Y, Takahashi S, Simons BD, Yoshida S (2019) Competition for mitogens regulates spermatogenic stem cell homeostasis in an open niche. Cell Stem Cell 24(1): 79–92. e76. https://doi.org/10.1016/j.stem.2018.11.013
32. Withers HR, Hunter N, Barkley HT, Reid BO (1974) Radiation survival and regeneration characteristics of spermatogenic stem cell of mouse testis. Radiat Res 57:88–103
33. Shetty G, Porter KL, Zhou W, Shao SH, Weng CC, Meistrich ML (2011) Androgen suppression-induced stimulation of spermatogonial differentiation in juvenile spermatogonial depletion mice acts by elevating the testicular temperature. Endocrinology 152(9): 3504–3514. https://doi.org/10.1210/en.2011-0251. en.2011-0251 [pii]
34. Porter KL, Shetty G, Meistrich ML (2006) Testicular edema is associated with spermatogonial arrest in irradiated rats. Endocrinology 147(3):1297–1305
35. Meistrich ML, Finch M, Lu CC, De Ruiter-Bootsma AL, de Rooij DG, Davids JAG (1984) Strain differences in the response of mouse testicular stem cells to fractionated radiation. Radiat Res 97:478–487
36. Abuelhija M, Weng CC, Shetty G, Meistrich ML (2012) Differences in radiation sensitivity of recovery of spermatogenesis between rat strains. Toxicol Sci 126(2):545–553. https://doi.org/10.1093/toxsci/kfs021. kfs021 [pii]
37. Kangasniemi M, Huhtaniemi I, Meistrich ML (1996) Failure of spermatogenesis to recover despite the presence of a spermatogonia in the irradiated $LBNF_1$ rat. Biol Reprod 54:1200–1208
38. Jahnukainen K, Ehmcke J, Quader MA, Saiful Huq M, Epperly MW, Hergenrother S, Nurmio M, Schlatt S (2011) Testicular recovery after irradiation differs in prepubertal and pubertal non-human primates, and can be enhanced by autologous germ cell

transplantation. Hum Reprod 26(8): 1945–1954. https://doi.org/10.1093/humrep/der160. der160 [pii]

39. Schlatt S, Foppiani L, Rolf C, Weinbauer GF, Nieschlag E (2002) Germ cell transplantation into X-irradiated monkey testes. Hum Reprod 17(1):55–62
40. Boekelheide K, Schoenfeld H, Hall SJ, Weng CCY, Shetty G, Leith J, Harper J, Sigman M, Hess DL, Meistrich ML (2005) Gonadotropin-releasing hormone antagonist (cetrorelix) therapy fails to protect non-human primates (macaca arctoides) from radiation-induced spermatogenic failure. J Androl 26(2):222–234
41. Wreford NG (1995) Theory and practice of stereological techniques applied to the estimation of cell number and nuclear volume in the testis. Microsc Res Tech 32:423–436
42. Noorafshan A (2014) Stereology as a valuable tool in the toolbox of testicular research. Ann Anat 196(1):57–66. https://doi.org/10.1016/j.aanat.2012.07.008
43. Kangasniemi M, Wilson G, Huhtaniemi I, Meistrich ML (1995) Protection against procarbazine-induced testicular damage by GnRH-agonist and antiandrogen treatment in the rat. Endocrinology 136(8):3677–3680
44. Ogawa T, Dobrinski I, Avarbock MR, Brinster RL (1998) Leuprolide, a gonadotropin-releasing hormone agonist, enhances colonization after spermatogonial transplantation into mouse testes. Tissue Cell 30(5):583–588
45. Shetty G, Wu Z, Lam TNA, Phan TT, Orwig KE, Meistrich ML (2018) Effect of hormone modulations on donor-derived spermatogenesis or colonization after syngeneic and xenotransplantation in mice. Andrology 7:257. https://doi.org/10.1111/andr.12571
46. Shetty G, Mitchell JM, Meyer JM, Wu Z, Lam TNA, Phan TT, Zhang J, Hill L, Tailor RC, Peters KA, Penedo MC, Hanna C, Orwig KE, Meistrich ML (2020) Restoration of functional sperm production in irradiated pubertal rhesus monkeys by spermatogonial stem cell transplantation. Andrology 8(5):1428–1441. https://doi.org/10.1111/andr.12807
47. de Rooij DG, Kramer MF (1968) Spermatogonial stem cell renewal in the rat, mouse and golden hamster. A study with the alkylating agent myleran. Z Zellforsch Mikrosk Anat 92(3):400–405
48. Kramer MF, de Rooij DG (1970) The effect of three alkylating agents on the seminiferous epithelium of rodents. II Cytotoxic effect. Virchows Arch B Cell Pathol 4(4):276–282
49. Kanatsu-Shinohara M, Takehashi M, Takashima S, Lee J, Morimoto H, Chuma S, Raducanu A, Nakatsuji N, Fassler R, Shinohara T (2008) Homing of mouse spermatogonial stem cells to germline niche depends on beta1-integrin. Cell Stem Cell 3(5):533–542

Chapter 15

Defining Gene Function in Spermatogonial Stem Cells Through Conditional Knockout Approaches

Julien M. D. Legrand and Robin M. Hobbs

Abstract

Mammalian male fertility is maintained throughout life by a population of self-renewing mitotic germ cells known as spermatogonial stem cells (SSCs). Much of our current understanding regarding the molecular mechanisms underlying SSC activity is derived from studies using conditional knockout mouse models. Here, we provide a guide for the selection and use of mouse strains to develop conditional knockout models for the study of SSCs, as well as their precursors and differentiation-committed progeny. We describe Cre recombinase-expressing strains, breeding strategies to generate experimental groups, and treatment regimens for inducible knockout models and provide advice for verifying and improving conditional knockout efficiency. This resource can be beneficial to those aiming to develop conditional knockout models for the study of SSC development and postnatal function.

Key words Spermatogonial stem cells, Spermatogonia, Conditional knockout, Mouse models, Cre, Knockout mice

1 Introduction

The mammalian male germline is maintained by a population of mitotic germ cells known collectively as undifferentiated spermatogonia, which comprise self-renewing spermatogonial stem cells (SSCs) and their differentiation-committed progeny. These cells enable the lifelong maintenance of male fertility through the process of spermatogenesis, which ultimately results in the production of haploid spermatozoa capable of fertilizing an ovum during sexual reproduction [1]. Similar to all tissue-resident stem cell systems, SSC maintenance and function is regulated by a complex network of cell-intrinsic and cell-extrinsic mechanisms, at the core of which is tightly coordinated gene regulation [2–6].

Since the groundbreaking development of the first targeted genetic knockout mice in the late 1980s [7–10], the study of mammalian gene function has continued to be revolutionized

Jon M. Oatley and Brian P. Hermann (eds.), *Spermatogonial Stem Cells: Methods and Protocols*, Methods in Molecular Biology, vol. 2656, https://doi.org/10.1007/978-1-0716-3139-3_15,

with advancements in genetic engineering technologies. Mammalian genetic knockout models, in particular knockout mice, are routinely used in laboratories worldwide, providing an indispensable tool for understanding fundamental biological processes and the underlying mechanisms of disease.

Organism-wide (or "conventional") knockout models continue to be used widely as they offer broad scope for understanding the role of a particular gene within all tissues. In this method, genetic knockout is achieved in embryonic stem cells (ESCs) following homologous recombination that replaces target genetic sequences with introduced vectors. In general, these vectors are designed to achieve the removal of exons, leading to nonsense-mediated decay of the resulting mRNA transcript or translation of a non-functional protein, and introduction of a gene that confers resistance to a drug used for positive selection of mutant cells. ESCs with the knockout allele are then injected into blastocysts to generate chimeric mice, which are used to transmit the null allele to offspring [11, 12]. More recently, the development of CRISPR/Cas9 gene editing techniques has enabled the rapid and highly efficient generation of knockout mice through direct manipulation of zygotes [13].

Given that gene disruption is evident within the embryo, conventional knockout models can provide insight into gene function from embryonic development to adult tissue homeostasis. However, disruption of genes that are critical for early embryonic development, organogenesis, or those that are widely expressed often result in embryonic lethality, precluding their study in specific cell or tissue types, or at later developmental time points. In addition, loss of gene function in multiple tissues can result in unexpected indirect consequences on the tissue of interest (e.g., genes important for endocrine or neural function). Furthermore, it has been suggested that conventional knockout models may result in the establishment of genetic compensatory networks early in development that circumvent consequences of gene inactivation [14]. These limitations were overcome following the identification and development of site-specific recombination technologies such as Cre-*loxP* [15–17] and Flp-*FRT* [18, 19]. By introducing site-specific recombinase recognition sequences around genetic regions of interest, gene knockout can be restricted to particular cell types with concurrent expression of an appropriate recombinase enzyme driven by a tissue-specific promoter. Such models are termed "conditional" knockout models and the myriad tissue-specific Cre and Flp mouse lines available are now central to the study of cell-specific gene function [20].

Besides cell type specificity, both an advantage and disadvantage of conditional knockout models is that recombinase expression is restricted to the developmental time point(s) at which the chosen promoter is activated. This may present challenges similar to those encountered using conventional knockout models when seeking to distinguish gene function at a precise developmental stage. These limitations can be circumvented by selection of a suitable promoter to drive spatiotemporally restricted recombinase expression; however, alternative strategies are required when none are available for the desired time point and the gene targeted for conditional knockout is expressed across developmental stages. To this end, technologies have been developed that allow inducible activation of recombinases to enable precise temporal control of gene knockout within specific tissues [20]. Accordingly, models employing this strategy are known as "inducible" conditional knockouts. Perhaps the most commonly used to date are chemically regulated Cre-recombinases, and in particular, those that are tamoxifen-induced. However, other well-established and more recently developed inducible recombinases include inducible Flp, progesterone- and trimethoprim-induced Cre, photoactivatable Cre, and doxycycline-regulated recombinases [20–23].

The continual development of gene targeting techniques and advent of technologies such as CRISPR/Cas9 gene editing now enable the development of transgenic and knockout animals with unprecedented ease and precision. Combinations of knockout models continue to be used universally to understand fundamental biological mechanisms and disease pathogenesis both in vivo and in vitro, and the study of SSCs is no exception (Fig. 1). Here, we provide a guide for the selection and use of conditional and inducible knockout mouse models for the investigation of SSC biology. We describe established mouse lines that have been used to achieve reliable gene knockout in SSCs (or their precursors) for the study of cell-intrinsic regulatory mechanisms at various stages of development. In addition, we describe mouse models that enable gene knockout in SSC niche cell populations in order to study extrinsic mechanisms that may dictate SSC function.

2 Materials

The mouse strains listed are readily available from commercial sources but may have been crossed to a genetic background that differs from the originally developed line. Please ensure that the genetic background is suitable for your study. Descriptions of each strain are provided in Subheading 3. All animal use must be conducted in accordance with legal and institutional animal ethics requirements. Materials required for inducible knockout experiments (*see* Subheadings 4.2 and 4.3) are also provided here.

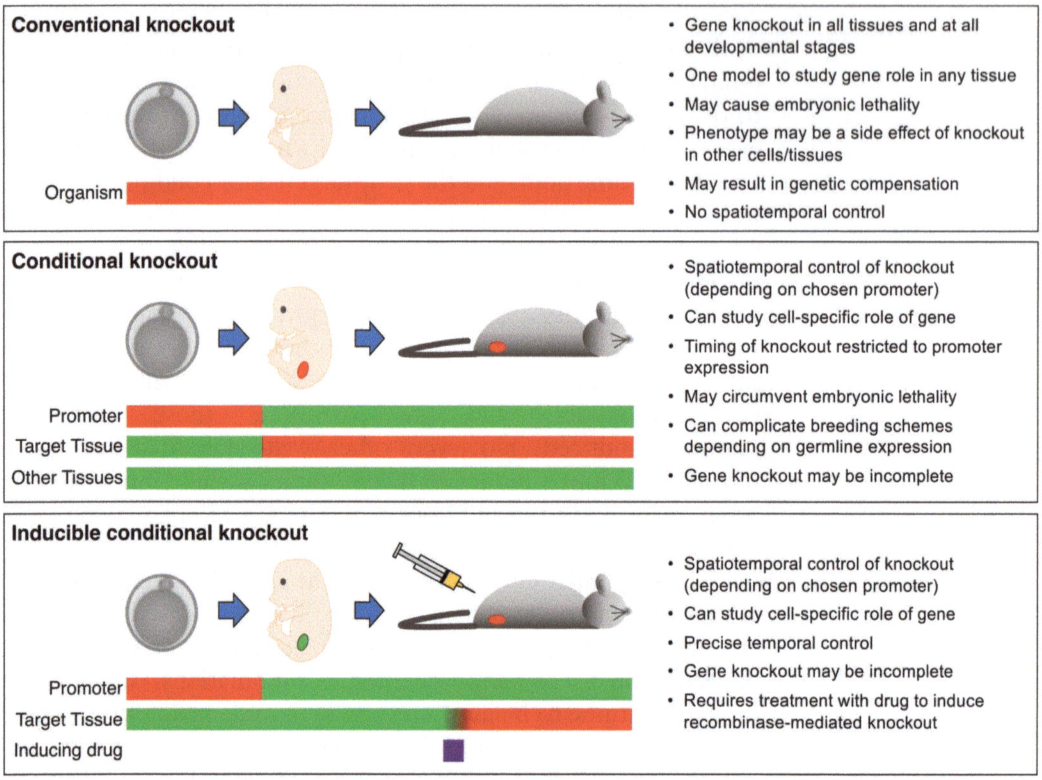

Fig. 1 *Comparison of knockout mouse models.* Conventional knockout mice are generated by disrupting target gene expression within embryonic stem cells, leading to organism-wide knockout. Conditional knockout models allow tissue- and time point-specific gene knockout (e.g., SSC-specific), requiring expression of recombinase enzyme under the control of a tissue-specific promoter and presence of a conditional allele (e.g., Cre/*loxP*). Inducible conditional knockout models express a recombinase enzyme that is only activated following treatment with a specific drug (e.g., tamoxifen in Cre^{ER}/*loxP* models). Target gene and promoter-driven recombinase expression indicated in red (not expressed) and green (expressed) for the entire organism, target tissue, and other tissues where recombinase is not expressed. Example target tissue (testis) shown in colored ovals. Timing of drug induction indicated in purple

2.1 Non-inducible Cre-Driver Strains

1. FVB-Tg(Ddx4-cre)1Dcas/J.
2. B6.Cg-Tg(Prdm1-cre)1Masu/J.
3. 129-*Alpl*$^{tm1(cre)Nagy}$/J.
4. STOCK Tg(Stra8-icre)1Reb/J.
5. B6.B6D2-Tg(Neurog3-cre)24Syos or B6.FVB(Cg)-Tg(Neurog3-cre)C1Able/J.

2.2 Inducible Cre-Driver Strains

1. B6.Cg-*Ndor1*$^{(Tg(UBC-cre/ERT2)1Ejb}$/J.
2. B6.129-*Gt(ROSA)26Sor*^{tm1}*(cre/ERT2)Tyj*/J.
3. B6(SJL)-*Pou5f1*$^{tm1.1(cre/Esr1*)Yseg}$/J.
4. FVB/N-Tg(Ddx4-cre/ERT2)1Dcas/J.

2.3 SSC Niche Targeting Cre-Driver Strains

1. 129S.FVB-Tg(Amh-cre)8815Reb/J.
2. B6;129S7-Amhr2$^{tm3(cre)Bhr}$/Mmnc.
3. STOCK Tg(Sox9-cre/ERT2)1Msan/J.
4. B6;SJL-Tg(Cyp17a1-icre)AJako/J.
5. STOCK Tg(Cx3cr1-cre)MW126Gsat/Mmucd.
6. B6.Cg-Tg(Myh11-cre,-EGFP)2Mik/J.

2.4 In Vivo Tamoxifen Treatment

1. Tamoxifen (Sigma-Aldrich, catalogue number: T5648).
2. Sesame oil (Sigma-Aldrich, catalogue number: S3547).
3. Absolute ethanol.
4. 1.5 mL microcentrifuge tube.
5. Vortex.
6. Heat block.
7. 1 mL syringes.
8. 25G 3/4-inch beveled needles.
9. 18G drawing up needles.

2.5 In Vitro 4-Hydroxytamoxifen Induction

1. 4-Hydroxytamoxifen (4-OHT) (Sigma-Aldrich, catalogue number: H6278).
2. Undifferentiated spermatogonia cell culture media (according to culture technique).
3. Absolute ethanol.

3 Selection of Mouse Strains for Conditional Knockout in SSCs

To date, a wide variety of conditional and inducible knockout mouse strains have been developed to study SSC biology. To select an appropriate recombinase-expressing mouse strain for generating conditional knockouts through site-specific gene recombination, consideration must be given to the stage of SSC development that will be assessed in the study (e.g., SSC specification, first wave of mouse spermatogenesis, or maintenance of adult SSC pool). SSC development has been comprehensively reviewed elsewhere [1, 24]. Selection of study time points may be guided by previous literature concerning the gene that will be targeted for ablation, e.g., whether it is known to be expressed in embryonic development or in other tissues. Similarly, these factors may also dictate decisions regarding selection of inducible vs. non-inducible conditional knockouts. Fortunately, several recombinase-expressing mouse strains have been developed and used successfully to achieve

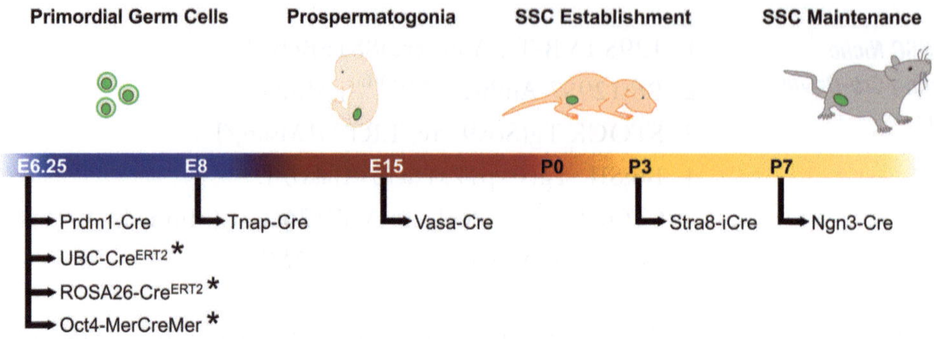

Fig. 2 *Mouse strains with Cre-recombinase expression in the male germline.* Timeline of Cre-expressing mouse strains used to generate conditional knockout models for the study of spermatogonial stem cells (SSCs) and their precursors. A developmental timeline is shown starting from primordial germ cell specification (E6.25) through to postnatal SSC maintenance. Cre-expressing strains are listed below the time point at which the transgene is first active within the male germline (or embryonic developing germline). Strains marked with an asterisk (*) express inducible Cre recombinase from that time point, but Cre activity only occurs following treatment with tamoxifen

targeted gene knockout in SSC studies covering a broad range of developmental stages in an inducible or non-inducible manner. The following sections describe widely used strains, many of which are available from commercial sources, to develop conditional knockout models for the study of SSCs (Fig. 2). However, this is not an exhaustive list, as new mouse models continue to be developed and may be made available by individual researchers (*see* Table 1).

3.1 Conditional Alleles

Conditional knockout models depend on expression of a recombinase such as Cre or Flp within the cells of interest, combined with the presence of recombinase recognition sites that flank the target gene (or a portion of that gene) in what is known as a "conditional allele" [20]. Within the *Cre-loxP* system, conditional alleles generated through insertion of *loxP* sites are described as "floxed," whereas loci flanked by *FRT* sites in the Flp-*FRT* system are described as "flrted." In general, conditional knockout is achieved in mice that are hemi- or heterozygous for recombinase expression and homozygous (or hemizygous for X-linked genes in males) for the conditional allele (*see* Subheading 4). In these mice, loss of gene function is achieved when recombination of conditional alleles prevents expression of mRNA and/or translation of functional protein (Fig. 3). If mice harboring conditional alleles suitable for knockout are not available for the target gene from commercial or collaborative sources, these will need to be developed de novo. Methods for generating conditional alleles have been described previously [25].

Table 1
Cre-expressing mouse strains used for the study of spermatogonia and their precursors

Common name	Allele/transgene name	Germline specific	Male germ cell expression	Reference
Non-inducible Cre-expressing strains				
Expressed during embryonic and fetal development				
Vasa-Cre (Ddx4-Cre)	Tg(Ddx4-Cre)1Dcas	Yes—male and female	Prospermatogonia	[28]
Prdm1-Cre (Blimp1-Cre)	Tg(Prdm1-Cre)1Masu	No	Primordial germ cells	[36]
Tnap-Cre (Alpl-Cre)	$Alpl^{tm1(cre)Nagy}$	No	Primordial germ cells	[48]
$Mvh^{Cre\text{-}mOrange}$	$Ddx4^{tm1.1(cre/mOrange)Dcp}$	Yes—male and female	Prospermatogonia	[209]
Nanos3-Cre	$Nanos3^{tm2.1(cre)Ysa}$	Yes—male and female embryo	Primordial germ cells	[210]
Expressed postnatally				
Stra8-iCre	Tg(Stra8-icre)1Reb	Yes—male only	Undifferentiated spermatogonia	[58]
Ngn3-Cre	Tg(Neurog3-cre)24Syos/ Tg(Neurog3-cre) C1Able	No	Undifferentiated spermatogonia	[68, 72]
Inducible Cre-expressing strains				
UBC-Cre^{ERT2}	$Ndor1^{Tg(UBC\text{-}cre/ERT2)1Ejb}$	No	Spermatogonia—Not reported in others	[88]
ROSA26-Cre^{ERT2}	Multiple – e.g. $Gt(ROSA)26Sor^{tm1(cre/ERT2)Tyj}$	No	Spermatogonia—Not reported in others	Examples [99–101]
Oct4-MerCreMer	$Pou5f1^{tm1.1(cre/Esr1*)Yseg}$	No	Primordial germ cells	[121]
Gfra1-Cre^{ERT2}	$Gfra1^{tm1(cre/ERT2)Sjn}$	No	Undifferentiated spermatogonia	[211]
Nanos2-MCM	Tg(Nanos2-cre/ERT) 1Ysa	Yes—male only	Undifferentiated spermatogonia—Not reported in others	[212]
Vasa-Cre^{ERT2}	Tg(Ddx4-cre/ERT2) 1Dcas	Yes—male and female	Spermatogonia, spermatocytes—Not reported in others	[124]

(continued)

Table 1 (continued)

Common name	Allele/transgene name	Germline specific	Male germ cell expression	Reference
Ngn3/ CreER™	Tg(Neurog3-cre/Esr1*) 1Dam	No	Undifferentiated spermatogonia	[213]
Pax7-Cre^{ERT2}	*Pax7$^{tm2.1(cre/ERT2)Fan}$*	No	Undifferentiated spermatogonia	[31]
PrP-Cre-ER^{T}	Tg(Prnp-cre/ERT) 28.8Ics	Yes—male only	Spermatogonia, spermatocytes	[214, 215]

A selection of mouse strains expressing Cre-recombinase that have been used for studies of the male (and embryonic) germline is listed. Note that some strains have only been employed for lineage tracing, i.e., not all strains have been used in conditional knockout models. For each strain, the commonly reported name and official allele/transgene name are included, as well as details regarding their germline expression specificity. For non-inducible Cre strains, the first germ cell populations where Cre is expressed are listed. For inducible Cre strains, the germ cell populations where Cre activity can be induced are listed. References listed are the first report describing generation of the strain. Not all strains are commercially available

3.2 Strains Expressing Non-inducible Recombinase During Embryonic and Fetal Germ Cell Development

The following strains express recombinase driven by promoters that are active within SSC precursors. These strains may be particularly useful for studying mechanisms involved in primordial germ cell (PGC) and pro-spermatogonia specification, as well as postnatal establishment of the foundational SSC pool (reviewed in [24, 26]). However, recombinase-expressing strains must be carefully selected to ensure that the timing of gene deletion is compatible with study aims. Importantly, when postnatal/adult SSC mechanisms are under investigation, gene knockout during embryonic or fetal stages may lead to unexpected consequences at later developmental stages, preventing unambiguous assessment of gene function in the established SSC pool. These strains can be used to study adult SSC mechanisms, providing that expression of the gene targeted for conditional knockout is well characterized and not known to be expressed or play an important functional role prior to establishment of the SSC pool.

3.2.1 Vasa-Cre

Ddx4 (also known as *Vasa* or *Mvh*, *mouse vasa homolog*) encodes a DEAD box RNA helicase that is highly conserved across invertebrate and vertebrate species, with expression restricted to the germline of both sexes [27, 28]. In mice, VASA is detected from embryonic day (E) 10.5–11.5 in PGCs that have migrated to the embryonic gonad, and expression is maintained in germ cells of both sexes into adulthood [29, 30]. In males, VASA is detectable in the cytoplasm of spermatogenic cells up to and including round spermatids and in females within the cytoplasm of oocytes up to their presence in antral follicles [30]. Genetic knockout of *Vasa* results in male infertility due to complete loss of post-meiotic germ

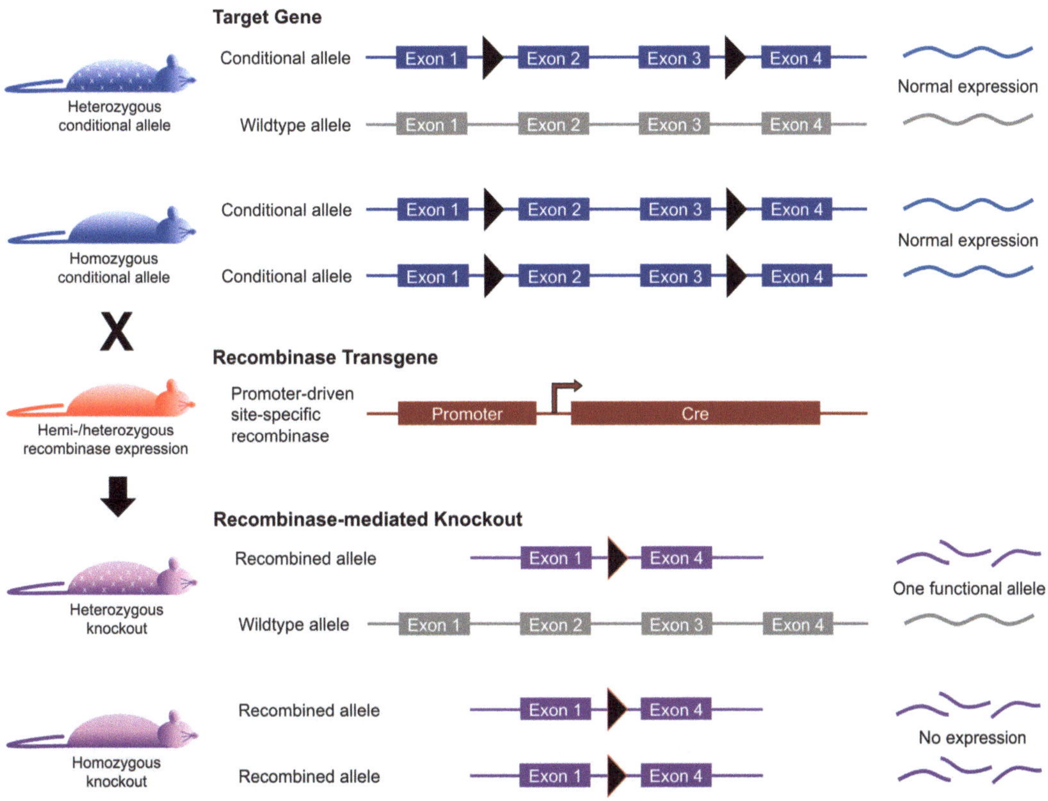

Fig. 3 *Site-specific recombination of conditional alleles.* Conditional gene knockout is achieved when a promoter-driven site-specific recombinase (SSR) is expressed in the presence of a conditional allele (s) containing SSR recognition sites (black triangles) flanking target gene regions. SSRs recognize these sites, and recombination results in excision of the flanked region (black triangle with red border). This leads to transcript decay (or translation of non-functional protein) and no gene expression from the recombined allele. To generate conditional knockout mouse models, mice expressing recombinase transgene (red mice) are crossed with mice containing a conditional allele for the target gene (blue mice). To achieve homozygous conditional knockout, at least two generations of breeding may be required to generate mice homozygous for conditional alleles that also express recombinase (purple mice). Heterozygous mice (purple with gray crosses) maintain a wild-type allele (gray) with functional gene expression and may not display any phenotype as a result

cells; however, females are fertile and capable of producing viable offspring [29].

Vasa-Cre mice were developed by random genomic integration of a DNA construct driving Cre-recombinase expression from a 5.6 kb fragment of the mouse *Vasa* promoter, with founder strains generated in FVB mice [28]. Vasa-Cre transgene expression occurs in embryonic gonads from E15, slightly later than expression of endogenous *Vasa*, with Cre-mediated recombination detected in over 97% of both male and female germ cells by postnatal day (P) 3 [28]. Thus, the Vasa-Cre strain serves as a highly efficient option for the generation of germ cell-specific conditional knockouts and has since been used widely for studies of SSCs (examples found in [31–34]). However, given that *Vasa* is expressed during germline development, studies investigating established SSCs should carefully consider the suitability of this line to avoid confounding developmental consequences. In addition, a recent study found that the genomic integration site of the Vasa-Cre transgene causes a 1.1 Mb deletion, resulting in knockout of two genes, *Neto1* and *Cbln2* [35]. Therefore, it is essential that Vasa-Cre mice are not bred to homozygosity to avoid potentially confounding phenotypes driven by loss of *Neto1* and *Cbln2* expression.

Importantly, Vasa-Cre expression occurs in oocytes and Cre protein is suggested to persist within zygotes [28]. Consequently, embryo-wide recombination of floxed alleles occurs in the presence of maternal Vasa-Cre, even in embryos that do not inherit the Cre transgene; therefore, where germline-specific conditional knockout is desired, breeding strategies where Vasa-Cre is paternally transmitted should be adopted (*see* Subheading 4).

3.2.2 Prdm1-Cre

Prdm1 (also known as *Blimp1*) is widely expressed in early embryonic development and encodes a transcriptional repressor that is essential for PGC specification [36, 37]. *Blimp1* is reported to be expressed from E6.25 in mice within a subset of cells in the proximal posterior epiblast. In these cells, BLIMP1 represses somatic gene expression programs, enabling their PGC lineage restriction. Loss of *Blimp1* leads to defects in PGC specification, proliferation, and migration [36].

Prdm1-Cre mice were generated using a bacterial artificial chromosome (BAC) construct containing the *Blimp1* locus where the coding region of *Blimp1* in exons 1 and 2 was replaced with Cre sequence followed by a transcriptional STOP cassette. The BAC was then microinjected into B6CBAF1 embryos, resulting in random integration of the construct into the mouse genome [36]. Thus, Cre expression in these mice is driven by *Blimp1* regulatory elements and corresponds with endogenous *Blimp1* expression [36]. Cre-mediated recombination efficiency was originally reported to be between 55% and 76% [36]; however, recombination efficiencies of up to 100% in PGCs at E13.5 have since been reported using this line [38, 39]. Although conditional knockout occurs in PGCs rather than established SSCs, the Prdm1-Cre line can be valuable tool for studies of SSC development [24].

It is important to note that *Blimp1* expression is not restricted to PGCs and is readily detected throughout the developing mouse embryo and in a variety of adult tissues [37]. *Blimp1* was originally identified as a transcription factor important for B lymphocyte maturation [40] and has since been described as a key regulator in multiple hematopoietic lineages [41, 42]. Besides its function in embryonic development of multiple tissues and the hematopoietic system, *Blimp1* is reported to play important roles in other tissues including the placenta, skin and its appendages, and intestinal epithelium [43, 44]. Therefore, conditional knockouts using the Prdm1-Cre strain to target PGCs may have inadvertent consequences within other tissues that should be carefully considered.

3.2.3 Tnap-Cre

Tissue non-specific alkaline phosphatase (*Tnap*, also known as *Alpl*) belongs to a family of genes encoding alkaline phosphatase isozymes within various tissues. *Tnap* is expressed broadly during embryonic development and in postnatal tissues and, similar to other family members, functions to catalyze dephosphorylation of a wide range of substrates [45, 46]. During embryonic development, *Tnap* expression is first detected in extraembryonic tissues at E4.5 and is then robustly detected from E7 throughout developing tissues including those that form the digestive, skeletal, and central nervous systems [47]. *Tnap* is reported to be expressed in PGC precursors from E7; however, it was found to be dispensable for PGC specification and migration [47].

Tnap-Cre mice were generated by targeted knock-in of Cre recombinase into the endogenous *Tnap* locus and established on the 129S genetic background [48]. By crossing with a reporter mouse line, Tnap-Cre expression was shown to initiate in PGCs between E8 and E9; however, Tnap-Cre activity was detected within the placenta, intestine, and neural tube after E10.5 [48]. At E13.5, Cre-mediated recombination efficiency in germ cells was originally determined to be approximately 60% [48], and similar efficiencies have since been reported in conditional knockouts targeting PGCs [49–52]. Since their development, Tnap-Cre mice have been used to study mechanisms of germ cell development [49–52], as well as postnatal SSC function [51, 53].

Given its broad expression during development and in adult tissues, use of the Tnap-Cre strain for conditional knockout in SSCs should be approached with caution. Spatiotemporal expression of the floxed allele should be well-defined in order to avoid unexpected effects from conditional knockout in other tissues. Notably, maternal Cre effects leading to ectopic Cre activity throughout the embryo have been described in this strain [48]; therefore, Tnap-Cre should be paternally transmitted (*see* Subheading 4).

3.3 Strains Expressing Non-inducible Recombinase at Postnatal SSC Developmental Stages

The following strains express recombinase driven by promoters that are active from postnatal stages of SSC development. Transgene expression in the strains described primarily occurs in differentiation-destined progenitors derived from SSCs; therefore, these strains are suited for generating conditional knockouts that target progenitor spermatogonia to study the establishment and maintenance of spermatogenesis. Given that these strains express non-inducible recombinase, issues regarding spatiotemporal expression of recombinase and conditional allele expression as described above should be considered.

3.3.1 *Stra8-iCre*

Retinoic acid promotes differentiation of spermatogonia and is required for subsequent meiotic initiation [54, 55]. During both male and female gametogenesis, retinoic acid induces expression of *Stimulated by retinoic acid 8* (*Stra8*), an essential regulator of meiotic induction [56]. *Stra8* expression is restricted to germ cells of both sexes and encodes a transcriptional activator responsible for directing gene expression programs required for the initiation and progression of meiosis [57]. In female mice, *Stra8* expression is restricted to premeiotic germ cells within the developing ovary between E12.5 and E16.5 and is not detected postnatally. In male mice, *Stra8* expression has been reported in spermatogonia from P2 and continues to be detected in the adult testis within spermatogonia at the onset of differentiation and preleptotene spermatocytes at meiotic induction [55].

Given its postnatal expression in the male germline, *Stra8* was identified as a unique candidate from which to drive Cre-recombinase expression within premeiotic germ cells of the postnatal testis [58]. To this end, Stra8-iCre (also referred to as Stra8-Cre) mice were developed on an FVB genetic background by random genomic integration of a construct driving codon-optimized Cre-recombinase [59] from a 1.4 kb fragment of the mouse *Stra8* promoter [58]. Transgene expression was detected by RT-PCR from P3 and restricted to the testis, as expected, with expression increasing at P7. Histological examination of adult testes using Cre antibody found that Stra8-Cre is expressed in Type A spermatogonia through to preleptotene spermatocytes, consistent with the known role of STRA8 [58]. Detection of Stra8-iCre activity by crossing with a reporter mouse line found Cre-mediated recombination occurred in the majority of undifferentiated spermatogonia marked by PLZF [58, 60, 61], confirming its activity in SSC-derived progenitors. Recombination efficiency was found to be over 95% when quantified between P3 and P5 [58]. Stra8-iCre was also shown to be active in juvenile mice within a substantial portion of undifferentiated spermatogonia positive for the SSC marker GFRα1, in addition to the majority of differentiation-primed spermatogonia marked by RARγ [33]. Importantly, despite known expression of *Stra8* in the developing ovary, Stra8-iCre expression was not detected in female germ cells and was concluded to be non-functional in female mice. It was suggested that the *Stra8* promoter fragment used to drive Cre-expression lacks key cis-regulatory elements required for expression in female germ cells [58].

Since its development, the Stra8-iCre strain has been used extensively to generate conditional knockout models to study adult spermatogonial maintenance and differentiation (examples in [32–34, 62]), as well as postnatal development and establishment of the spermatogonial pool (examples in [63, 64]).

3.3.2 Ngn3-Cre

Neurogenin 3 (*Ngn3*) is a basic helix–loop–helix transcription factor originally found to be expressed in the developing central nervous system and pancreas [65]. Besides its role in these systems, *Ngn3* is expressed in postnatal undifferentiated spermatogonia from P3 and primarily marks differentiation-primed cells that contribute variably to the SSC population [66–68]. Lineage tracing studies have shown that NGN3+ spermatogonia give rise to all differentiating, meiotic, and post-meiotic male germ cells and contribute to regeneration of the SSC pool following germline damage [68–70].

Two distinct non-inducible Ngn3-Cre strains have been used for the study of SSCs and spermatogenesis. As part of the original study identifying *Ngn3* expression in the male germline, an Ngn3-Cre strain was developed using a BAC construct containing a Cre-recombinase cassette inserted within a 23 kb region of the mouse *Ngn3* locus. This construct was randomly integrated into the mouse genome after microinjection into B6D2F1 embryos, and the resultant transgenic mice were backcrossed to C57BL/6 [68]. After crossing with a LacZ reporter strain, robust Cre activity was detected in seminiferous tubules by P7, corresponding to the time point at which upregulation of endogenous *Ngn3* occurs [68]. Recombination efficiency using this strain has been reported to be high, with near complete recombination observed in juvenile and adult testes [67, 68, 71]. However, it is important to note that only a fraction of germ cells express *Ngn3* during initial postnatal stages [67, 68]; therefore, the high proportion of germ cells labelled following Ngn3-Cre-mediated recombination in juvenile and adult testes is likely derived from this early postnatal NGN3+ population.

An alternative Ngn3-Cre strain that is similarly used for studies of the male germline was developed in an investigation of enteroendocrine cells. To generate this strain, a Cre-recombinase cassette was inserted by homologous recombination immediately following the start codon of mouse *Ngn3* contained within a BAC construct. Transgenic mice were obtained following nuclear injection and random integration of this construct and backcrossed on to a CD1 genetic background [72]. In studies of spermatogenesis, transgene expression within this Ngn3-Cre strain has been reported to correspond to endogenous *Ngn3* expression, with Cre-mediated recombination occurring from P7 [73–75]. Accordingly, this strain has been used to investigate postnatal germline development [76], as well as adult SSC function [34, 77, 78].

Given its role in embryonic central nervous system and pancreatic development, as well as adult enteroendocrine cells [72], similar precautions for other more widely expressed Cre transgenes should be considered (as described herein and previously [79]).

3.4 Strains Expressing Inducible Recombinase Active in Spermatogonia

The development of recombinase-expressing strains with the capacity to induce recombinase activity at precise time points allows the circumvention of experimental limitations presented by developmentally active transgene expression. Temporal control of recombinase activity enables conditional knockout in specific cell populations or at particular developmental stages where the recombinase driver or conditional allele being used is broadly expressed.

The strains described in this section employ a modified version of Cre-recombinase, CreER^{T2}, where Cre is fused to a mutated ligand-binding domain of the human estrogen receptor. CreER^{T2} contains a triple mutation in the estrogen-binding domain that prevents binding of physiological 17β-estradiol but allows it to be bound by synthetic 4-hydroxytamoxifen (4-OHT), a metabolite of the selective estrogen receptor modulator, tamoxifen [80]. In the absence of 4-OHT, CreER^{T2} is sequestered to the cytoplasm, rendering it inactive; however, upon binding 4-OHT, CreER^{T2} is translocated to the nucleus where it can perform its site-specific recombinase activity [81]. CreER^{T2} was developed shortly after the originally developed CreER^{T}, which has a single mutation of the estrogen-binding domain [82]. CreER^{T2} is favored for mouse model development, as the additional mutations present in its ligand-binding domain confer at least fourfold higher sensitivity to 4-OHT compared to CreER^{T} [80], thereby requiring less in vivo tamoxifen administration to induce efficient recombination [83].

Here, we describe tamoxifen-inducible Cre expressing strains that have been used extensively for studies of SSCs and spermatogenesis. The included strains drive recombinase expression from promoters that are ubiquitously expressed and result in high recombination efficiency within the male germline. In addition, we describe a strain that expresses a distinct version of inducible Cre-recombinase under the control of a tissue-restricted promoter.

3.4.1 UBC-CreER^{T2}

Ubiquitin is a highly evolutionarily conserved protein expressed throughout multiple tissues that play essential roles in diverse cellular processes [84]. *Ubiquitin C* (*UBC*) is one of four mammalian genes encoding ubiquitin, and although originally described as stress-responsive, it is now appreciated to have broad roles in cellular homeostasis [85]. In mice, *Ubc* homozygous knockout results in embryonic lethality from E11.5 due to defects in fetal liver development [86].

Earlier studies demonstrated that the human *UBC* promoter was able to drive efficient and ubiquitous transgene expression in mice from at least E10.5 [87]. This promoter was subsequently used to drive expression of CreER^{T2} in a lentiviral construct used to generate the UBC-CreER^{T2} strain. Transgenic mice were obtained by microinjection of lentivirus containing this construct into the perivitelline space of 129S × C57BL/6 zygotes [88]. Commercially available UBC-CreER^{T2} strains have since been backcrossed to C57BL/6 for at least seven generations. The technique used to generate this strain results in presumably random transgene integration, with a later study finding this transgene was integrated within the *Ndor1* locus resulting in a 5 bp deletion [35]. The consequences of this deletion in *Ndor1* remains unclear; however, UBC-CreER^{T2} mice are viable and fertile [88].

The efficiency of conditional knockout driven by UBC--CreER^{T2} in adult mice (8–12 weeks) was originally reported to be over 70% at 1-week post-induction in all tissues examined [88]. A later study found that efficient UBC-CreER^{T2}-mediated recombination could be induced within embryos from E5.5 [89], highlighting the versatility of this strain for use in conditional knockout models.

In studies of spermatogenesis, the UBC-CreER^{T2} strain has been used to drive efficient conditional knockout in adult spermatogonia, while testis somatic cells and meiotic/post-meiotic germ cells remain essentially unaffected [90–93]. In one study, UBC--CreER^{T2}-mediated gene knockout was observed in almost all spermatogonia at 2 days following tamoxifen administration in adult mice, with no detectable knockout in Sertoli cells [93]. Similarly, later studies have shown efficient knockout in all spermatogonia from 5 to 7 days post-tamoxifen treatment in adult mice, with no Cre-mediated knockout observed in Sertoli cells or spermatocytes [90–92]. In contrast, knockout in testis somatic cells has been reported using this strain when tamoxifen was administered to 4-week-old juvenile males [94, 95], suggesting differences in UBC-CreER^{T2} activity in the testis according to developmental stage. Importantly, although conditional knockout using this strain is highly efficient, it is often incomplete resulting in mosaic deletion that is common to tamoxifen-inducible Cre-expressing strains [88, 90, 96]. This can be beneficial for lineage-tracing studies but may be a limitation in knockout models where complete ablation is required.

Although UBC-CreER^{T2} can be used to drive efficient deletion in spermatogonia, use of this strain should be approached with care. Given its widespread expression, cell autonomous effects of gene knockout in SSCs should be confirmed to exclude phenotypes that may occur due to disruption of gene function in other cell types. To this end, induction of UBC-CreER^{T2}-mediated knockout in in vitro cultured spermatogonia can be used to assess cell intrinsic effects [90–92] (*see* Subheading 4). In addition, spontaneous CreER^{T2} activity in the absence of tamoxifen has been reported for this strain [97]; therefore, appropriate controls must be included to account for any background Cre activity when analyzing experiments using these mice (*see* Subheading 4).

3.4.2 *ROSA26-CreER^{T2}*

The *ROSA26* locus is a region of the mouse genome that has been used widely for targeted transgene insertion. *ROSA26* is ubiquitously expressed in both embryonic and adult tissues; however, it does not encode products that have known biological significance. Therefore, the *ROSA26* locus has been described as a "safe" region for transgene insertion and has been exploited for the generation of over 130 transgenic mouse strains [98].

To date, multiple ROSA26-CreER^{T2} strains have been developed by independent research groups [99–101], all of which were generated by targeted knock-in of CreER^{T2} to the *ROSA26* locus. Similar to the UBC-CreER^{T2} strain, ROSA26-CreER^{T2} has been used to drive conditional gene knockout in the juvenile and adult male germline. Following tamoxifen administration at 8 and 12 weeks of age, analysis of 16-week-old mice found near complete ROSA26-CreER^{T2}-mediated gene knockout in spermatogonia while recombination did not appear to occur in Sertoli cells [102]. Besides this study, limited analysis of ROSA26-CreER^{T2} activity within testis cell types has been performed. However, several studies have reported efficient gene knockout in spermatogonia following tamoxifen treatment in adults, juveniles, and neonates using these strains [102–105].

Like the UBC-CreER^{T2} strain, use of the ROSA26-CreER^{T2} for studies of SSCs and spermatogenesis must be carefully controlled, and cell autonomous effects should be confirmed using independent analyses. For example, independent Cre-driver strains can be used to confirm germ cell-specific effects of gene knockout [104] and cultured spermatogonia derived from ROSA26-CreER^{T2}-based models can be assessed for SSC-intrinsic effects [102].

3.4.3 *Oct4-MerCreMer*

POU class 5 homeobox 1 (*Pou5f1*, also known as *Oct4*) is a transcription factor well known for its role in maintenance of embryonic stem cell pluripotency [106] and as one of four key factors required for the reprogramming of somatic cells to induce pluripotent stem cells [107, 108]. OCT4 was originally found to be expressed in embryonic stem cells from pre-implantation stages, as well as in PGCs and oocytes [109–111]. In the developing embryo, OCT4 is abundant and ubiquitously expressed by the eight-cell stage, with its localization becoming restricted to cells of the inner cell mass and eventually to PGCs upon their specification [110, 112, 113]. In male germ cells, OCT4 remains detectable throughout embryonic development and within postnatal undifferentiated spermatogonia, where it continues to be expressed throughout adult life [114]. Later studies have suggested that OCT4 is a marker of postnatal SSCs and is required for their self-renewal

[115–117]; however, it has more recently been demonstrated that OCT4 is a progenitor-associated marker in adult undifferentiated spermatogonia during steady-state spermatogenesis [118]. Given selective expression within undifferentiated spermatogonia and their precursors, *Oct4* is an attractive candidate to drive recombinase expression for SSC-targeted conditional knockouts.

Accordingly, a strain of mice known as Oct4-MerCreMer that express tamoxifen-inducible Cre from endogenous *Oct4* have been used to develop conditional knockout models for studies of SSCs and germ cell development [70, 119, 120]. Oct4-MerCreMer mice were developed by targeted knock-in of a construct containing tamoxifen-inducible Cre linked to an internal ribosome entry site into the 3′-untranslated region of the endogenous *Oct4* locus [121]. The inducible Cre-recombinase expressed in this strain, MerCreMer, was developed by fusion of a mutant hormone binding domain of the mouse estrogen receptor to each side of the Cre enzyme [122]. Like Cre^{T2}, MerCreMer is not bound by endogenous 17β-estradiol and is more stringently regulated by 4-OHT compared to similar versions of this recombinase [122].

Oct4-MerCreMer mice displayed expected expression patterns of endogenous *Oct4*, including within adult spermatogonia [121]. Rapid and efficient Oct4-MerCreMer-mediated recombination has been demonstrated in SSC-enriched populations marked by GFRα1 and ID4 following tamoxifen treatment at P8 and analysis 8 h later [70]. Subsequently, conditional knockout using Oct4-MerCreMer resulted in target gene ablation within all subsets of undifferentiated spermatogonia (A_s, A_{pr}, and A_{al}) by 4 days' post-tamoxifen treatment [70]. In separate studies using Oct4-MerCreMer mice, tamoxifen treatment of pregnant females from E10.5 was shown to result in near complete conditional knockout in PGCs within the gonads of developing embryos by E13.5 [119, 120].

The Oct4-MerCreMer strain allows induction of conditional knockout at precise SSC developmental time points, thereby offering flexibility for various research questions. However, although restricted to PGCs upon their specification, *Oct4* expression has been reported in myriad postnatal tissues, including somatic stem and progenitor cells within the bone marrow, pancreas, skin, kidney, intestine, and liver [123]. Therefore, Oct4-MerCreMer may be active in these tissues, and, similar to the strains previously described herein, conditional knockouts should be carefully assessed to confirm specific cell-intrinsic effects in SSCs.

3.4.4 Vasa-Cre^{ERT2}

Following development of the highly efficient Vasa-Cre strain for germline-specific conditional knockout (*see* Subheading 3.2.1), the Vasa-Cre^{ERT2} strain (also known as Ddx4-Cre^{ERT2}) was subsequently generated [124]. In this strain, the same *Vasa* promoter fragment used to drive recombinase expression in Vasa-Cre mice [28] was used to promote expression of CreER^{T2} [80] in a plasmid construct that was subsequently microinjected into FVB oocytes, resulting in random genomic integration [124]. To date, no functional disruption of endogenous genetic loci has been identified in the Vasa-Cre^{ERT2} strain, unlike that which was discovered in later analyses of Vasa-Cre mice [35].

Originally reported in a study investigating oocyte development, Vasa-Cre^{ERT2} was shown using reporter mice to result in 90–95% oocyte-specific recombination following tamoxifen treatment at 6 weeks of age [124]. Similar recombination efficiency was reported in the testis after treatment between 3 and 6 weeks of age, with specific Cre activity found in germ cells and not interstitial cells [124]. However, quantitative analysis of recombination in specific germ cell populations within the testis was not performed.

To date, the Vasa-Cre^{ERT2} strain has seen limited use in studies of the male germline despite its initially reported efficiency. Studies reporting use of this strain to study male germ cells are primarily focused on meiosis [125–127], with limited data available regarding conditional knockout efficiency in SSCs. Previous studies have indirectly suggested efficient recombination in undifferentiated spermatogonia using the Vasa-Cre^{ERT2} strain [124, 125]; however, further studies are required to validate its use for inducible conditional knockout in SSCs.

3.5 Strains Expressing Recombinase in SSC Niche Cell Populations

Regulation of SSC self-renewal and differentiation depends on extrinsic factors derived from supporting somatic cells in the testis microenvironment, or SSC "niche." Growth factors that promote SSC self-renewal such as GDNF are derived from Sertoli and peritubular myoid cells, and CSF1, another growth factor promoting self-renewal, is secreted by Leydig cells [3, 128–130]. Leydig cells are known to stimulate Sertoli cell activity and together with testis endothelial cells, peritubular myoid cells, and macrophages, function in a complex network to regulate spermatogenesis [3, 128–132].

Given the role of niche cell populations in the regulation of SSC activity, in this section we describe examples of Cre strains that can be used for conditional gene knockout in testis somatic cells. These models can be valuable for studying extrinsic mechanisms regulating SSC function.

3.5.1 Sertoli Cell-Expressed Cre Strains

In males, *anti-Müllerian hormone* (*Amh*) is specifically expressed in developing Sertoli cells (from E11.5 in mice) and its expression declines until it is no longer detectable by puberty [133]. To generate a Sertoli cell-specific Cre strain, the promoter region of mouse *Amh* was used to drive non-inducible Cre expression in a construct that was introduced into FVB zygotes by microinjection. The Amh-Cre construct randomly integrated into the mouse genome and resulting founder mice were subsequently backcrossed onto a 129S background for several generations [134]. Efficient conditional knockout in Sertoli cells using Amh-Cre was reported to occur by E14.5 [134], and later studies confirmed that Amh-Cre is expressed from E11.5 [135], corresponding to endogenous *Amh* expression. Since its development, this strain has been widely used for efficient and restricted conditional gene knockout in Sertoli cells (examples in [33, 135–137]).

In addition to the Amh-Cre strain, Amhr2-Cre mice have been used successfully in several studies to drive gene recombination in Sertoli cells (examples in [138–142]). In males, *anti-Müllerian hormone receptor type 2* (*Amhr2*) is expressed in the developing testis and persists postnatally within Sertoli and Leydig cells [143, 144]. Amhr2-Cre mice were developed by targeted knock-in of Cre recombinase into the endogenous *Amhr2* locus [143]. Using reporter mice, Amhr2-Cre activity has been detected in Sertoli cells from E14, as well as a proportion of Leydig cells [141]. Recombination efficiency in Sertoli cells is reported to be high in conditional knockout models using this strain, with minimal recombination observed in Leydig cells [138–142].

To our knowledge, inducible conditional knockout models to study Sertoli cell function have not been reported to date. To this end, the Sox9CreER^{T2} strain may be suitable given the strong expression of *Sox9* in Sertoli cells [145–147]. However, it is important to note that *Sox9* is broadly expressed during development and in adult tissues [148, 149], and Sox9CreER^{T2} mice have been used widely to target various cell types [146, 150–152]. Therefore, despite expression in the testis that is restricted to Sertoli cells, models employing Sox9CreER^{T2} for conditional knockout in this population should be carefully analyzed to exclude any indirect consequences arising from knockout in other tissues.

3.5.2 Cyp17-iCre

Cytochrome P450 17A1 (*Cyp17a1*, also known simply as *Cyp17*) encodes a cytochrome P450 enzyme essential for sex steroid synthesis in males and females [153]. In the testis, *Cyp17a1* expression is restricted to fetal and adult Leydig cells [154, 155]; thus, it presents as an ideal driver of recombinase expression for conditional knockout in these cells. The Cyp17-iCre strain was developed by microinjection of a DNA fragment containing 3.34 kb of the *Cyp17* promoter preceding a codon-optimized Cre-recombinase sequence into the pronuclei of C57BL/6xSJL zygotes [156]. Cyp17-iCre expression was detected in fetal Leydig cells at E16.5, and conditional knockout using this strain was subsequently confirmed [156]. It is important to note that *Cyp17* expression occurs in other steroidogenic tissues, and Cyp17-iCre activity in males was also detected in the kidneys, adrenal glands, spleen, brain, lungs, and liver [156].

3.5.3 Cx3cr1-Cre

Testicular macrophages have been reported to play important roles in the SSC niche, providing a source of CSF1 to regulate SSC function. In addition, testicular macrophages produce enzymes involved in retinoic acid biosynthesis, and their loss has been reported to impair spermatogonial differentiation. Testicular macrophages are found in close association with Leydig cells and testis vasculature, as well as within the peritubular region of seminiferous tubules [131]. *Cxc3 chemokine receptor 1* (*Cx3cr1*) is expressed in monocytes and tissue-resident macrophages throughout the mouse, including within all testicular macrophages [131, 157].

Cx3cr1-Cre mice were generated as part of the GENSAT project by random integration of a BAC construct driving Cre expression from the mouse *Cx3cr1* promoter [158, 159]. Although not used for conditional knockout per se, Cx3cr1-Cre has been used to induce conditional diphtheria toxin receptor expression in testis macrophages for their ablation. Cx3cr1-Cre-mediated recombination was found to occur in approximately 95% of testis macrophages, demonstrating its effectiveness in targeting this population [131]. The broad expression of *Cx3cr1* in monocytes and macrophages in other tissues should be considered when using this strain for testis studies.

3.5.4 Myh11-Cre

Peritubular myoid cells are smooth muscle cells that surround the seminiferous tubule to provide important structural support. In addition, they produce growth factors and extracellular matrix that supports spermatogenesis and the function of other testis somatic cells [3, 160]. Peritubular myoid cells produce GDNF in response to testosterone stimulation, and it is reported that conditional knockout of *Gdnf* in these cells results in progressive loss of undifferentiated spermatogonia [161, 162].

The Myh11-Cre strain has been used to generate conditional knockout models targeting peritubular myoid cells [162, 163]. This strain was developed through random genomic integration of a construct containing Cre recombinase and EGFP downstream of a 16 kb mouse *myosin heavy chain 11* (*Myh11*) promoter fragment [164]. The Myh11-Cre/EGFP transgene was found to be expressed in all vascular and non-vascular smooth muscle from E12.5 [164]. In the testis, Myh11-Cre is robustly detected in peritubular myoid cells, as well as testis vasculature [162, 165]; thus, broader activity of this transgene should be considered when assessing effects within peritubular myoid cells. Importantly, Myh11-Cre is reported to be active in the germline of both sexes [166, 167] and maternal Cre transmission results in embryo-wide deletion as *Myh11* is expressed in oocytes and persists in zygotes [168]. Therefore, the Myh11-Cre transgene should be paternally transmitted (*see* Subheading 4.1.1).

4 Methods for Using Conditional Knockout Lines to Study SSCs

The following sections provide practical advice for the development and use of conditional knockout models to study SSCs and spermatogenesis. We include breeding schemes suitable for various modes of transgene inheritance, examples of treatment regimens for inducible knockouts, suggestions for confirming and improving conditional knockouts, and other in vivo and in vitro contexts where conditional alleles can be used. Given that only Cre-expressing strains have been described in Subheading 3, we have focused our descriptions on methods used for Cre/*loxP* models. However, the principles for breeding, treatment, and analysis are similar for other conditional knockout systems, e.g., Flp/FRT.

4.1 Breeding Strategies

The choice of breeding strategy is often dictated by spatiotemporal expression of the selected transgenes. For example, whether expression of the Cre-driver is restricted to the male germline or is expressed in oocytes and maternally derived Cre persists during early embryonic development determines whether Cre should be paternally transmitted. In contrast, inducible Cre activity may occur ectopically in SSCs; therefore, it may be preferable for Cre to be transmitted through the maternal side. Furthermore, different breeding strategies are required where Cre is non-inducible vs. inducible, to ensure an appropriate control group is generated. Here, we provide breeding schemes that cover the various strains introduced in Subheading 3. Variations to these schemes necessitated by knockout efficiency or strain-specific features are described in later sections.

In general, regardless of the required breeding strategy, Cre-expressing mice are maintained as hemizygous (for random transgene insertions) or heterozygous (for knock-in mice) at all stages of breeding. Hemi- or heterozygous Cre expression is typically sufficient to achieve efficient recombination of floxed alleles, as exemplified by the strains described herein. Importantly, methods used to generate Cre strains often result in random transgene integration and may disrupt endogenous gene expression, leading to unexpected consequences when bred to homozygosity (examples in [169]). Cre recombinase itself can be toxic to cells when expressed at high levels (known as "Cre toxicity"), as a result of DNA damage that may in part be due to cryptic *loxP* sites present within the genome [170]. Therefore, when developing a new conditional knockout model, it is important to determine the effects of Cre expression alone using appropriate control groups.

4.1.1 Paternal Transmission of Non-inducible Cre

The following breeding strategy is suitable for generating conditional knockout models using non-inducible Cre strains, where the Cre-driver is expressed in oocytes and maternally derived Cre

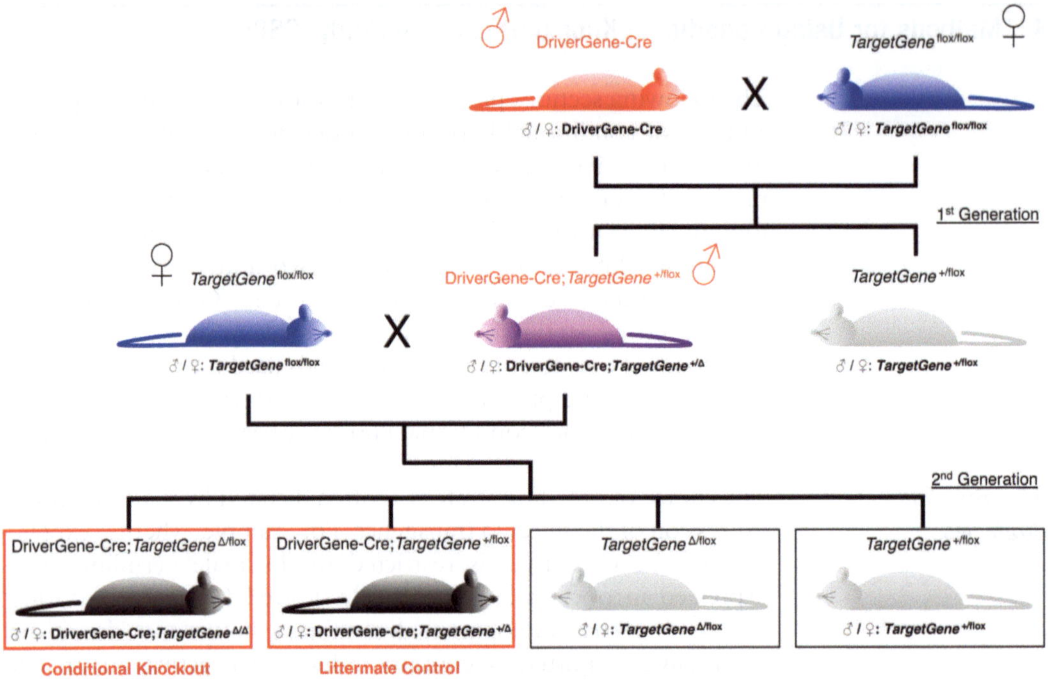

Fig. 4 *Conditional knockout model breeding scheme for paternal transmission of a non-inducible Cre transgene.* This breeding scheme describes paternal transmission of a non-inducible Cre transgene that is active within the male and female germlines (e.g., Vasa-Cre). Male transmission is required to avoid embryo-wide recombination. Male (♂) mice carrying the Cre transgene (DriverGene-Cre; red text; red-colored mice) are first crossed with female (♀) mice carrying a conditional (flox) allele (*TargetGene*$^{flox/flox}$; blue-colored mice). At the first generation, male DriverGene-Cre;*TargetGene*$^{flox/+}$ progeny (red text; purple-colored mice) are backcrossed to female *TargetGene*$^{flox/flox}$ mice. First-generation progeny of *TargetGene*$^{flox/+}$ genotype (gray mice) are not required for further breeding. Second-generation mice of genotype DriverGene-Cre; *TargetGene*$^{\Delta/flox}$ (black-colored mice; red text "Conditional Knockout") and DriverGene-Cre;*TargetGene*$^{+/flox}$ (black-colored mice; red text "Littermate Control") are experimental groups (red boxes), whereas progeny of other genotypes (black boxes) are not required but can be used as alternative control groups. Germline genotypes (following Cre-mediated recombination in germ cells) are listed in bold black text underneath mice. "Δ" denotes a recombined allele. "+" denotes a wild-type allele

persists during early embryonic development (Fig. 4). Cre expression in zygotes and the early embryo will lead to ubiquitous recombination and organism-wide genetic knockout. Examples of strains described herein that require paternally transmitted Cre to generate tissue-specific conditional knockouts include Vasa-Cre, Tnap-Cre, and Myh11-Cre.

We denote Cre-expressing strains as "DriverGene-Cre," in which the transgene is always maintained in hemi- or heterozygosity. Mice homozygous for floxed alleles are described as "*TargetGene*$^{flox/flox}$." Compound heterozygotes where one null allele and one floxed allele has been inherited are indicated as "*TargetGene*$^{\Delta/flox}$," where "Δ" denotes the recombined allele.

Breeding procedure:

1. Mate *male DriverGene-Cre* mice with *female TargetGene$^{flox/flox}$* mice to generate **DriverGene-Cre;*TargetGene*$^{+/flox}$** progeny. All progeny in this generation that inherit the Cre allele (expected 50%) will recombine the inherited floxed allele (expected 100%) within their germline; therefore, their germline genotype will be **DriverGene-Cre;*TargetGene*$^{+/\Delta}$**. The remaining (expected 50%) progeny will be *TargetGene*$^{+/flox}$ and negative for the Cre-expressing allele and no recombination will occur in the germline.
2. Mate *male DriverGene-Cre;TargetGene*$^{+/flox}$ mice (whose germline genotypes are DriverGene-Cre;*TargetGene*$^{+/\Delta}$) with *female TargetGene*$^{flox/flox}$ mice to generate **DriverGene-Cre; *TargetGene*$^{\Delta/flox}$** progeny (conditional knockout group; expected 25%) and **DriverGene-Cre;*TargetGene*$^{+/flox}$** progeny (control group; expected 25%). This generation will also include *TargetGene*$^{\Delta/flox}$ (expected 25%) and *TargetGene*$^{+/flox}$ (expected 25%) progeny that are generally not required for experimental purposes using this strategy.
3. All progeny of the cross in **step 2** will have the following genotypes (expected germline genotypes in parentheses):
   - DriverGene-Cre;*TargetGene*$^{\Delta/flox}$ (**DriverGene-Cre; *TargetGene*$^{\Delta/\Delta}$**)—conditional knockout mice.
   - DriverGene-Cre;*TargetGene*$^{+/flox}$ (**DriverGene-Cre;*TargetGene*$^{+/\Delta}$**)—littermate control group.
   - *TargetGene*$^{\Delta/flox}$—not required for experiments but can be alternative control.
   - *TargetGene*$^{+/flox}$—not required for experiments but can be alternative control.

Control group mice in these experiments are heterozygous for the target gene in the germline and express Cre recombinase. This is an appropriate control to account for any concerns regarding Cre toxicity. Where there are concerns regarding haploinsufficiency (reviewed in [171–173]), *TargetGene*$^{\Delta/flox}$ and *TargetGene*$^{+/flox}$ may be analyzed as alternative control groups, as *TargetGene*$^{+/flox}$ mice are expected to have two functional target gene alleles. However, all four genotypes obtained in **step 3** cannot be obtained in male mice when the target gene is hemizygous (e.g., X-linked) and a modified breeding strategy is required (*see* Subheading 4.1.4).

4.1.2 Maternal Transmission of Non-inducible Cre

If Cre activity does not occur in the female germline, it may be preferable to transmit a Cre transgene via the maternal lineage when generating conditional knockout models. For example, first-generation crosses that result in heterozygous germline mutants may lead to impaired male fertility as a result of haploinsufficiency [174–176]. Consequently, difficulties may arise when

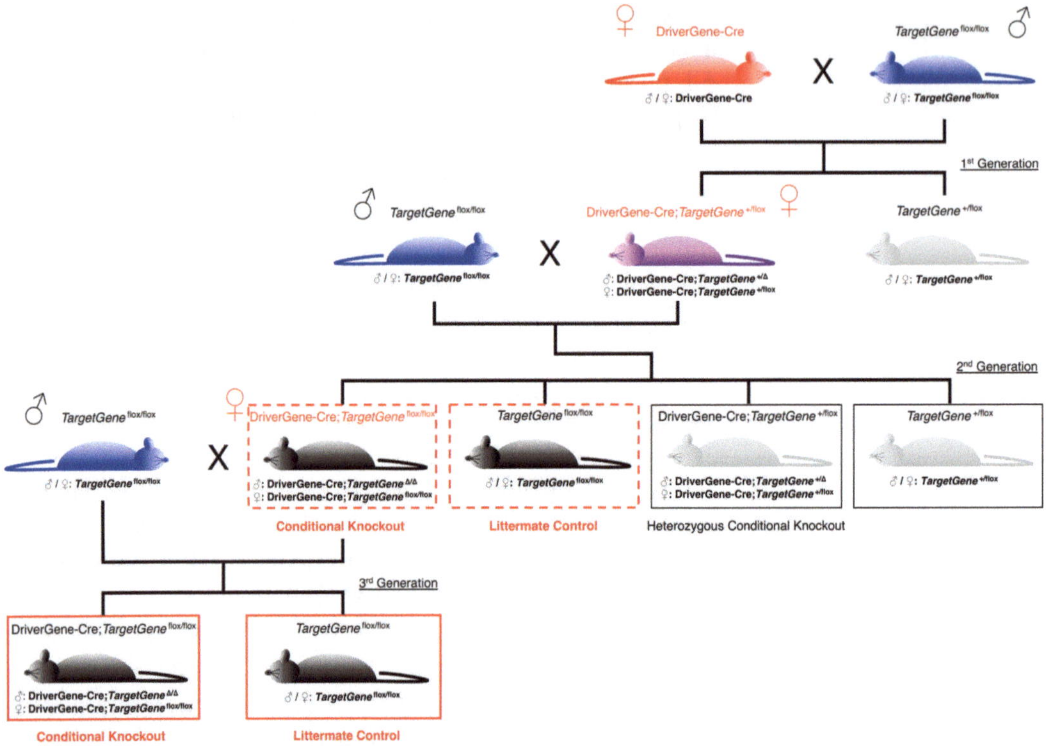

Fig. 5 *Conditional knockout model breeding scheme for maternal transmission of a non-inducible Cre transgene.* This breeding scheme describes maternal transmission of a non-inducible Cre transgene that is active within the male germline, but not active within the female germline or early embryonic development (e.g., Stra8-iCre). Female (♀) mice carrying the Cre transgene (DriverGene-Cre; red text; red-colored mice) are first crossed with male (♂) mice carrying a conditional (flox) allele (*TargetGene*$^{flox/flox}$; blue-colored mice). At the first generation, female DriverGene-Cre;*TargetGene*$^{flox/+}$ progeny (red text; purple-colored mice) are backcrossed to male *TargetGene*$^{flox/flox}$ mice. First-generation progeny of *TargetGene*$^{flox/+}$ genotype (gray mice) are not required for further breeding. Second-generation mice of genotype DriverGene-Cre;*TargetGene*$^{flox/flox}$ (black-colored mice; red dashed box; red text "Conditional Knockout") and *TargetGene*$^{flox/flox}$ (black-colored mice; red dashed box; red text "Littermate Control") can be used as experimental groups, whereas progeny of other genotypes (black boxes) are not required but can be used as alternative control groups. However, an additional backcross of DriverGene-Cre;*TargetGene*$^{flox/flox}$ females to *TargetGene*$^{flox/flox}$ males is recommended to generate final experimental groups (third generation, red boxes). Germline genotypes for males and females are listed in bold black text underneath mice. "Δ" denotes a recombined allele. "+" denotes a wild-type allele

breeding males to generate homozygous conditional knockouts in the following generation.

The following breeding scheme describes maternal non-inducible Cre transmission of a transgene that is active within the male germline to generate male germline conditional knockouts (e.g., models using Stra8-iCre) (Fig. 5). Where non-inducible Cre activity occurs in both male and female germlines (e.g., Vasa-Cre), breeding schemes with paternal Cre transmission should be used (*see* Subheading 4.1.1). Note that the DriverGene-Cre transgene is always maintained in hemi- or heterozygosity.

Breeding procedure:

1. Mate *female DriverGene-Cre* mice with *male TargetGene*$^{flox/flox}$ mice to generate **DriverGene-Cre;*TargetGene***$^{+/\mathbf{flox}}$ progeny. Male progeny in this generation that inherit the Cre allele (expected 50% of males; 25% of all progeny) will recombine the inherited floxed allele (expected 100%) within their germline; therefore, their germline genotype will be DriverGene-Cre;*TargetGene*$^{+/\Delta}$. Cre is not active in the female germline; therefore, female progeny in this generation that inherit the Cre allele (expected 50% of females; 25% of all progeny) will not recombine the inherited floxed allele and their germline genotype will remain **DriverGene-Cre;*TargetGene***$^{+/\mathbf{flox}}$. In theory, the Cre allele may also be paternally transmitted for this first-generation cross.
2. Mate *female DriverGene-Cre;TargetGene*$^{+/flox}$ mice with *male TargetGene*$^{flox/flox}$ mice to generate progeny with the following genotypes (expected 25% each):
   - **DriverGene-Cre;*TargetGene***$^{\mathbf{flox/flox}}$.
   - **DriverGene-Cre;*TargetGene***$^{+/\mathbf{flox}}$.
   - ***TargetGene***$^{\mathbf{flox/flox}}$.
   - ***TargetGene***$^{+/\mathbf{flox}}$.
3. Male mice obtained in this generation will have the following germline genotypes:
   - **DriverGene-Cre;*TargetGene***$^{\boldsymbol{\Delta/\Delta}}$—homozygous conditional knockout group.
   - **DriverGene-Cre;*TargetGene***$^{+/\boldsymbol{\Delta}}$—heterozygous conditional knockout group.
   - ***TargetGene***$^{\mathbf{flox/flox}}$—Cre-negative littermate control group.
   - ***TargetGene***$^{+/\mathbf{flox}}$ —not required for experiments but can be alternative control.

In this breeding scheme, we have designated *TargetGene*$^{flox/flox}$ littermates obtained in the second generation of breeding as control group mice, as is commonly reported in previous studies [32–34, 177, 178]. However, as previously suggested herein, mice carrying only the Cre transgene should be assessed to exclude any potential effects of Cre expression alone. Where there are no concerns regarding haploinsufficiency, DriverGene-Cre;*TargetGene*$^{+/flox}$ (heterozygous conditional knockout group) may also be used as a control group to account for Cre effects.

Where it is determined that second-generation DriverGene-Cre;*TargetGene*$^{+/\Delta}$ and *TargetGene*$^{+/flox}$ mice are not experimentally required, an additional cross can be performed to eliminate the production of these genotypes for ethical reasons as follows:

4. Mate *female DriverGene-Cre;TargetGene*$^{flox/flox}$ mice with *male TargetGene*$^{flox/flox}$ mice to generate progeny with the following genotypes (expected 50% each):
   - *DriverGene-Cre;TargetGene*$^{flox/flox}$.
   - *TargetGene*$^{flox/flox}$.

Male mice obtained within this third generation that inherit the Cre allele (expected 50% of males; 25% of all progeny) will have germline genotype **DriverGene-Cre;*TargetGene***$^{\Delta/\Delta}$ (homozygous conditional knockout group). Genotype of the remaining male mice will be ***TargetGene***$^{flox/flox}$ and serve as littermate controls. At this generation, all animals can be used for experimental purposes, eliminating the euthanasia of mice due solely to incorrect genotype. Female DriverGene-Cre;*TargetGene*$^{flox/flox}$ mice produced in this generation can continue to be crossed to male *TargetGene*$^{flox/flox}$ mice to maintain the production of experimental groups, provided that potential effects of genetic drift are managed appropriately if breeding for several generations [179].

4.1.3 Inducible Conditional Knockout Models

The use of inducible Cre strains to generate conditional knockout models offers flexibility with choice of breeding strategy. Given that Cre is only active following tamoxifen treatment, whether transgenes are transmitted paternally or maternally generally has no bearing on generation of experimental groups.

A typical breeding scheme is described below, with Cre transmission not restricted to male or female breeders (Fig. 6). The inducible Cre allele is designated "CreER" for the purposes of this description and is always maintained in hemi- or heterozygosity.

Breeding procedure:

1. Mate *DriverGene-CreER* mice with *TargetGene*$^{flox/flox}$ mice to generate **DriverGene-CreER;*TargetGene***$^{+/flox}$ progeny (expected 50%). The remaining progeny will be *TargetGene*$^{+/flox}$ and are not generally required for further breeding.
2. Mate *DriverGene-CreER;TargetGene*$^{+/flox}$ mice with *TargetGene*$^{flox/flox}$ mice to generate progeny with the following genotypes (expected 25% each):
   - **DriverGene-CreER;*TargetGene***$^{flox/flox}$.
   - **DriverGene-CreER;*TargetGene***$^{+/flox}$.
   - ***TargetGene***$^{flox/flox}$.
   - ***TargetGene***$^{+/flox}$.

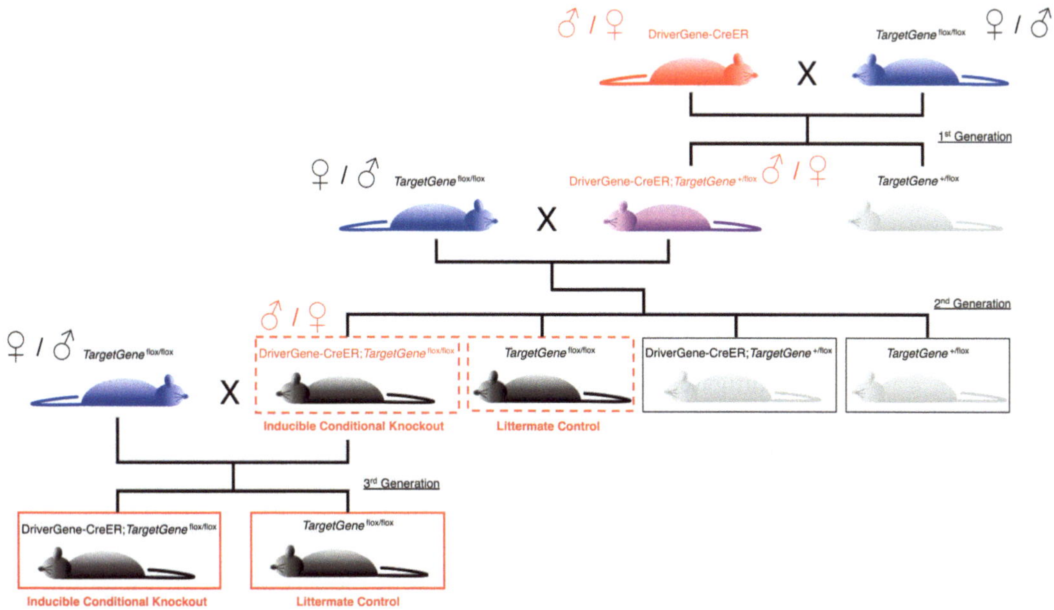

Fig. 6 *Inducible conditional knockout model breeding scheme.* This breeding scheme describes the generation of experimental groups using a Cre-driver strain carrying an inducible Cre transgene (CreER). CreER can be transmitted via female (♀) mice or male (♂) mice, as it is only activated following treatment by tamoxifen. Mice carrying the CreER transgene (DriverGene-CreER; red text; red-colored mice) are first crossed with mice carrying a conditional (flox) allele (*TargetGene*$^{flox/flox}$; blue-colored mice). At the first generation, DriverGene-CreER;*TargetGene*$^{flox/+}$ progeny (red text; purple-colored mice) are backcrossed to *TargetGene*$^{flox/flox}$ mice. First-generation progeny of *TargetGene*$^{flox/+}$ genotype (gray mice) are not required for further breeding. Second-generation mice of genotype DriverGene-CreER;*TargetGene*$^{flox/flox}$ (black-colored mice; red dashed box; red text "Conditional Knockout") and *TargetGene*$^{flox/flox}$ (black-colored mice; red dashed box; red text "Littermate Control") can be used as experimental groups, whereas progeny of other genotypes (black boxes) are not required but can be used as alternative control groups. However, an additional backcross of DriverGene-CreER;*TargetGene*$^{flox/flox}$ mice to *TargetGene*$^{flox/flox}$ mice is recommended to generate final experimental groups (third generation, red boxes). "+" denotes a wild-type allele

At this step, DriverGene-CreER;*TargetGene*$^{flox/flox}$ mice can be used as the inducible conditional knockout group, and *TargetGene*$^{flox/flox}$ littermates serve as control mice. For ethical reasons, in order to reduce the production of mice with genotypes not suitable for experimental purposes the following breeding step is recommended.

3. Mate *DriverGene-CreER;TargetGene*$^{flox/flox}$ mice with *TargetGene*$^{flox/flox}$ mice to generate progeny with the following genotypes (expected 50% each):
   - **DriverGene-CreER;*TargetGene*$^{flox/flox}$**—inducible conditional knockout group.
   - ***TargetGene*$^{flox/flox}$**—littermate control group.

Similar to non-inducible Cre strains, mice expressing Cre only can be used as an additional control group. However, all experimental mice must be treated with tamoxifen to maintain an appropriately controlled study design (*see* Subheading 4.2).

Importantly, spontaneous activity of inducible Cre (or "leaky Cre") within a number of separate strains has been reported [97, 180, 181]. Although the rate of spontaneous recombination is generally low, unexpected recombination in the germline of breeding animals can be problematic and may generate conventional knockout offspring. Therefore, it is important to maintain separate breeding colonies of mice that only carry floxed alleles to mitigate potential consequences of leaky Cre activity. Where there are concerns regarding spontaneous recombination in experimental mice, an experimental group where DriverGene-CreER;*TargetGene*$^{flox/flox}$ mice are treated with vehicle can be used as an additional control (*see* Subheading 4.2).

4.1.4 Considerations for Sex Chromosome-Linked Conditional Alleles

To generate conditional knockout models for the investigation of sex chromosome-linked genes in spermatogenesis, breeding strategies and experimental designs used for studies of autosomal genes must be altered. Fortunately, breeding schemes are simpler as experimental groups can be achieved in a single generation given that males are hemizygous for sex chromosome genes.

The breeding scheme to generate male conditional knockouts for X chromosome genes is described here and is applicable regardless of Cre activity (i.e., inducible vs. non-inducible; male or male and female germline active). A floxed allele present on the X chromosome is denoted as "X(flox)," whereas wild-type X and Y chromosomes are simply depicted as "X" or "Y."

Breeding procedure:

1. Mate *male DriverGene-Cre* mice with *female TargetGene*$^{X(flox)/X(flox)}$ mice to generate progeny of the following genotypes according to sex (expected 25% each):
   - **Male DriverGene-Cre;*TargetGene*$^{X(flox)/Y}$**—conditional knockout group.
   - **Male *TargetGene*$^{X(flox)/Y}$**—Cre-negative littermate control group.
   - Female *DriverGene-Cre;TargetGene*$^{X(flox)/X}$—heterozygous conditional knockout.
   - Female *TargetGene*$^{X(flox)/X}$—heterozygous floxed allele.

   Male mice from this single generation of breeding are suitable for experimental use. Control littermates are Cre-negative; therefore, males expressing Cre transgene only should be used if there are concerns regarding Cre toxicity.

To our knowledge, conditional knockout models targeting Y chromosome genes have not been reported but are theoretically possible. A breeding scheme to generate such models is described here and is applicable regardless of Cre activity. A floxed allele present on the Y chromosome is denoted as "Y(flox)," and wild-type X chromosomes are simply depicted as "X."

Breeding procedure:

1. Mate *female DriverGene-Cre* mice with *male TargetGene*$^{X/Y(flox)}$ mice to generate progeny of the following genotypes according to sex (expected 25% each):
   - **Male DriverGene-Cre;*TargetGene*$^{\mathbf{X/Y(flox)}}$**—conditional knockout group.
   - **Male *TargetGene*$^{\mathbf{X/Y(flox)}}$**—Cre-negative littermate control group.
   - Female *DriverGene-Cre*—Cre-expressing only as females do not have a Y chromosome.
   - Female *wild-types*—produced due to hemi−/heterozygous Cre and no Y chromosome.

   Like X chromosome gene conditional knockouts, male mice from this generation of breeding are suitable for experimental use. Similarly, control littermates are Cre-negative and males expressing Cre transgene only can be used to exclude potential effects of Cre toxicity.

4.2 Treatment Regimens for Inducible Conditional Knockouts

Tamoxifen treatment protocols to induce Cre activity can vary according to a number of factors including age of mice; whether pregnant mice are being treated to induce recombination in embryos; expression level of Cre recombinase; and efficiency of recombination in the target tissue. Here, we describe our standard tamoxifen treatment protocol to induce UBC-CreER^{T2}-mediated conditional knockout in adult mice (6–8 weeks old) for the study of spermatogonia [90–92]. All animal experimentation must be conducted in accordance with legal and institutional animal ethics requirements.

4.2.1 Tamoxifen Preparation

The following protocol will prepare a final solution of 10 mg/ml tamoxifen in 15% ethanol (v/v)/85% sesame oil (v/v) vehicle. The quantities provided are enough to prepare 5 ml tamoxifen solution (25 doses for a 20 g mouse). Tamoxifen is light-sensitive and should be protected from light. Preparation of tamoxifen should be performed using sterile reagents and sterile technique.

1. Preheat a heat block to 37 °C.
2. Prepare 5 × 1.5 mL centrifuge tubes containing 850 μL sesame oil per tube and incubate at 37 °C.
3. Place 50 mg tamoxifen into a microcentrifuge tube.

4. Add 750 μL absolute ethanol and vortex until dissolved (can take several minutes).
5. Once tamoxifen is dissolved in ethanol, add 150 μL of this solution to each of the tubes containing 850 μL preheated sesame oil.
6. Vortex well to combine.
7. Once combined, incubate at 37 °C for an additional 10 min.
8. Protect from light and proceed to tamoxifen treatment of mice.

For optimal results, prepare fresh tamoxifen solution on each day of treatment. However, tamoxifen solution can be stored at 4 ° C protected from light for up to 5 days. Where vehicle-treated controls are required, the vehicle solution is prepared according to the same protocol, omitting tamoxifen (final solution of 15% ethanol (v/v)/85% sesame oil (v/v) only).

4.2.2 Tamoxifen Treatment

1. Preheat tamoxifen (or vehicle-only) solution to 37 °C and ensure mixture is homogenous. If precipitate is visible, the solution should be discarded.
2. Determine tamoxifen dosage based on mouse mass, using 100 mg/kg of body mass. For a 20 g mouse, this is 2 mg tamoxifen (200 μL of prepared solution). For vehicle-only treatment, administer an equivalent volume of solution.
3. Draw up tamoxifen into a 1 ml syringe using an 18 G drawing up needle.
4. Remove the drawing up needle and fit a 25G 3/4″ beveled needle to the syringe.
5. Sanitize the injection site with 80% ethanol (v/v).
6. Administer tamoxifen by intraperitoneal injection.
7. Dispose of all sharps and tamoxifen waste according to institutional requirements.

Mice are treated with tamoxifen *each day* (24 h apart) for *two consecutive days*. We deem the second day of tamoxifen treatment as experimental day (D) 0. Mice should be monitored each day according to approved ethics protocols for any adverse reactions. For initial analyses, we collect testes from treated mice at D7 and D14 (*see* Subheading 4.4).

4.3 In Vitro SSC Conditional Knockout

In order to confirm cell-autonomous effects of a conditional knockout in SSCs or perform more in-depth biochemical analyses that require large numbers of cells, it may be desirable to generate in vitro cultures. Established methods exist for the isolation of undifferentiated spermatogonia and their long-term in vitro propagation [182–186]. For conditional knockouts generated using a non-inducible model, cultures can be generated from separate control and knockout animals for subsequent analysis. However, gene knockout may affect spermatogonial function and their ability to establish colonies in vitro. Therefore, it may be necessary to generate cultures using inducible knockout models. These cultures can be derived from untreated DriverGene-CreER;*TargetGene*$^{flox/flox}$ mice and treated with vehicle (control group) or tamoxifen (conditional knockout group) in vitro. An advantage of this method is the ability to control the timing of gene knockout, circumventing difficulties that may arise when using non-inducible models. In addition, this method eliminates the requirement to establish cultures from mice of different genotypes for control purposes. However, if there are concerns regarding the effects of tamoxifen, cultures can also be established from *TargetGene*$^{flox/flox}$ mice for treatment.

Here, we describe a basic protocol used for in vitro tamoxifen treatment of cultured undifferentiated spermatogonia to induce conditional knockout in our previous studies [90–92]. For in vitro treatment, 4-hydroxytamoxifen (4-OHT) must be used to induce conditional knockout. 4-OHT is an active metabolite of tamoxifen generated in vivo after oxidation by cytochrome P450 in the liver and has higher affinity for estrogen receptors compared to tamoxifen [187, 188].

4.3.1 4-OHT Treatment

1. Culture DriverGene-CreER;*TargetGene*$^{flox/flox}$ undifferentiated spermatogonia to approximately 60–70% confluency. Two treatment groups, vehicle and 4-OHT, are required per experiment. An additional group of 4-OHT-treated *TargetGene*$^{flox/flox}$ undifferentiated spermatogonia may be included if there are concerns regarding the effects of 4-OHT.
2. Prepare a 5 mM solution of 4-OHT dissolved in absolute ethanol. Aliquot and store unused solution at −20 °C for up to 1 month.
3. On the first day of treatment, prepare vehicle and 4-OHT treatment media in separate tubes. For the 4-OHT media, add 5 mM 4-OHT solution (prepared in **step 2**) to fresh cell culture media to a final concentration of 200 nM. For the vehicle media, add an equivalent volume of absolute ethanol to fresh cell culture media. The final amount of absolute ethanol within cell culture media should not exceed 0.1% (v/v).
4. Replace culture media with vehicle or 4-OHT treatment media and return cells to incubator.
5. After 48 h, replace treatment media with freshly prepared treatment media (according to **step 3**) and return cells to incubator.
6. After another 48 h, replace treatment media with standard culture media. We deem this time point to be D1.

7. For initial analyses, collect cells at D4. Until this time, maintain cells according to standard culture protocols.

This protocol should be used as a basic guide and may need to be optimized depending on growth rate of cultured cells (e.g., adjusting starting confluency), efficiency of Cre recombination (e.g., adjusting 4-OHT dosage and timing), and consequence of genetic knockout (e.g., adjusting analysis time point).

4.4 Confirming and Improving Conditional Knockout Efficiency

Key to the success of any conditional knockout model is the efficiency of recombination that is achieved. Although several factors can contribute to incomplete recombination including Cre expression level, floxed allele design, or treatment regimens, the strains we have described have been reported to achieve sufficiently high rates of knockout to realize experimental aims.

In this section, we provide an outline of steps that can be taken to confirm and improve conditional knockout efficiency.

4.4.1 Genotyping

When working with conditional knockout models, it is essential that accurate and reliable genotyping protocols are maintained to ensure efficiency in breeding and experimental group production. Several established protocols exist for genotyping mouse samples (examples in [189–191]), and recommended primer sequences to detect transgenes are generally provided by commercial sources and original sources describing strains. However, understanding the structure of chosen transgenes is important in order to design primers to detect Cre expression, as well as wild-type, floxed, and recombined alleles for the target gene, if required. Ideal primers for floxed alleles are able to distinguish between wild-type, heterozygous, and homozygous genotypes. For Cre transgenes, primers may only be able to detect their presence, rather than zygosity, if the strain was generated using methods that result in random genomic insertion.

When developing a new conditional knockout model, each generation should be genotyped for all transgenic alleles regardless of expected breeding outcome (i.e., even following crosses where all progeny are expected to be hetero- or homozygous). This prudent approach ensures that any unexpected breeding difficulties, genotyping inaccuracies, or breeder pairing errors are detected early and rectified before additional mice are generated.

For experimental animals, Cre-mediated recombination should be confirmed at least within whole testis tissue and ideally in isolated SSCs (or other cells being targeted) using magnetic- or fluorescence-based sorting. Recombined alleles can be detected by PCR using genomic DNA and suitable primers, followed by gel electrophoresis. For inducible Cre models, this should be performed on Cre-expressing mice following tamoxifen treatment, as well as mice from this group that have not been treated with tamoxifen. Although largely qualitative, these steps can confirm successful recombination prior to further analysis.

4.4.2 Confirming Conditional Knockout Efficiency Using RNA- and Protein-Based Methods

To quantify the efficiency of conditional knockout in the male germline, a combination of RNA- and protein-based methods are commonly used. For in vitro cultured spermatogonia, obtaining material to perform molecular biology techniques to confirm and quantify knockout is relatively straightforward compared to in vivo tissues. For in vivo conditional knockout, additional steps may be required to isolate target cells, although histological techniques on tissue samples can be used to provide valuable spatial information regarding knockout specificity and efficiency.

Although recombination of floxed coding regions is anticipated to result in nonsense-mediated decay of any resulting transcript, truncated gene products may still be generated [192, 193]. Therefore, RT-qPCR should be used as a quantitative method to assess expression of mRNA from the target gene. Ideally, primers spanning junctions of the targeted exon(s) can be used to determine relative levels of target gene expression, as well as primers upstream and downstream of these regions to determine whether mRNA is still being transcribed from a recombined allele.

Ultimately, the absence of protein translated from the target gene is an important indicator of conditional knockout success. Using isolated cells for techniques such as Western blot and flow cytometry enables reliable quantification of protein remaining after knockout, provided that suitable antibodies are available. Given that truncated proteins may be produced following recombination, antibodies targeting different epitopes can be used to verify knockout efficiency. Antibodies can also be used to detect target proteins using histological techniques, and testis cross-sections and whole mount samples are commonly used to demonstrate successful conditional knockout in SSCs and other testis cell populations (examples in [90–92]).

4.4.3 Improving Conditional Knockout Efficiency

The observed efficiency of conditional knockout can be affected by a number of factors related to the transgenes being used or experimental protocols. If the Cre-driver being used is not sufficiently expressed within the cells of interest, recombination efficiency may be low. In this case, an alternative strain where Cre is expressed via a more strongly expressed promoter may ultimately be required. Cre expression in the tissue or cells of interest can be analyzed by RT-qPCR for *Cre*, or Cre reporter strains can be used to visualize Cre-mediated recombination in situ [194] (*see* Subheading 4.5). Recombination efficiency can also be affected by factors related to conditional allele design, including genetic distance between *loxP* sites, sequences flanking these sites, and their location within chromosomes [181, 195–197]. Where recombination levels are not sufficient for experimental purposes, there may be no choice but to redesign the floxed allele.

Redesigning a mouse model can be laborious and costly and may not be necessary to improve experimental outcomes. For inducible knockout models, optimization of tamoxifen treatment regimens by increasing dose and/or duration of treatment can improve recombination efficiency. In addition, tamoxifen treatment of mice at younger ages has been suggested to increase recombination efficiency [198]. For non-inducible Cre models, generating compound heterozygotes within the final experimental group (i.e., DriverGene-Cre;*TargetGene*$^{\Delta/\mathrm{flox}}$, *see* Subheading 4.1.1) can improve knockout efficiency, as only one allele is required to be recombined by Cre. This breeding strategy is commonly reported in previous studies of SSCs and spermatogenesis (examples in [32, 34, 62, 199]). Timing of analysis post-knockout may also need to be optimized, as collecting samples prematurely may not allow sufficient time for recombination and protein turnover.

4.5 Other Uses for Conditional Alleles in the Study of SSCs

In this chapter, we have described the use of conditional alleles in the context of generating mouse models with targeted gene knockout. However, Cre/*loxP* and other recombinase systems have broader applicability beyond their use in conditional knockout models. Perhaps the most prominent use of recombinase systems, and in particular Cre/*loxP*, is in genetic lineage tracing studies (*see* Chapter 13). In these experiments, tissue-specific Cre strains are crossed with reporter strains that ubiquitously express a transgene containing a label (e.g., fluorescent protein or β-galactosidase) preceded by a "STOP" cassette containing transcription terminating signals. The STOP cassette is flanked by *loxP* sites and therefore, upon Cre-mediated recombination, is excised, and expression of a fluorescent protein or β-galactosidase is possible [200]. As a result, the Cre-expressing cells are labelled, and their fate can be traced at later time points. Since its development, this technique has revolutionized the study of stem and progenitor cells and has been used extensively in studies of SSCs. Lineage tracing studies have provided critical insight into the molecular characteristics and dynamics of SSCs (examples in [31, 66, 69, 118, 201]). Besides the tracing of cell fate, this system is also routinely used when establishing new recombinase-expressing strains and developing novel conditional knockout models. In general, samples from reporter models are collected shortly after the onset of recombinase activity and analyzed to determine cell specificity and efficiency of recombination [194]. These reporter models are essential for the validation of new strains, providing critical data regarding their suitability to target specific cell populations for conditional gene knockout prior to the generation of experimental cohorts.

Similar to inducing expression of a reporter gene, recombinase systems have also been used in studies of the male germline to induce expression of gain-of-function [202, 203] and dominant-negative [204] alleles, as well as overexpression of genes encoding transcription factors [205], microRNAs [206], and RNA-binding proteins [71]. In addition, conditional models have been used to express toxin or toxin receptors (e.g., diphtheria toxin and its receptor) to enable ablation of specific cell populations [207]. In testis studies, these models have been used to ablate testicular macrophages, Sertoli cells, and germ cells [131, 208]. This demonstrates the versatility of conditional alleles for use in the study of SSCs and spermatogenesis and highlights the advantages of having reliable Cre-driver strains to target the male germline.

5 Conclusions

The development and use of conditional knockout models are indispensable to biomedical research. These tools have provided unprecedented insight into the mechanisms regulating SSC function and will undoubtedly continue to be a foundation of studies for the foreseeable future. In this chapter, we have provided readily available strains to drive conditional knockout in the male germline and other testis cell populations, as well as breeding strategies to generate appropriate experimental groups. In addition, we have provided advice for verifying and improving recombination efficiency.

Future studies will inevitably generate novel and improved methods for genetic manipulation that can be harnessed in studies of SSCs and spermatogenesis. Already, cutting-edge technologies such as CRISPR/Cas9 have enabled more rapid development of transgenic and knockout animals. Undoubtedly, knockout animals will continue to play a central role in our understanding of SSCs, reproductive biology, and stem cells in general.

References

1. Kanatsu-Shinohara M, Shinohara T (2013) Spermatogonial stem cell self-renewal and development. Ann Rev Cell Dev Biol 29: 163–187. https://doi.org/10.1146/annurev-cellbio-101512-122353
2. Hermann BP, Cheng K, Singh A, Roa-De La Cruz L, Mutoji KN, Chen IC, Gildersleeve H, Lehle JD, Mayo M, Westernstroer B, Law NC, Oatley MJ, Velte EK, Niedenberger BA, Fritze D, Silber S, Geyer CB, Oatley JM, McCarrey JR (2018) The mammalian spermatogenesis single-cell transcriptome, from spermatogonial stem cells to spermatids. Cell Rep 25(6):1650–1667. e1658. https://doi.org/10.1016/j.celrep.2018.10.026
3. La HM, Hobbs RM (2019) Mechanisms regulating mammalian spermatogenesis and fertility recovery following germ cell depletion. Cell Mol Life Sci 76(20):4071–4102. https://doi.org/10.1007/s00018-019-03201-6
4. Legrand JMD, Hobbs RM (2018) RNA processing in the male germline: mechanisms and implications for fertility. Semin Cell Dev Biol 79:80–91. https://doi.org/10.1016/j.semcdb.2017.10.006

5. Oatley JM, Brinster RL (2008) Regulation of spermatogonial stem cell self-renewal in mammals. Ann Rev Cell Dev Biol 24:263–286. https://doi.org/10.1146/annurev.cellbio.24.110707.175355
6. Yoshida S (2012) Elucidating the identity and behavior of spermatogenic stem cells in the mouse testis. Reproduction 144(3): 293–302. https://doi.org/10.1530/REP-11-0320
7. Koller BH, Hagemann LJ, Doetschman T, Hagaman JR, Huang S, Williams PJ, First NL, Maeda N, Smithies O (1989) Germ-line transmission of a planned alteration made in a hypoxanthine phosphoribosyltransferase gene by homologous recombination in embryonic stem cells. Proc Natl Acad Sci U S A 86(22): 8927–8931. https://doi.org/10.1073/pnas.86.22.8927
8. Thomas KR, Capecchi MR (1990) Targeted disruption of the murine int-1 proto-oncogene resulting in severe abnormalities in midbrain and cerebellar development. Nature 346(6287):847–850. https://doi.org/10.1038/346847a0
9. Thompson S, Clarke AR, Pow AM, Hooper ML, Melton DW (1989) Germ line transmission and expression of a corrected HPRT gene produced by gene targeting in embryonic stem cells. Cell 56(2):313–321. https://doi.org/10.1016/0092-8674(89)90905-7
10. Zijlstra M, Li E, Sajjadi F, Subramani S, Jaenisch R (1989) Germ-line transmission of a disrupted beta 2-microglobulin gene produced by homologous recombination in embryonic stem cells. Nature 342(6248): 435–438. https://doi.org/10.1038/342435a0
11. Melton DW (2002) Gene-targeting strategies. Methods Mol Biol 180:151–173. https://doi.org/10.1385/1-59259-178-7:151
12. Hall B, Limaye A, Kulkarni AB (2009) Overview: generation of gene knockout mice. Curr Protoc Cell Biol 44:Unit 19.12 19.12.1. https://doi.org/10.1002/0471143030.cb1912s44
13. Wefers B, Bashir S, Rossius J, Wurst W, Kuhn R (2017) Gene editing in mouse zygotes using the CRISPR/Cas9 system. Methods 121–122:55–67. https://doi.org/10.1016/j.ymeth.2017.02.008
14. El-Brolosy MA, Stainier DYR (2017) Genetic compensation: a phenomenon in search of mechanisms. PLoS Genet 13(7):e1006780. https://doi.org/10.1371/journal.pgen.1006780
15. Sternberg N, Hamilton D (1981) Bacteriophage P1 site-specific recombination. I. Recombination between loxP sites. J Mol Biol 150(4):467–486. https://doi.org/10.1016/0022-2836(81)90375-2
16. Austin S, Ziese M, Sternberg N (1981) A novel role for site-specific recombination in maintenance of bacterial replicons. Cell 25(3):729–736. https://doi.org/10.1016/0092-8674(81)90180-x
17. Sauer B, Henderson N (1988) Site-specific DNA recombination in mammalian cells by the Cre recombinase of bacteriophage P1. Proc Natl Acad Sci U S A 85(14): 5166–5170. https://doi.org/10.1073/pnas.85.14.5166
18. McLeod M, Craft S, Broach JR (1986) Identification of the crossover site during FLP-mediated recombination in the Saccharomyces cerevisiae plasmid 2 microns circle. Mol Cell Biol 6(10):3357–3367. https://doi.org/10.1128/mcb.6.10.3357
19. Broach JR, Guarascio VR, Jayaram M (1982) Recombination within the yeast plasmid 2mu circle is site-specific. Cell 29(1):227–234. https://doi.org/10.1016/0092-8674(82)90107-6
20. Branda CS, Dymecki SM (2004) Talking about a revolution: the impact of site-specific recombinases on genetic analyses in mice. Dev Cell 6(1):7–28. https://doi.org/10.1016/s1534-5807(03)00399-x
21. Kellendonk C, Tronche F, Monaghan AP, Angrand PO, Stewart F, Schutz G (1996) Regulation of Cre recombinase activity by the synthetic steroid RU 486. Nucleic Acids Res 24(8):1404–1411. https://doi.org/10.1093/nar/24.8.1404
22. Morikawa K, Furuhashi K, de Sena-Tomas C, Garcia-Garcia AL, Bekdash R, Klein AD, Gallerani N, Yamamoto HE, Park SE, Collins GS, Kawano F, Sato M, Lin CS, Targoff KL, Au E, Salling MC, Yazawa M (2020) Photoactivatable Cre recombinase 3.0 for in vivo mouse applications. Nat Commun 11(1): 2141. https://doi.org/10.1038/s41467-020-16030-0
23. Sando R 3rd, Baumgaertel K, Pieraut S, Torabi-Rander N, Wandless TJ, Mayford M, Maximov A (2013) Inducible control of gene expression with destabilized Cre. Nat Methods 10(11):1085–1088. https://doi.org/10.1038/nmeth.2640
24. Law NC, Oatley JM (2020) Developmental underpinnings of spermatogonial stem cell establishment. Andrology 8(4):852–861. https://doi.org/10.1111/andr.12810

25. Friedel RH, Wurst W, Wefers B, Kuhn R (2011) Generating conditional knockout mice. Methods Mol Biol 693:205–231. https://doi.org/10.1007/978-1-60761-974-1_12
26. Saitou M, Yamaji M (2012) Primordial germ cells in mice. Cold Spring Harb Perspect Biol 4(11):a008375. https://doi.org/10.1101/cshperspect.a008375
27. Castrillon DH, Quade BJ, Wang TY, Quigley C, Crum CP (2000) The human VASA gene is specifically expressed in the germ cell lineage. Proc Natl Acad Sci U S A 97(17):9585–9590. https://doi.org/10.1073/pnas.160274797
28. Gallardo T, Shirley L, John GB, Castrillon DH (2007) Generation of a germ cell-specific mouse transgenic Cre line, vasa-Cre. Genesis 45(6):413–417. https://doi.org/10.1002/dvg.20310
29. Tanaka SS, Toyooka Y, Akasu R, Katoh-Fukui Y, Nakahara Y, Suzuki R, Yokoyama M, Noce T (2000) The mouse homolog of drosophila vasa is required for the development of male germ cells. Genes Dev 14(7):841–853
30. Toyooka Y, Tsunekawa N, Takahashi Y, Matsui Y, Satoh M, Noce T (2000) Expression and intracellular localization of mouse vasa-homologue protein during germ cell development. Mech Dev 93(1–2):139–149. https://doi.org/10.1016/s0925-4773(00)00283-5
31. Aloisio GM, Nakada Y, Saatcioglu HD, Pena CG, Baker MD, Tarnawa ED, Mukherjee J, Manjunath H, Bugde A, Sengupta AL, Amatruda JF, Cuevas I, Hamra FK, Castrillon DH (2014) PAX7 expression defines germline stem cells in the adult testis. J Clin Invest 124(9):3929–3944. https://doi.org/10.1172/JCI75943
32. Hobbs RM, Fagoonee S, Papa A, Webster K, Altruda F, Nishinakamura R, Chai L, Pandolfi PP (2012) Functional antagonism between Sall4 and Plzf defines germline progenitors. Cell Stem Cell 10(3):284–298. https://doi.org/10.1016/j.stem.2012.02.004
33. Hobbs RM, La HM, Makela JA, Kobayashi T, Noda T, Pandolfi PP (2015) Distinct germline progenitor subsets defined through Tsc2-mTORC1 signaling. EMBO Rep 16(4):467–480. https://doi.org/10.15252/embr.201439379
34. Yang QE, Gwost I, Oatley MJ, Oatley JM (2013) Retinoblastoma protein (RB1) controls fate determination in stem cells and progenitors of the mouse male germline. Biol Reprod 89(5):113. https://doi.org/10.1095/biolreprod.113.113159
35. Goodwin LO, Splinter E, Davis TL, Urban R, He H, Braun RE, Chesler EJ, Kumar V, van Min M, Ndukum J, Philip VM, Reinholdt LG, Svenson K, White JK, Sasner M, Lutz C, Murray SA (2019) Large-scale discovery of mouse transgenic integration sites reveals frequent structural variation and insertional mutagenesis. Genome Res 29(3):494–505. https://doi.org/10.1101/gr.233866.117
36. Ohinata Y, Payer B, O'Carroll D, Ancelin K, Ono Y, Sano M, Barton SC, Obukhanych T, Nussenzweig M, Tarakhovsky A, Saitou M, Surani MA (2005) Blimp1 is a critical determinant of the germ cell lineage in mice. Nature 436(7048):207–213. https://doi.org/10.1038/nature03813
37. Chang DH, Cattoretti G, Calame KL (2002) The dynamic expression pattern of B lymphocyte induced maturation protein-1 (Blimp-1) during mouse embryonic development. Mech Dev 117(1–2):305–309. https://doi.org/10.1016/s0925-4773(02)00189-2
38. Li Z, Yu J, Hosohama L, Nee K, Gkountela S, Chaudhari S, Cass AA, Xiao X, Clark AT (2015) The Sm protein methyltransferase PRMT5 is not required for primordial germ cell specification in mice. EMBO J 34(6):748–758. https://doi.org/10.15252/embj.201489319
39. Kim S, Gunesdogan U, Zylicz JJ, Hackett JA, Cougot D, Bao S, Lee C, Dietmann S, Allen GE, Sengupta R, Surani MA (2014) PRMT5 protects genomic integrity during global DNA demethylation in primordial germ cells and preimplantation embryos. Mol Cell 56(4):564–579. https://doi.org/10.1016/j.molcel.2014.10.003
40. Turner CA Jr, Mack DH, Davis MM (1994) Blimp-1, a novel zinc finger-containing protein that can drive the maturation of B lymphocytes into immunoglobulin-secreting cells. Cell 77(2):297–306. https://doi.org/10.1016/0092-8674(94)90321-2
41. Kim SJ (2015) Immunological function of Blimp-1 in dendritic cells and relevance to autoimmune diseases. Immunol Res 63(1–3):113–120. https://doi.org/10.1007/s12026-015-8694-5
42. Martins G, Calame K (2008) Regulation and functions of Blimp-1 in T and B lymphocytes. Annu Rev Immunol 26:133–169. https://doi.org/10.1146/annurev.immunol.26.021607.090241
43. John SA, Garrett-Sinha LA (2009) Blimp1: a conserved transcriptional repressor critical for

differentiation of many tissues. Exp Cell Res 315(7):1077–1084. https://doi.org/10.1016/j.yexcr.2008.11.015

44. Muncan V, Heijmans J, Krasinski SD, Buller NV, Wildenberg ME, Meisner S, Radonjic M, Stapleton KA, Lamers WH, Biemond I, van den Bergh Weerman MA, O'Carroll D, Hardwick JC, Hommes DW, van den Brink GR (2011) Blimp1 regulates the transition of neonatal to adult intestinal epithelium. Nat Commun 2:452. https://doi.org/10.1038/ncomms1463
45. Hoshi K, Amizuka N, Oda K, Ikehara Y, Ozawa H (1997) Immunolocalization of tissue non-specific alkaline phosphatase in mice. Histochem Cell Biol 107(3):183–191. https://doi.org/10.1007/s004180050103
46. Sebastian-Serrano A, de Diego-Garcia L, Martinez-Frailes C, Avila J, Zimmermann H, Millan JL, Miras-Portugal MT, Diaz-Hernandez M (2015) Tissue-nonspecific alkaline phosphatase regulates purinergic transmission in the central nervous system during development and disease. Comput Struct Biotechnol J 13:95–100. https://doi.org/10.1016/j.csbj.2014.12.004
47. MacGregor GR, Zambrowicz BP, Soriano P (1995) Tissue non-specific alkaline phosphatase is expressed in both embryonic and extraembryonic lineages during mouse embryogenesis but is not required for migration of primordial germ cells. Development 121(5):1487–1496
48. Lomeli H, Ramos-Mejia V, Gertsenstein M, Lobe CG, Nagy A (2000) Targeted insertion of Cre recombinase into the TNAP gene: excision in primordial germ cells. Genesis 26(2):116–117
49. Elliott AM, de Miguel MP, Rebel VI, Donovan PJ (2007) Identifying genes differentially expressed between PGCs and ES cells reveals a role for CREB-binding protein in germ cell survival. Dev Biol 311(2):347–358. https://doi.org/10.1016/j.ydbio.2007.08.029
50. Yamaguchi YL, Tanaka SS, Kumagai M, Fujimoto Y, Terabayashi T, Matsui Y, Nishinakamura R (2015) Sall4 is essential for mouse primordial germ cell specification by suppressing somatic cell program genes. Stem Cells 33(1):289–300. https://doi.org/10.1002/stem.1853
51. Hayashi K, Chuva de Sousa Lopes SM, Kaneda M, Tang F, Hajkova P, Lao K, O'Carroll D, Das PP, Tarakhovsky A, Miska EA, Surani MA (2008) MicroRNA biogenesis is required for mouse primordial germ cell development and spermatogenesis. PLoS One 3(3):e1738. https://doi.org/10.1371/journal.pone.0001738
52. Kimura T, Suzuki A, Fujita Y, Yomogida K, Lomeli H, Asada N, Ikeuchi M, Nagy A, Mak TW, Nakano T (2003) Conditional loss of PTEN leads to testicular teratoma and enhances embryonic germ cell production. Development 130(8):1691–1700. https://doi.org/10.1242/dev.00392
53. Kim S, Bardwell VJ, Zarkower D (2007) Cell type-autonomous and non-autonomous requirements for Dmrt1 in postnatal testis differentiation. Dev Biol 307(2):314–327. https://doi.org/10.1016/j.ydbio.2007.04.046
54. Busada JT, Geyer CB (2016) The role of retinoic acid (RA) in Spermatogonial differentiation. Biol Reprod 94(1):10. https://doi.org/10.1095/biolreprod.115.135145
55. Endo T, Mikedis MM, Nicholls PK, Page DC, de Rooij DG (2019) Retinoic acid and germ cell development in the ovary and testis. Biomol Ther 9(12). https://doi.org/10.3390/biom9120775
56. Anderson EL, Baltus AE, Roepers-Gajadien HL, Hassold TJ, de Rooij DG, van Pelt AM, Page DC (2008) Stra8 and its inducer, retinoic acid, regulate meiotic initiation in both spermatogenesis and oogenesis in mice. Proc Natl Acad Sci U S A 105(39):14976–14980. https://doi.org/10.1073/pnas.0807297105
57. Kojima ML, de Rooij DG, Page DC (2019) Amplification of a broad transcriptional program by a common factor triggers the meiotic cell cycle in mice. elife 8. https://doi.org/10.7554/eLife.43738
58. Sadate-Ngatchou PI, Payne CJ, Dearth AT, Braun RE (2008) Cre recombinase activity specific to postnatal, premeiotic male germ cells in transgenic mice. Genesis 46(12):738–742. https://doi.org/10.1002/dvg.20437
59. Shimshek DR, Kim J, Hubner MR, Spergel DJ, Buchholz F, Casanova E, Stewart AF, Seeburg PH, Sprengel R (2002) Codon-improved Cre recombinase (iCre) expression in the mouse. Genesis 32(1):19–26. https://doi.org/10.1002/gene.10023
60. Buaas FW, Kirsh AL, Sharma M, McLean DJ, Morris JL, Griswold MD, de Rooij DG, Braun RE (2004) Plzf is required in adult male germ cells for stem cell self-renewal. Nat Genet 36(6):647–652. https://doi.org/10.1038/ng1366
61. Costoya JA, Hobbs RM, Barna M, Cattoretti G, Manova K, Sukhwani M,

Orwig KE, Wolgemuth DJ, Pandolfi PP (2004) Essential role of Plzf in maintenance of spermatogonial stem cells. Nat Genet 36(6):653–659. https://doi.org/10.1038/ng1367

62. Gely-Pernot A, Raverdeau M, Teletin M, Vernet N, Feret B, Klopfenstein M, Dennefeld C, Davidson I, Benoit G, Mark M, Ghyselinck NB (2015) Retinoic acid receptors control Spermatogonia cell-fate and induce expression of the SALL4A transcription factor. PLoS Genet 11(10): e1005501. https://doi.org/10.1371/journal.pgen.1005501
63. Bao J, Zhang Y, Schuster AS, Ortogero N, Nilsson EE, Skinner MK, Yan W (2014) Conditional inactivation of Miwi2 reveals that MIWI2 is only essential for prospermatogonial development in mice. Cell Death Differ 21(5):783–796. https://doi.org/10.1038/cdd.2014.5
64. Lores P, Vernet N, Kurosaki T, Van de Putte T, Huylebroeck D, Hikida M, Gacon G, Toure A (2014) Deletion of MgcRacGAP in the male germ cells impairs spermatogenesis and causes male sterility in the mouse. Dev Biol 386(2):419–427. https://doi.org/10.1016/j.ydbio.2013.12.006
65. Sommer L, Ma Q, Anderson DJ (1996) Neurogenins, a novel family of atonal-related bHLH transcription factors, are putative mammalian neuronal determination genes that reveal progenitor cell heterogeneity in the developing CNS and PNS. Mol Cell Neurosci 8(4):221–241. https://doi.org/10.1006/mcne.1996.0060
66. Nakagawa T, Nabeshima Y, Yoshida S (2007) Functional identification of the actual and potential stem cell compartments in mouse spermatogenesis. Dev Cell 12(2):195–206. https://doi.org/10.1016/j.devcel.2007.01.002
67. Yoshida S, Sukeno M, Nakagawa T, Ohbo K, Nagamatsu G, Suda T, Nabeshima Y (2006) The first round of mouse spermatogenesis is a distinctive program that lacks the self-renewing spermatogonia stage. Development 133(8):1495–1505. https://doi.org/10.1242/dev.02316
68. Yoshida S, Takakura A, Ohbo K, Abe K, Wakabayashi J, Yamamoto M, Suda T, Nabeshima Y (2004) Neurogenin3 delineates the earliest stages of spermatogenesis in the mouse testis. Dev Biol 269(2):447–458. https://doi.org/10.1016/j.ydbio.2004.01.036
69. Nakagawa T, Sharma M, Nabeshima Y, Braun RE, Yoshida S (2010) Functional hierarchy and reversibility within the murine spermatogenic stem cell compartment. Science 328(5974):62–67. https://doi.org/10.1126/science.1182868
70. Zhang T, Oatley J, Bardwell VJ, Zarkower D (2016) DMRT1 is required for mouse Spermatogonial stem cell maintenance and replenishment. PLoS Genet 12(9):e1006293. https://doi.org/10.1371/journal.pgen.1006293
71. Sada A, Hasegawa K, Pin PH, Saga Y (2012) NANOS2 acts downstream of glial cell line-derived neurotrophic factor signaling to suppress differentiation of spermatogonial stem cells. Stem Cells 30(2):280–291. https://doi.org/10.1002/stem.790
72. Schonhoff SE, Giel-Moloney M, Leiter AB (2004) Neurogenin 3-expressing progenitor cells in the gastrointestinal tract differentiate into both endocrine and non-endocrine cell types. Dev Biol 270(2):443–454. https://doi.org/10.1016/j.ydbio.2004.03.013
73. Korhonen HM, Meikar O, Yadav RP, Papaioannou MD, Romero Y, Da Ros M, Herrera PL, Toppari J, Nef S, Kotaja N (2011) Dicer is required for haploid male germ cell differentiation in mice. PLoS One 6(9):e24821. https://doi.org/10.1371/journal.pone.0024821
74. Ellnati E, Russell HR, Ojarikre OA, Sangrithi M, Hirota T, de Rooij DG, McKinnon PJ, Turner JMA (2017) DNA damage response protein TOPBP1 regulates X chromosome silencing in the mammalian germ line. Proc Natl Acad Sci U S A 114(47): 12536–12541. https://doi.org/10.1073/pnas.1712530114
75. Zheng K, Wang PJ (2012) Blockade of pachytene piRNA biogenesis reveals a novel requirement for maintaining post-meiotic germline genome integrity. PLoS Genet 8(11):e1003038. https://doi.org/10.1371/journal.pgen.1003038
76. Evans E, Hogarth C, Mitchell D, Griswold M (2014) Riding the spermatogenic wave: profiling gene expression within neonatal germ and Sertoli cells during a synchronized initial wave of spermatogenesis in mice. Biol Reprod 90(5):108. https://doi.org/10.1095/biolreprod.114.118034
77. Bai S, Cheng L, Zhang Y, Zhu C, Zhu Z, Zhu R, Cheng CY, Ye L, Zheng K (2018) A germline-specific role for the mTORC2 component Rictor in maintaining spermatogonial differentiation and intercellular adhesion in mouse testis. Mol Hum Reprod 24(5):

244–259. https://doi.org/10.1093/molehr/gay009
78. Jin C, Zhang Y, Wang ZP, Wang XX, Sun TC, Li XY, Tang JX, Cheng JM, Li J, Chen SR, Deng SL, Liu YX (2017) EZH2 deletion promotes spermatogonial differentiation and apoptosis. Reproduction 154(5):615–625. https://doi.org/10.1530/REP-17-0302
79. Magnuson MA, Osipovich AB (2013) Pancreas-specific Cre driver lines and considerations for their prudent use. Cell Metab 18(1):9–20. https://doi.org/10.1016/j.cmet.2013.06.011
80. Feil R, Wagner J, Metzger D, Chambon P (1997) Regulation of Cre recombinase activity by mutated estrogen receptor ligand-binding domains. Biochem Biophys Res Commun 237(3):752–757. https://doi.org/10.1006/bbrc.1997.7124
81. Hayashi S, McMahon AP (2002) Efficient recombination in diverse tissues by a tamoxifen-inducible form of Cre: a tool for temporally regulated gene activation/inactivation in the mouse. Dev Biol 244(2):305–318. https://doi.org/10.1006/dbio.2002.0597
82. Feil R, Brocard J, Mascrez B, LeMeur M, Metzger D, Chambon P (1996) Ligand-activated site-specific recombination in mice. Proc Natl Acad Sci U S A 93(20):10887–10890. https://doi.org/10.1073/pnas.93.20.10887
83. Indra AK, Warot X, Brocard J, Bornert JM, Xiao JH, Chambon P, Metzger D (1999) Temporally-controlled site-specific mutagenesis in the basal layer of the epidermis: comparison of the recombinase activity of the tamoxifen-inducible Cre-ER(T) and Cre-ER(T2) recombinases. Nucleic Acids Res 27(22):4324–4327. https://doi.org/10.1093/nar/27.22.4324
84. Hershko A, Ciechanover A (1998) The ubiquitin system. Annu Rev Biochem 67:425–479. https://doi.org/10.1146/annurev.biochem.67.1.425
85. Radici L, Bianchi M, Crinelli R, Magnani M (2013) Ubiquitin C gene: structure, function, and transcriptional regulation. Adv Biosci Biotechnol 04(12):6. https://doi.org/10.4236/abb.2013.412141
86. Ryu KY, Maehr R, Gilchrist CA, Long MA, Bouley DM, Mueller B, Ploegh HL, Kopito RR (2007) The mouse polyubiquitin gene UbC is essential for fetal liver development, cell-cycle progression and stress tolerance. EMBO J 26(11):2693–2706. https://doi.org/10.1038/sj.emboj.7601722
87. Schorpp M, Jager R, Schellander K, Schenkel J, Wagner EF, Weiher H, Angel P (1996) The human ubiquitin C promoter directs high ubiquitous expression of transgenes in mice. Nucleic Acids Res 24(9):1787–1788. https://doi.org/10.1093/nar/24.9.1787
88. Ruzankina Y, Pinzon-Guzman C, Asare A, Ong T, Pontano L, Cotsarelis G, Zediak VP, Velez M, Bhandoola A, Brown EJ (2007) Deletion of the developmentally essential gene ATR in adult mice leads to age-related phenotypes and stem cell loss. Cell Stem Cell 1(1):113–126. https://doi.org/10.1016/j.stem.2007.03.002
89. Loebel DA, Plageman TF Jr, Tang TL, Jones VJ, Muccioli M, Tam PP (2016) Thyroid bud morphogenesis requires CDC42- and SHROOM3-dependent apical constriction. Biol Open 5(2):130–139. https://doi.org/10.1242/bio.014415
90. Chan AL, La HM, Legrand JMD, Makela JA, Eichenlaub M, De Seram M, Ramialison M, Hobbs RM (2017) Germline stem cell activity is sustained by SALL4-dependent silencing of distinct tumor suppressor genes. Stem Cell Reports 9(3):956–971. https://doi.org/10.1016/j.stemcr.2017.08.001
91. La HM, Chan AL, Legrand JMD, Rossello FJ, Gangemi CG, Papa A, Cheng Q, Morand EF, Hobbs RM (2018) GILZ-dependent modulation of mTORC1 regulates spermatogonial maintenance. Development 145(18). https://doi.org/10.1242/dev.165324
92. Legrand JMD, Chan AL, La HM, Rossello FJ, Anko ML, Fuller-Pace FV, Hobbs RM (2019) DDX5 plays essential transcriptional and post-transcriptional roles in the maintenance and function of spermatogonia. Nat Commun 10(1):2278. https://doi.org/10.1038/s41467-019-09972-7
93. Matson CK, Murphy MW, Griswold MD, Yoshida S, Bardwell VJ, Zarkower D (2010) The mammalian doublesex homolog DMRT1 is a transcriptional gatekeeper that controls the mitosis versus meiosis decision in male germ cells. Dev Cell 19(4):612–624. https://doi.org/10.1016/j.devcel.2010.09.010
94. Ferreira M, Boens S, Winkler C, Szeker K, Verbinnen I, Van Eynde A, Fardilha M, Bollen M (2017) The protein phosphatase 1 regulator NIPP1 is essential for mammalian spermatogenesis. Sci Rep 7(1):13364. https://doi.org/10.1038/s41598-017-13809-y
95. Lopez IP, Rodriguez-de la Rosa L, Pais RS, Pineiro-Hermida S, Torrens R, Contreras J, Varela-Nieto I, Pichel JG (2015) Differential

organ phenotypes after postnatal Igf1r gene conditional deletion induced by tamoxifen in UBC-CreERT2; Igf1r fl/fl double transgenic mice. Transgenic Res 24(2):279–294. https://doi.org/10.1007/s11248-014-9837-5

96. Puri P, Phillips BT, Suzuki H, Orwig KE, Rajkovic A, Lapinski PE, King PD, Feng GS, Walker WH (2014) The transition from stem cell to progenitor spermatogonia and male fertility requires the SHP2 protein tyrosine phosphatase. Stem Cells 32(3):741–753. https://doi.org/10.1002/stem.1572
97. Kristianto J, Johnson MG, Zastrow RK, Radcliff AB, Blank RD (2017) Spontaneous recombinase activity of Cre-ERT2 in vivo. Transgenic Res 26(3):411–417. https://doi.org/10.1007/s11248-017-0018-1
98. Casola S (2010) Mouse models for miRNA expression: the ROSA26 locus. Methods Mol Biol 667:145–163. https://doi.org/10.1007/978-1-60761-811-9_10
99. de Luca C, Kowalski TJ, Zhang Y, Elmquist JK, Lee C, Kilimann MW, Ludwig T, Liu SM, Chua SC Jr (2005) Complete rescue of obesity, diabetes, and infertility in db/db mice by neuron-specific LEPR-B transgenes. J Clin Invest 115(12):3484–3493. https://doi.org/10.1172/JCI24059
100. Seibler J, Zevnik B, Kuter-Luks B, Andreas S, Kern H, Hennek T, Rode A, Heimann C, Faust N, Kauselmann G, Schoor M, Jaenisch R, Rajewsky K, Kuhn R, Schwenk F (2003) Rapid generation of inducible mouse mutants. Nucleic Acids Res 31(4):12e. https://doi.org/10.1093/nar/gng012
101. Ventura A, Kirsch DG, McLaughlin ME, Tuveson DA, Grimm J, Lintault L, Newman J, Reczek EE, Weissleder R, Jacks T (2007) Restoration of p53 function leads to tumour regression in vivo. Nature 445(7128):661–665. https://doi.org/10.1038/nature05541
102. Tomizawa SI, Kobayashi Y, Shirakawa T, Watanabe K, Mizoguchi K, Hoshi I, Nakajima K, Nakabayashi J, Singh S, Dahl A, Alexopoulou D, Seki M, Suzuki Y, Royo H, Peters A, Anastassiadis K, Stewart AF, Ohbo K (2018) Kmt2b conveys monovalent and bivalent H3K4me3 in mouse spermatogonial stem cells at germline and embryonic promoters. Development 145(23). https://doi.org/10.1242/dev.169102
103. Glaser S, Lubitz S, Loveland KL, Ohbo K, Robb L, Schwenk F, Seibler J, Roellig D, Kranz A, Anastassiadis K, Stewart AF (2009) The histone 3 lysine 4 methyltransferase, Mll2, is only required briefly in development and spermatogenesis. Epigenetics Chromatin 2(1):5. https://doi.org/10.1186/1756-8935-2-5
104. Kawai Y, Oda A, Kanai Y, Goitsuka R (2018) Germ cell-intrinsic requirement for the homeodomain transcription factor PKnox1/Prep1 in adult spermatogenesis. PLoS One 13(1):e0190702. https://doi.org/10.1371/journal.pone.0190702
105. Shirakawa T, Yaman-Deveci R, Tomizawa S, Kamizato Y, Nakajima K, Sone H, Sato Y, Sharif J, Yamashita A, Takada-Horisawa Y, Yoshida S, Ura K, Muto M, Koseki H, Suda T, Ohbo K (2013) An epigenetic switch is crucial for spermatogonia to exit the undifferentiated state toward a kit-positive identity. Development 140(17):3565–3576. https://doi.org/10.1242/dev.094045
106. Shi G, Jin Y (2010) Role of Oct4 in maintaining and regaining stem cell pluripotency. Stem Cell Res Ther 1(5):39. https://doi.org/10.1186/scrt39
107. Takahashi K, Tanabe K, Ohnuki M, Narita M, Ichisaka T, Tomoda K, Yamanaka S (2007) Induction of pluripotent stem cells from adult human fibroblasts by defined factors. Cell 131(5):861–872. https://doi.org/10.1016/j.cell.2007.11.019
108. Takahashi K, Yamanaka S (2006) Induction of pluripotent stem cells from mouse embryonic and adult fibroblast cultures by defined factors. Cell 126(4):663–676. https://doi.org/10.1016/j.cell.2006.07.024
109. Rosner MH, Vigano MA, Ozato K, Timmons PM, Poirier F, Rigby PW, Staudt LM (1990) A POU-domain transcription factor in early stem cells and germ cells of the mammalian embryo. Nature 345(6277):686–692. https://doi.org/10.1038/345686a0
110. Scholer HR, Hatzopoulos AK, Balling R, Suzuki N, Gruss P (1989) A family of octamer-specific proteins present during mouse embryogenesis: evidence for germline-specific expression of an Oct factor. EMBO J 8(9):2543–2550
111. Scholer HR, Ruppert S, Suzuki N, Chowdhury K, Gruss P (1990) New type of POU domain in germ line-specific protein Oct-4. Nature 344(6265):435–439. https://doi.org/10.1038/344435a0
112. Palmieri SL, Peter W, Hess H, Scholer HR (1994) Oct-4 transcription factor is differentially expressed in the mouse embryo during establishment of the first two extraembryonic cell lineages involved in implantation. Dev Biol 166(1):259–267. https://doi.org/10.1006/dbio.1994.1312

113. Yeom YI, Fuhrmann G, Ovitt CE, Brehm A, Ohbo K, Gross M, Hubner K, Scholer HR (1996) Germline regulatory element of Oct-4 specific for the totipotent cycle of embryonal cells. Development 122(3):881–894
114. Pesce M, Wang X, Wolgemuth DJ, Scholer H (1998) Differential expression of the Oct-4 transcription factor during mouse germ cell differentiation. Mech Dev 71(1–2):89–98. https://doi.org/10.1016/s0925-4773(98)00002-1
115. Dann CT, Alvarado AL, Molyneux LA, Denard BS, Garbers DL, Porteus MH (2008) Spermatogonial stem cell self-renewal requires OCT4, a factor downregulated during retinoic acid-induced differentiation. Stem Cells 26(11):2928–2937. https://doi.org/10.1634/stemcells.2008-0134
116. Filipponi D, Hobbs RM, Ottolenghi S, Rossi P, Jannini EA, Pandolfi PP, Dolci S (2007) Repression of kit expression by Plzf in germ cells. Mol Cell Biol 27(19):6770–6781. https://doi.org/10.1128/MCB.00479-07
117. Hammoud SS, Low DH, Yi C, Lee CL, Oatley JM, Payne CJ, Carrell DT, Guccione E, Cairns BR (2015) Transcription and imprinting dynamics in developing postnatal male germline stem cells. Genes Dev 29(21):2312–2324. https://doi.org/10.1101/gad.261925.115
118. La HM, Mäkelä JA, Chan AL, Rossello F, Nefzger C, Legrand JMD, De Seram M, Polo JM, Hobbs RM (2018) Identification of dynamic undifferentiated cell states within the male germline. Nature Commun 9:2819
119. Piprek RP, Kolasa M, Podkowa D, Kloc M, Kubiak JZ (2019) Tissue-specific knockout of E-cadherin (Cdh1) in developing mouse gonads causes germ cells loss. Reproduction 158(2):147–157. https://doi.org/10.1530/REP-18-0621
120. Piprek RP, Kolasa M, Podkowa D, Kloc M, Kubiak JZ (2019) N-cadherin is critical for the survival of germ cells, the formation of steroidogenic cells, and the architecture of developing mouse gonads. Cell 8(12). https://doi.org/10.3390/cells8121610
121. Greder LV, Gupta S, Li S, Abedin MJ, Sajini A, Segal Y, Slack JM, Dutton JR (2012) Analysis of endogenous Oct4 activation during induced pluripotent stem cell reprogramming using an inducible Oct4 lineage label. Stem Cells 30(11):2596–2601. https://doi.org/10.1002/stem.1216
122. Zhang Y, Riesterer C, Ayrall AM, Sablitzky F, Littlewood TD, Reth M (1996) Inducible site-directed recombination in mouse embryonic stem cells. Nucleic Acids Res 24(4):543–548. https://doi.org/10.1093/nar/24.4.543
123. Lengner CJ, Camargo FD, Hochedlinger K, Welstead GG, Zaidi S, Gokhale S, Scholer HR, Tomilin A, Jaenisch R (2007) Oct4 expression is not required for mouse somatic stem cell self-renewal. Cell Stem Cell 1(4):403–415. https://doi.org/10.1016/j.stem.2007.07.020
124. John GB, Gallardo TD, Shirley LJ, Castrillon DH (2008) Foxo3 is a PI3K-dependent molecular switch controlling the initiation of oocyte growth. Dev Biol 321(1):197–204. https://doi.org/10.1016/j.ydbio.2008.06.017
125. Abe H, Alavattam KG, Kato Y, Castrillon DH, Pang Q, Andreassen PR, Namekawa SH (2018) CHEK1 coordinates DNA damage signaling and meiotic progression in the male germline of mice. Hum Mol Genet 27(7):1136–1149. https://doi.org/10.1093/hmg/ddy022
126. Guan Y, Leu NA, Ma J, Chmatal L, Ruthel G, Bloom JC, Lampson MA, Schimenti JC, Luo M, Wang PJ (2020) SKP1 drives the prophase I to metaphase I transition during male meiosis. Sci Adv 6(13):eaaz2129. https://doi.org/10.1126/sciadv.aaz2129
127. Zhang J, Tu Z, Watanabe Y, Shibuya H (2017) Distinct TERB1 domains regulate different protein interactions in meiotic telomere movement. Cell Rep 21(7):1715–1726. https://doi.org/10.1016/j.celrep.2017.10.061
128. de Rooij DG (2009) The spermatogonial stem cell niche. Microsc Res Tech 72(8):580–585. https://doi.org/10.1002/jemt.20699
129. Oatley JM, Brinster RL (2012) The germline stem cell niche unit in mammalian testes. Physiol Rev 92(2):577–595. https://doi.org/10.1152/physrev.00025.2011
130. Oatley JM, Oatley MJ, Avarbock MR, Tobias JW, Brinster RL (2009) Colony stimulating factor 1 is an extrinsic stimulator of mouse spermatogonial stem cell self-renewal. Development 136(7):1191–1199. https://doi.org/10.1242/dev.032243
131. DeFalco T, Potter SJ, Williams AV, Waller B, Kan MJ, Capel B (2015) Macrophages contribute to the Spermatogonial niche in the adult testis. Cell Rep 12(7):1107–1119. https://doi.org/10.1016/j.celrep.2015.07.015
132. Yoshida S, Sukeno M, Nabeshima Y (2007) A vasculature-associated niche for

undifferentiated spermatogonia in the mouse testis. Science 317(5845):1722–1726. https://doi.org/10.1126/science.1144885

133. Beau C, Vivian N, Munsterberg A, Dresser DW, Lovell-Badge R, Guerrier D (2001) In vivo analysis of the regulation of the anti-Mullerian hormone, as a marker of Sertoli cell differentiation during testicular development, reveals a multi-step process. Mol Reprod Dev 59(3):256–264. https://doi.org/10.1002/mrd.1030
134. Holdcraft RW, Braun RE (2004) Androgen receptor function is required in Sertoli cells for the terminal differentiation of haploid spermatids. Development 131(2):459–467. https://doi.org/10.1242/dev.00957
135. Payne CJ, Gallagher SJ, Foreman O, Dannenberg JH, Depinho RA, Braun RE (2010) Sin3a is required by Sertoli cells to establish a niche for undifferentiated spermatogonia, germ cell tumors, and spermatid elongation. Stem Cells 28(8):1424–1434. https://doi.org/10.1002/stem.464
136. Elliott MR, Zheng S, Park D, Woodson RI, Reardon MA, Juncadella IJ, Kinchen JM, Zhang J, Lysiak JJ, Ravichandran KS (2010) Unexpected requirement for ELMO1 in clearance of apoptotic germ cells in vivo. Nature 467(7313):333–337. https://doi.org/10.1038/nature09356
137. Nalam RL, Andreu-Vieyra C, Braun RE, Akiyama H, Matzuk MM (2009) Retinoblastoma protein plays multiple essential roles in the terminal differentiation of Sertoli cells. Mol Endocrinol 23(11):1900–1913. https://doi.org/10.1210/me.2009-0184
138. Boyer A, Girard M, Thimmanahalli DS, Levasseur A, Celeste C, Paquet M, Duggavathi R, Boerboom D (2016) mTOR regulates gap junction Alpha-1 protein trafficking in Sertoli cells and is required for the maintenance of spermatogenesis in mice. Biol Reprod 95(1):13. https://doi.org/10.1095/biolreprod.115.138016
139. Boyer A, Hermo L, Paquet M, Robaire B, Boerboom D (2008) Seminiferous tubule degeneration and infertility in mice with sustained activation of WNT/CTNNB1 signaling in sertoli cells. Biol Reprod 79(3): 475–485. https://doi.org/10.1095/biolreprod.108.068627
140. Kyronlahti A, Euler R, Bielinska M, Schoeller EL, Moley KH, Toppari J, Heikinheimo M, Wilson DB (2011) GATA4 regulates Sertoli cell function and fertility in adult male mice. Mol Cell Endocrinol 333(1):85–95. https://doi.org/10.1016/j.mce.2010.12.019
141. Tanwar PS, Kaneko-Tarui T, Zhang L, Rani P, Taketo MM, Teixeira J (2010) Constitutive WNT/beta-catenin signaling in murine Sertoli cells disrupts their differentiation and ability to support spermatogenesis. Biol Reprod 82(2):422–432. https://doi.org/10.1095/biolreprod.109.079335
142. Tanwar PS, Kaneko-Tarui T, Zhang L, Teixeira JM (2012) Altered LKB1/AMPK/TSC1/TSC2/mTOR signaling causes disruption of Sertoli cell polarity and spermatogenesis. Hum Mol Genet 21(20): 4394–4405. https://doi.org/10.1093/hmg/dds272
143. Jamin SP, Arango NA, Mishina Y, Hanks MC, Behringer RR (2002) Requirement of Bmpr1a for Mullerian duct regression during male sexual development. Nat Genet 32(3): 408–410. https://doi.org/10.1038/ng1003
144. Jeyasuria P, Ikeda Y, Jamin SP, Zhao L, De Rooij DG, Themmen AP, Behringer RR, Parker KL (2004) Cell-specific knockout of steroidogenic factor 1 reveals its essential roles in gonadal function. Mol Endocrinol 18(7):1610–1619. https://doi.org/10.1210/me.2003-0404
145. Frojdman K, Harley VR, Pelliniemi LJ (2000) Sox9 protein in rat sertoli cells is age and stage dependent. Histochem Cell Biol 113(1): 31–36. https://doi.org/10.1007/s004180050004
146. Kopp JL, Dubois CL, Schaffer AE, Hao E, Shih HP, Seymour PA, Ma J, Sander M (2011) Sox9+ ductal cells are multipotent progenitors throughout development but do not produce new endocrine cells in the normal or injured adult pancreas. Development 138(4):653–665. https://doi.org/10.1242/dev.056499
147. Yokonishi T, McKey J, Ide S, Capel B (2020) Sertoli cell ablation and replacement of the spermatogonial niche in mouse. Nat Commun 11(1):40. https://doi.org/10.1038/s41467-019-13879-8
148. Symon A, Harley V (2017) SOX9: a genomic view of tissue specific expression and action. Int J Biochem Cell Biol 87:18–22. https://doi.org/10.1016/j.biocel.2017.03.005
149. Garside VC, Cullum R, Alder O, Lu DY, Vander Werff R, Bilenky M, Zhao Y, Jones SJ, Marra MA, Underhill TM, Hoodless PA (2015) SOX9 modulates the expression of key transcription factors required for heart valve development. Development 142(24): 4340–4350. https://doi.org/10.1242/dev.125252
150. Font-Burgada J, Shalapour S, Ramaswamy S, Hsueh B, Rossell D, Umemura A,

Taniguchi K, Nakagawa H, Valasek MA, Ye L, Kopp JL, Sander M, Carter H, Deisseroth K, Verma IM, Karin M (2015) Hybrid Periportal hepatocytes regenerate the injured liver without giving rise to cancer. Cell 162(4): 766–779. https://doi.org/10.1016/j.cell.2015.07.026
151. Kopp JL, von Figura G, Mayes E, Liu FF, Dubois CL, Morris JP, Pan FC, Akiyama H, Wright CV, Jensen K, Hebrok M, Sander M (2012) Identification of Sox9-dependent acinar-to-ductal reprogramming as the principal mechanism for initiation of pancreatic ductal adenocarcinoma. Cancer Cell 22(6): 737–750. https://doi.org/10.1016/j.ccr.2012.10.025
152. Malhotra GK, Zhao X, Edwards E, Kopp JL, Naramura M, Sander M, Band H, Band V (2014) The role of Sox9 in mouse mammary gland development and maintenance of mammary stem and luminal progenitor cells. BMC Dev Biol 14:47. https://doi.org/10.1186/s12861-014-0047-4
153. Porubek D (2013) CYP17A1: a biochemistry, chemistry, and clinical review. Curr Top Med Chem 13(12):1364–1384. https://doi.org/10.2174/15680266113139990002
154. Liu C, Rodriguez K, Yao HH (2016) Mapping lineage progression of somatic progenitor cells in the mouse fetal testis. Development 143(20):3700–3710. https://doi.org/10.1242/dev.135756
155. Shima Y, Miyabayashi K, Haraguchi S, Arakawa T, Otake H, Baba T, Matsuzaki S, Shishido Y, Akiyama H, Tachibana T, Tsutsui K, Morohashi K (2013) Contribution of Leydig and Sertoli cells to testosterone production in mouse fetal testes. Mol Endocrinol 27(1):63–73. https://doi.org/10.1210/me.2012-1256
156. Bridges PJ, Koo Y, Kang DW, Hudgins-Spivey S, Lan ZJ, Xu X, DeMayo F, Cooney A, Ko C (2008) Generation of Cyp17iCre transgenic mice and their application to conditionally delete estrogen receptor alpha (Esr1) from the ovary and testis. Genesis 46(9):499–505. https://doi.org/10.1002/dvg.20428
157. Jung S, Aliberti J, Graemmel P, Sunshine MJ, Kreutzberg GW, Sher A, Littman DR (2000) Analysis of fractalkine receptor CX(3)CR1 function by targeted deletion and green fluorescent protein reporter gene insertion. Mol Cell Biol 20(11):4106–4114. https://doi.org/10.1128/mcb.20.11.4106-4114.2000
158. Gong S, Doughty M, Harbaugh CR, Cummins A, Hatten ME, Heintz N, Gerfen CR (2007) Targeting Cre recombinase to specific neuron populations with bacterial artificial chromosome constructs. J Neurosci 27(37):9817–9823. https://doi.org/10.1523/JNEUROSCI.2707-07.2007
159. Gong S, Zheng C, Doughty ML, Losos K, Didkovsky N, Schambra UB, Nowak NJ, Joyner A, Leblanc G, Hatten ME, Heintz N (2003) A gene expression atlas of the central nervous system based on bacterial artificial chromosomes. Nature 425(6961):917–925. https://doi.org/10.1038/nature02033
160. Maekawa M, Kamimura K, Nagano T (1996) Peritubular myoid cells in the testis: their structure and function. Arch Histol Cytol 59(1):1–13. https://doi.org/10.1679/aohc.59.1
161. Chen LY, Brown PR, Willis WB, Eddy EM (2014) Peritubular myoid cells participate in male mouse spermatogonial stem cell maintenance. Endocrinology 155(12):4964–4974. https://doi.org/10.1210/en.2014-1406
162. Chen LY, Willis WD, Eddy EM (2016) Targeting the Gdnf gene in peritubular myoid cells disrupts undifferentiated spermatogonial cell development. Proc Natl Acad Sci U S A 113(7):1829–1834. https://doi.org/10.1073/pnas.1517994113
163. Welsh M, Saunders PT, Atanassova N, Sharpe RM, Smith LB (2009) Androgen action via testicular peritubular myoid cells is essential for male fertility. FASEB J 23(12): 4218–4230. https://doi.org/10.1096/fj.09-138347
164. Xin HB, Deng KY, Rishniw M, Ji G, Kotlikoff MI (2002) Smooth muscle expression of Cre recombinase and eGFP in transgenic mice. Physiol Genomics 10(3):211–215. https://doi.org/10.1152/physiolgenomics.00054.2002
165. Chen SR, Liu YX (2016) Myh11-Cre is not limited to peritubular myoid cells and interaction between Sertoli and peritubular myoid cells needs investigation. Proc Natl Acad Sci U S A 113(17):E2352. https://doi.org/10.1073/pnas.1602873113
166. Chakraborty R, Saddouk FZ, Carrao AC, Krause DS, Greif DM, Martin KA (2019) Promoters to study vascular smooth muscle. Arterioscler Thromb Vasc Biol 39(4): 603–612. https://doi.org/10.1161/ATVBAHA.119.312449
167. de Lange WJ, Halabi CM, Beyer AM, Sigmund CD (2008) Germ line activation of the Tie2 and SMMHC promoters causes noncell-specific deletion of floxed alleles. Physiol Genomics 35(1):1–4. https://doi.org/10.1152/physiolgenomics.90284.2008

168. Fang J, Meng D, Huo J, Zhang Y, Peng H (2020) Expression and localization of MYH11 in mouse tissues, oocytes and preimplantation embryos. Biologia 75(7): 961–967. https://doi.org/10.2478/s11756-019-00353-5

169. Cox BC, Liu Z, Lagarde MM, Zuo J (2012) Conditional gene expression in the mouse inner ear using Cre-loxP. J Assoc Res Otolaryngol 13(3):295–322. https://doi.org/10.1007/s10162-012-0324-5

170. Schmidt-Supprian M, Rajewsky K (2007) Vagaries of conditional gene targeting. Nat Immunol 8(7):665–668. https://doi.org/10.1038/ni0707-665

171. Veitia RA, Birchler JA (2010) Dominance and gene dosage balance in health and disease: why levels matter! J Pathol 220(2):174–185. https://doi.org/10.1002/path.2623

172. Veitia RA, Bottani S, Birchler JA (2008) Cellular reactions to gene dosage imbalance: genomic, transcriptomic and proteomic effects. Trends Genet 24(8):390–397. https://doi.org/10.1016/j.tig.2008.05.005

173. Veitia RA, Bottani S, Birchler JA (2013) Gene dosage effects: nonlinearities, genetic interactions, and dosage compensation. Trends Genet 29(7):385–393. https://doi.org/10.1016/j.tig.2013.04.004

174. Burton KA, McDermott DA, Wilkes D, Poulsen MN, Nolan MA, Goldstein M, Basson CT, McKnight GS (2006) Haploinsufficiency at the protein kinase a RI alpha gene locus leads to fertility defects in male mice and men. Mol Endocrinol 20(10):2504–2513. https://doi.org/10.1210/me.2006-0060

175. Cho C, Willis WD, Goulding EH, Jung-Ha H, Choi YC, Hecht NB, Eddy EM (2001) Haploinsufficiency of protamine-1 or −2 causes infertility in mice. Nat Genet 28(1):82–86. https://doi.org/10.1038/ng0501-82

176. Yan W, Ma L, Burns KH, Matzuk MM (2004) Haploinsufficiency of kelch-like protein homolog 10 causes infertility in male mice. Proc Natl Acad Sci U S A 101(20): 7793–7798. https://doi.org/10.1073/pnas.0308025101

177. Tong M-H, Yang Q-E, Davis JC, Griswold MD (2013) Retinol dehydrogenase 10 is indispensible for spermatogenesis in juvenile males. Proc Natl Acad Sci 110(2):543. https://doi.org/10.1073/pnas.1214883110

178. Zhang H, Chen F, Dong H, Xie M, Zhang H, Chen Y, Liu H, Bai X, Li X, Chen Z (2019) Loss of Fbxw7 in Sertoli cells impairs testis development and causes infertility in mice†. Biol Reprod 102(4):963–974. https://doi.org/10.1093/biolre/ioz230

179. Taft RA, Davisson M, Wiles MV (2006) Know thy mouse. Trends Genet 22(12): 649–653. https://doi.org/10.1016/j.tig.2006.09.010

180. Liu Y, Suckale J, Masjkur J, Magro MG, Steffen A, Anastassiadis K, Solimena M (2010) Tamoxifen-independent recombination in the RIP-CreER mouse. PLoS One 5(10):e13533. https://doi.org/10.1371/journal.pone.0013533

181. Vooijs M, Jonkers J, Berns A (2001) A highly efficient ligand-regulated Cre recombinase mouse line shows that LoxP recombination is position dependent. EMBO Rep 2(4): 292–297. https://doi.org/10.1093/embo-reports/kve064

182. Kanatsu-Shinohara M, Inoue K, Lee J, Miki H, Ogonuki N, Toyokuni S, Ogura A, Shinohara T (2006) Anchorage-independent growth of mouse male germline stem cells in vitro. Biol Reprod 74(3):522–529. https://doi.org/10.1095/biolreprod.105.046441

183. Kanatsu-Shinohara M, Miki H, Inoue K, Ogonuki N, Toyokuni S, Ogura A, Shinohara T (2005) Long-term culture of mouse male germline stem cells under serum-or feeder-free conditions. Biol Reprod 72(4): 985–991. https://doi.org/10.1095/biolreprod.104.036400

184. Kanatsu-Shinohara M, Ogonuki N, Inoue K, Miki H, Ogura A, Toyokuni S, Shinohara T (2003) Long-term proliferation in culture and germline transmission of mouse male germline stem cells. Biol Reprod 69(2): 612–616. https://doi.org/10.1095/biolreprod.103.017012

185. Kubota H, Avarbock MR, Brinster RL (2004) Culture conditions and single growth factors affect fate determination of mouse spermatogonial stem cells. Biol Reprod 71(3): 722–731. https://doi.org/10.1095/biolreprod.104.029207

186. Kubota H, Avarbock MR, Brinster RL (2004) Growth factors essential for self-renewal and expansion of mouse spermatogonial stem cells. Proc Natl Acad Sci U S A 101(47): 16489–16494. https://doi.org/10.1073/pnas.0407063101

187. Borgna JL, Rochefort H (1981) Hydroxylated metabolites of tamoxifen are formed in vivo and bound to estrogen receptor in target tissues. J Biol Chem 256(2):859–868

188. Jahn HM, Kasakow CV, Helfer A, Michely J, Verkhratsky A, Maurer HH, Scheller A, Kirchhoff F (2018) Refined protocols of tamoxifen injection for inducible DNA recombination in mouse astroglia. Sci Rep 8(1):5913. https://doi.org/10.1038/s41598-018-24085-9
189. Jacquot S, Chartoire N, Piguet F, Herault Y, Pavlovic G (2019) Optimizing PCR for mouse genotyping: recommendations for reliable, rapid, cost effective, robust and adaptable to high-throughput genotyping protocol for any type of mutation. Curr Protoc Mouse Biol 9(4):e65. https://doi.org/10.1002/cpmo.65
190. Feil S, Valtcheva N, Feil R (2009) Inducible Cre mice. Methods Mol Biol 530:343–363. https://doi.org/10.1007/978-1-59745-471-1_18
191. Lopez ME (2012) A quick, no frills approach to mouse genotyping. Bio Protoc 2(15). https://doi.org/10.21769/bioprotoc.244
192. Byers PH (2002) Killing the messenger: new insights into nonsense-mediated mRNA decay. J Clin Invest 109(1):3–6. https://doi.org/10.1172/JCI14841
193. Hentze MW, Kulozik AE (1999) A perfect message: RNA surveillance and nonsense-mediated decay. Cell 96(3):307–310
194. Gofflot F, Wendling O, Chartoire N, Birling MC, Warot X, Auwerx J (2011) Characterization and validation of Cre-driver mouse lines. Curr Protoc Mouse Biol 1(1):1–15. https://doi.org/10.1002/9780470942390.mo100103
195. Coppoolse ER, de Vroomen MJ, van Gennip F, Hersmus BJ, van Haaren MJ (2005) Size does matter: cre-mediated somatic deletion efficiency depends on the distance between the target lox-sites. Plant Mol Biol 58(5):687–698. https://doi.org/10.1007/s11103-005-7705-7
196. Liu J, Willet SG, Bankaitis ED, Xu Y, Wright CV, Gu G (2013) Non-parallel recombination limits Cre-LoxP-based reporters as precise indicators of conditional genetic manipulation. Genesis 51(6):436–442. https://doi.org/10.1002/dvg.22384
197. Zheng B, Sage M, Sheppeard EA, Jurecic V, Bradley A (2000) Engineering mouse chromosomes with Cre-loxP: range, efficiency, and somatic applications. Mol Cell Biol 20(2):648–655. https://doi.org/10.1128/mcb.20.2.648-655.2000
198. Monvoisin A, Alva JA, Hofmann JJ, Zovein AC, Lane TF, Iruela-Arispe ML (2006) VE-cadherin-CreERT2 transgenic mouse: a model for inducible recombination in the endothelium. Dev Dyn 235(12):3413–3422. https://doi.org/10.1002/dvdy.20982
199. Wang C, Wang Z, Xiong Z, Dai H, Zou Z, Jia C, Bai X, Chen Z (2016) mTORC1 activation promotes Spermatogonial differentiation and causes subfertility in mice. Biol Reprod 95(5):97. https://doi.org/10.1095/biolreprod.116.140947
200. Kretzschmar K, Watt FM (2012) Lineage tracing. Cell 148(1–2):33–45. https://doi.org/10.1016/j.cell.2012.01.002
201. Sun F, Xu Q, Zhao D, Degui Chen C (2015) Id4 Marks Spermatogonial stem cells in the mouse testis. Sci Rep 5:17594. https://doi.org/10.1038/srep17594
202. Cantu AV, Altshuler-Keylin S, Laird DJ (2016) Discrete somatic niches coordinate proliferation and migration of primordial germ cells via Wnt signaling. J Cell Biol 214(2):215–229. https://doi.org/10.1083/jcb.201511061
203. Yamada M, Cai W, Martin LA, N'Tumba-Byn T, Seandel M (2019) Functional robustness of adult spermatogonial stem cells after induction of hyperactive Hras. PLoS Genet 15(5):e1008139. https://doi.org/10.1371/journal.pgen.1008139
204. Chen Y, Ma L, Hogarth C, Wei G, Griswold MD, Tong MH (2016) Retinoid signaling controls spermatogonial differentiation by regulating expression of replication-dependent core histone genes. Development 143(9):1502–1511. https://doi.org/10.1242/dev.135939
205. Helsel AR, Yang QE, Oatley MJ, Lord T, Sablitzky F, Oatley JM (2017) ID4 levels dictate the stem cell state in mouse spermatogonia. Development 144(4):624–634. https://doi.org/10.1242/dev.146928
206. Gao H, Wen H, Cao C, Dong D, Yang C, Xie S, Zhang J, Huang X, Huang X, Yuan S, Dong W (2019) Overexpression of MicroRNA-10a in germ cells causes male infertility by targeting Rad51 in mouse and human. Front Physiol 10:765. https://doi.org/10.3389/fphys.2019.00765
207. Buch T, Heppner FL, Tertilt C, Heinen TJ, Kremer M, Wunderlich FT, Jung S, Waisman A (2005) A Cre-inducible diphtheria toxin receptor mediates cell lineage ablation after toxin administration. Nat Methods 2(6):419–426. https://doi.org/10.1038/nmeth762
208. Rebourcet D, Darbey A, Curley M, O'Shaughnessy P, Smith LB (2018)

Testicular cell selective ablation using diphtheria toxin receptor transgenic mice. Methods Mol Biol 1748:203–228. https://doi.org/10.1007/978-1-4939-7698-0_15

209. Hu YC, de Rooij DG, Page DC (2013) Tumor suppressor gene Rb is required for self-renewal of spermatogonial stem cells in mice. Proc Natl Acad Sci U S A 110(31): 12685–12690. https://doi.org/10.1073/pnas.1311548110

210. Suzuki H, Tsuda M, Kiso M, Saga Y (2008) Nanos3 maintains the germ cell lineage in the mouse by suppressing both Bax-dependent and -independent apoptotic pathways. Dev Biol 318(1):133–142. https://doi.org/10.1016/j.ydbio.2008.03.020

211. Keefe Davis T, Hoshi M, Jain S (2013) Stage specific requirement of Gfralpha1 in the ureteric epithelium during kidney development. Mech Dev 130(9–10):506–518. https://doi.org/10.1016/j.mod.2013.03.001

212. Sada A, Suzuki A, Suzuki H, Saga Y (2009) The RNA-binding protein NANOS2 is required to maintain murine spermatogonial stem cells. Science 325(5946):1394–1398. https://doi.org/10.1126/science.1172645

213. Gu G, Dubauskaite J, Melton DA (2002) Direct evidence for the pancreatic lineage: NGN3+ cells are islet progenitors and are distinct from duct progenitors. Development 129(10):2447–2457

214. Weber P, Metzger D, Chambon P (2001) Temporally controlled targeted somatic mutagenesis in the mouse brain. Eur J Neurosci 14(11):1777–1783. https://doi.org/10.1046/j.0953-816x.2001.01803.x

215. Weber P, Schuler M, Gerard C, Mark M, Metzger D, Chambon P (2003) Temporally controlled site-specific mutagenesis in the germ cell lineage of the mouse testis. Biol Reprod 68(2):553–559. https://doi.org/10.1095/biolreprod.102.005801

Chapter 16

Lineage Tracing of Spermatogonial Stem Cells Within the Male Germline

Nathan C. Law

Abstract

Spermatogonial stem cells (SSCs) are the fundamental units from which continuous spermatogenesis arises. Although our knowledge regarding the basic properties of SSCs has grown, driven primarily through the advancement of techniques and technologies to study SSCs, the mechanisms controlling their fate remain largely unknown. Among the modern strategies to evaluate SSCs, lineage tracing is among the few established approaches that allow for functional assessment of stem cell capacity. As a result, lineage tracing continues to forge new discoveries underlying the basic attributes of SSCs as well as the molecular factors that govern SSC function. Traditional approaches to lineage tracing with dyes or radioactive labels suffer from progressive loss after successive cell divisions or unintentional label transfer to neighboring cells. To address these limitations, genetic approaches primarily leveraging transgenic technologies have prevailed as the preferred avenue for modern lineage tracing. This chapter will discuss current protocols for effective genetic lineage tracing and address applications of this technology, considerations when designing lineage tracing experiments, and the methods involved in utilizing lineage tracing to study SSCs and other cell populations.

Key words Spermatogonial stem cell, Lineage tracing, Tamoxifen, Germline

1 Introduction

Lineage tracing is a powerful approach to expand our understanding of cell lineages during development, homeostasis, and disease states. In essence, lineage tracing involves the marking of a founder cell in such a way that all progeny derived from the founder inherit the mark, thus forming multicellular populations labeled with a single mark, known as a clone. This mark can only be transmitted through direct inheritance, must be preserved through cell division, and cannot be transiently passed to neighboring cells. Analyzing the mark through time, development, in response to stimuli, or under different physiological conditions allows investigators to stitch together, or "trace," the pathways involved in cellular differentiation. This technique has become increasingly utilized in the

Jon M. Oatley and Brian P. Hermann (eds.), *Spermatogonial Stem Cells: Methods and Protocols*, Methods in Molecular Biology, vol. 2656, https://doi.org/10.1007/978-1-0716-3139-3_16,

study of stem cells, including SSCs, for which a wealth of information can be discerned from labeling SSCs and their progeny by lineage tracing. For example, lineage tracing can provide information about population size, differentiation potential, and spatial localization within the intact tissue. Furthermore, lineage tracing can effectively assess stem cell fate either through development or under homeostatic, disease, or stressed conditions.

Continuous spermatogenesis is supported by a population of undifferentiated spermatogonia within the testis. At periodic intervals within the seminiferous epithelium, a major portion of the undifferentiated population responds to retinoic acid forming differentiating spermatogonia and initiating a "round" of spermatogenesis that culminates in the formation of sperm within the testis. The undifferentiated spermatogonia that remain are tasked with regenerating the undifferentiated pool in preparation for the next pulse of retinoic acid. The same regenerative capacity enables spermatogenesis to recover after insult or stress and following transplantation into a suitable recipient. Undifferentiated spermatogonia bestowed with this regenerative capacity are commonly referred to as SSCs. Given that SSCs are responsible for both complete and sustained spermatogenesis, lineage tracing of SSCs can be identified through labeling of all germ cell types and retention of all labeled cell types through several rounds of spermatogenesis. Thus, lineage tracing is one of the few techniques to functionally evaluate stem cell capacity in the male germline.

Much of what we understand about SSC biology has been shaped from discoveries utilizing lineage tracing. For decades, morphological studies sought to define SSCs based on physical appearance with limited success. As modern gene discovery methods became available, molecular markers of the SSC population have been identified and functionally validated through techniques such as lineage tracing. While a definitive "SSC-only" marker remains elusive, a number of genes have been localized to the SSC population through lineage tracing [1–10]. Studies of the RNA binding protein NANOS2 were among the first applications of lineage tracing in the SSC population [3]. Since then, lineage tracing studies have revealed intriguing qualities of the SSC population across different developmental stages. For example, studies of precursor prospermatogonia prior to formation of the SSC population discovered that the propensity to form an SSC may be preprogrammed within a subpopulation of the germline based on lineage tracing approaches [2, 7, 11]. More recently, lineage tracing studies have revealed heterogeneity within the mature SSC population that may coincide with interconversion among different states [1]. In addition, lineage tracing studies suggest that spermatogonia that have taken the initial step toward differentiation may retain some level of self-renewal potential [12]. Finally, lineage tracing has been used to estimate the number of SSCs in adult mice. Studies utilizing transplantation estimated that ~3000 SSCs are present in each testis [13], while lineage tracing approaches estimated ~4800 SSCs per testis [8]. Ultimately, as more information regarding the factors expressed by SSCs emerges, our knowledge of basic SSC function and the population as a whole will continue to advance.

While lineage tracing is rooted in classic studies of developmental biology, modern approaches that utilize transgenic technologies have become invaluable in our understanding of stem cell biology, including that of SSCs. Recombination approaches to lineage tracing were first developed in the 1990s and have become the standard in a variety of model organisms. Two of the most common recombination approaches used in lineage tracing are the bacteriophage P1 Cre recombinase (Cre) and *S. cerevisiae* flipase (FLP) systems. Both of these proteins fall within the tyrosine recombinase enzyme family and utilize the same basic strategy; that is, recombinases, either Cre or FLP, recognize their associated DNA sequences, either locus of x-over P1 (*loxP*) or FLP recombinase recognition target (*FRT*), respectively, and catalyze recombination between paired recognition sequences. Based on directionality of the inverted repeats within each recognition sequence, the intervening DNA sequence is either excised or inverted if the repeats are orientated in the same or opposite directions, respectively. This approach is commonly used in conditional gene knockout studies, where a portion of the target gene is excised following recombination in an effort to ablate functional gene expression. In contrast, lineage tracing technologies aim to activate or alter the expression of reporter constructs. Activation of reporter expression is typically achieved through the excision of a STOP cassette that is flanked by recognition sites appropriately matched to the recombinase system (Fig. 1). Alternatively, more sophisticated reporters switch expression from one cassette to another. For example, reporter expression in the mT/mG mouse line changes from constitutive tdTOMATO to eGFP following Cre-mediate recombination (discussed more below). The array of different reporter constructs will be briefly discussed in more detail below. Between the two recombinase systems widely used for lineage tracing, Cre-*loxP* is most commonly used in transgenic rodent applications because Cre is significantly more efficient at 37 °C [14]. In contrast, FLP-*FRT* is more common to non-mammalian systems such as *Drosophila melanogaster*. For the purpose of this chapter, we will primarily focus on transgenic rodent applications of lineage tracing technologies using Cre-*loxP* systems. For more information on lineage tracing in model organisms, readers can refer to [15].

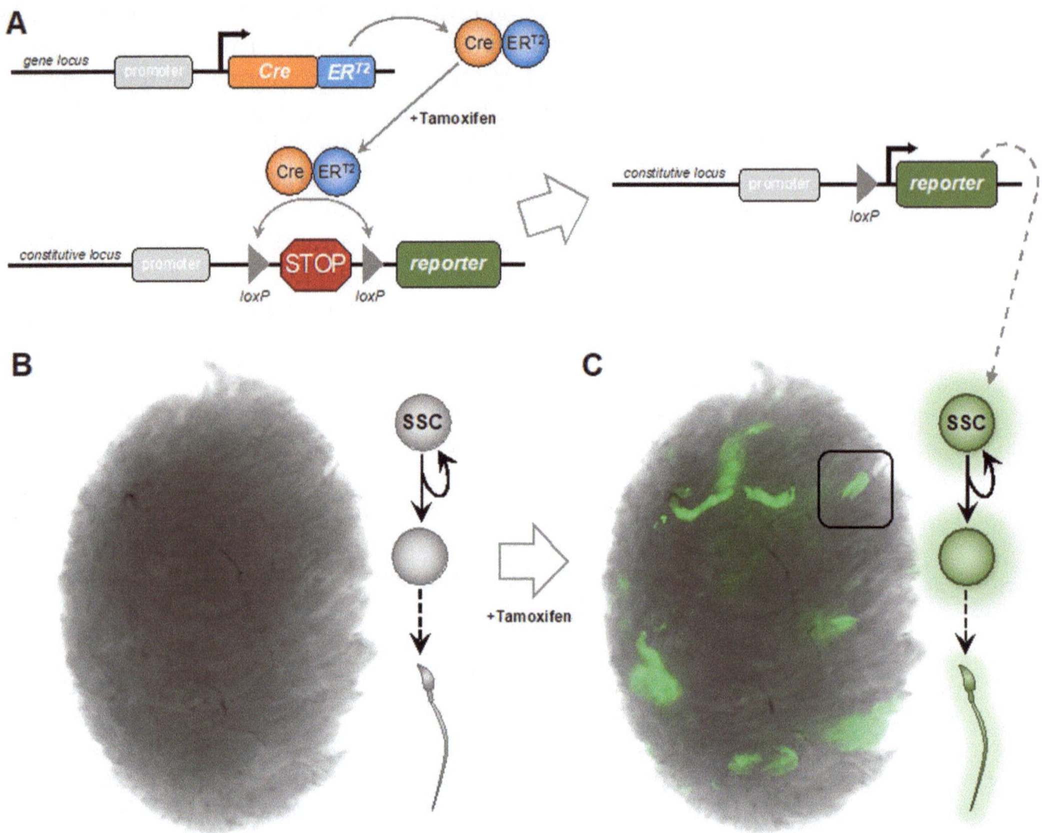

Fig. 1 Lineage tracing of SSCs within the male germline using transgenic mouse technologies. (**a**) Schematic representation of inducible reporter expression. Dual-transgenic male mice positive for both $CreER^{T2}$ and the reporter cassette are generated by standard breeding crosses. In the absence of tamoxifen, a STOP sequence located upstream the constitutive and ubiquitous reporter cassette blocks expression of the fluorescent (or biochemical) reporter. When exposed to tamoxifen, $CreER^{T2}$ mediates recombination at paired *loxP* sites flanking the STOP sequence, thereby permitting constitutive expression from the reporter cassette within cells expressing $CreER^{T2}$ and all progeny derived from targeted cells. (**b**, **c**) Effective lineage tracing of the SSC population can be visualized as reporter expression (i.e., labeling) of all cells within the germline and retention of labeled germ cells after several rounds of spermatogenesis

2 Materials

To achieve cell type-specific labeling, either the recombinase or reporter (or both) must be under the control of gene regulatory elements that confer specific expression. Currently, due to the widespread use of conditional gene knockout strategies, cell type-specific Cre recombinases are widely available. Therefore, most lineage tracing experiments to date have been performed with cell type-specific Cre's and constitutive reporters with global expression. The proper combination of Cre drivers and reporters are important to consider when planning a lineage tracing experiment.

Furthermore, the types of Cre's and reporters that are available will guide the potential outcomes and limitations of each lineage tracing experiment.

2.1 Cre Recombinases

Cre-*loxP* technology has become the standard in gene ablation approaches when global gene knockout strategies are not feasible. This cell-specific control has further enabled effective and highly coordinated applications of lineage tracing. Cell- or tissue-specific expression of Cre recombinase is typically accomplished through either methods of transgenesis or targeted knock-in approaches. Transgenesis involves random genomic insertion of an expression cassette that contains regulatory elements from a cell- or tissue-specific gene that ultimately drive expression of Cre. Alternatively, modern advances in CRISPR/Cas9 genome editing have greatly advanced the ease to which Cre expression elements can be targeted to particular gene loci that exhibit cell- or tissue-specific expression. Standard mating crosses between the Cre mouse line and an appropriate reporter line can then achieve labeling of the target population and any progeny derived from that population. These approaches accomplish spatial labeling at generally high efficiency depending on the level of Cre expression and do not require exogenous treatments or artificial forms of induction. However, a common limitation of Cre transgenic lines is that expression from the promoter of interest often occurs at an earlier stage of development. Furthermore, to study the temporal dynamics of cell populations, including SSCs, dynamic control of Cre activity is advantageous.

To address this demand, inducible Cre recombination systems have been developed that permit temporal control of Cre activity (not expression) and resulting recombination-based expression from Cre reporters. The most common approach involves the fusion of Cre recombinase to the estrogen receptor (ER). In the absence of ER ligands such as 17β-estradiol, the anti-estrogenic tamoxifen, or its active metabolite 4-hydroxy-tamoxifen (4-OHT), heat-shock proteins sequester ER and the fused Cre recombinase to the cytoplasm, thereby preventing recombination within the nucleus. Upon ligand binding, ER changes conformation, releases bound heat-shock proteins, and transits to the nucleus carrying the Cre recombinase in tow. Once in the nucleus, Cre can initiate recombination at target *loxP* sites [16]. Early iterations of this technology represented a major leap in advancement for lineage tracing but presented significant limitations, particularly that the presence of endogenous estradiol would initiate recombination. This is especially problematic in reproductive tissues. To overcome this issue, researchers developed mutated forms of either mouse or human ER that preferentially bind tamoxifen as opposed to estrogens ($CreER^{Tam}$ or $CreER^{T}$) [17, 18]. Unfortunately, high levels of tamoxifen were necessary to achieve adequate recombination, so a second-generation isoform was developed (CreER^{T2}) to enhance tamoxifen sensitivity and further depress activity induced by endogenous estrogen [19].

The potential for unintended recombination of reporter constructs, or "leaky" activity of inducible Cre isoforms, is an important consideration with lineage tracing experiments. Leaky recombination in some tissues has been attributed to endogenous estrogens [20, 21], which is particularly crucial to address in SSC biology considering the high levels of local estrogens in reproductive organs. In males, peripheral blood concentrations of estrogen typically fall well below the sub-nanomolar range, but intratesticular concentrations of estrogen are far higher [22]. In rodents, estrogen concentration measured within rete testis fluid is approximately 250 pg/mL (or approximately 917.75 pM or 0.918 nM) [23]. During the development of CreER^{T2}, recombination activity measured in vitro remained at background levels even in the presence of 25 and 250 nM estradiol [19], far exceeding intratesticular estrogen concentrations. In our experience, and consistent with published observations [7], unintended recombination has not been observed in the testis in vivo. Thus, leaky recombination due to endogenous estrogens is, at most, limited. However, the use of a no treatment control (i.e., inducible Cre+, tamoxifen-) can easily evaluate the level of background recombination that may occur and should be included with every lineage tracing experiment.

In addition to ER, forms of Cre recombinase conjugated to the progesterone receptor (PR) have been developed (CrePR). Similar to ER, early iterations of CrePR suffered from leaky activation in the presence of endogenous progesterone, and subsequent mutant lines were generated to overcome this limitation [24, 25]. As with CreER, CrePR leakiness is a critical consideration in every experiment, and an untreated control is necessary to assess the level of leaky activity relative to induced activity. To date, utilization of CrePR for lineage tracing has not been used in the study of the male germline for lineage tracing.

Finally, an alternative approach to temporal control of recombinase activity is through inducible control of Cre protein expression which can be accomplished via tetracycline (Tet)-inducible systems [26, 27]. Tet systems consist of two elements, a Tet-controlled transactivator (tTA) and a Tet-responsive promoter element (TRE). The tTA is composed of the Tet repressor DNA-binding protein (TetR) and transactivating domain VP16 from Herpes simplex virus. The TRE is composed of Tet operon (tetO) concatemers fused to a promoter element that collectively control downstream expression. For inducible Cre systems, the tTA or TRE are driven by cell- or tissue-specific gene promoter elements (commonly the tTA) and Cre expression is driven from the TRE. Tet systems are available into two different iterations, Tet-OFF or Tet-ON, in which Cre expression is either inactive or active, respectively, in the presence of Tet or its derivative doxycycline (Dox). With the Tet-OFF system, tTA binds to the TRE and induces expression of Cre in the absence of Tet/Dox. In the presence of Tet/Dox, the tTA cannot bind TRE and Cre expression is inactive. On the other hand, with the Tet-ON system a reversible form of tTA (rtTA) can only bind to a TRE in the presence of Tet/Dox, thus activating Cre expression. Thus, in the absence of Tet/Dox, the Tet-ON system lacks Cre expression. Notably, rtTA responds better to Dox than Tet. Collectively, both Tet-OFF and Tet-ON systems permit temporal control of Cre expression and Tet/Dox can be easily administered through the drinking water. To date, Tet inducible Cre systems have not been widely used for lineage tracing in the male germline.

2.2 Tamoxifen Administration

Lineage tracing approaches utilizing CreER^{T2} are among the most common given the temporal control of recombination. However, there are important considerations to be made in terms of the route to which temporal control is achieved. Modern lineage tracing approaches utilize either tamoxifen or its active metabolite 4-OHT.

Tamoxifen is more stable and can be administered to mice through either direct injection (intraperitoneal or subcutaneous) or through the digestive tract (food containing tamoxifen or oral gavage with tamoxifen solutions). Each approach has strengths and weaknesses. Direct injection provides more repeatable measures of lineage labeling, and the precise dosage can be both controlled and normalized for differences in body weight, particularly as body mass changes during development. In circumstances when repeated treatments are necessary, oral gavage or chow with tamoxifen are more convenient approaches. Unfortunately, digestive routes can be variable among animals given differences in factors such as appetite and absorption. Thus, it is important to consider the preferred route of administration when using tamoxifen.

As mentioned above, tamoxifen is metabolized to form 4-OHT via cytochrome P450 enzymes primarily in the liver [28]. While both tamoxifen and 4-OHT bind ER^{T2}, 4-OHT does so with a binding affinity of approximately 100-fold greater than that of tamoxifen [29]. Thus, direct administration of 4-OHT remains an alternative to the use of tamoxifen. Typically, lower concentrations of 4-OHT are necessary to achieve similar levels of recombination compared to tamoxifen in vivo. Furthermore, due the timeline involved in tamoxifen metabolism, direct application of 4-OHT can be kinetically favorable depending on the goals of each lineage tracing experiment. For example, if the experimental goal is to analyze lineage-labeled clones within hours of injection, 4-OHT may be favorable. The kinetics of tamoxifen- and 4-OHT-based lineage tracing in the germline have not been studied in detail.

Unfortunately, 4-OHT is significantly more costly than tamoxifen, which is why tamoxifen is often favored in academic research settings. Another limitation of 4-OHT is that the compound commonly undergoes conversion from *trans* to *cis* isoforms when in common laboratory solvents [30]. *Trans* isoforms exhibit a 300-fold greater affinity for ER compared to *cis* [30]; thus, *trans* is favorable for optimal control of recombination while mitigating any potential negative side effects. Powdered 4-OHT is far more stable than pure solutions of *trans*-4-OHT which can undergo *cis* conversion. Thus, it is recommended to make fresh solutions of 4-OHT for lineage tracing experiments.

After selecting the preferred ER^{T2} agonist and route of administration, dosage optimization is necessary to achieve optimal and repeatable lineage labeling. Preliminary studies should include a range of tamoxifen/4-OHT doses followed by assessment of labeling efficiency (outlined below) and evaluation of any health-related secondary effects. It is important to consider that acute exposure to tamoxifen is generally well-tolerated, but studies suggest that repeated or chronic exposure may cause liver toxicity in mice [31, 32]. Thus, maximal dosages should be determined for each experiment. For reference, previous lineage tracing studies of the male germline have utilized either single administrations or sequential injections (3–5 daily injections) of either tamoxifen or 4-OHT at doses as high as 2 mg per injection in adult mice [1, 2, 5, 12].

Limited studies have evaluated the extent to which tamoxifen or its derivatives impact SSCs or spermatogenesis in general. Previous studies demonstrated no significant difference in the level of apoptosis 12–24 h post-tamoxifen injection within seminiferous tubule cross sections of adult mice treated with tamoxifen at a dose of 225 ug/g body weight [8]. Furthermore, no apparent difference in testis weight or morphology have been observed several months after acute tamoxifen administration [7], suggesting that short-term applications of tamoxifen/4-OHT used in lineage tracing approaches do not appear to influence gross metrics of male reproduction. However, chronic exposure to tamoxifen or its derivatives can influence fertility parameters in rodents [33], and estrogenic compounds such as bisphenol A have documented impacts on the SSC population [34, 35]. Collectively, tamoxifen/4-OHT treatment controls are critical when evaluating cellular phenotypes with lineage tracing experiments.

2.3 Reporter Constructs

The measurable outcome of lineage tracing stems from the reporter construct. Numerous reporters have been generated, many of which are commonly available from commercial repositories such as the Jackson Laboratory (Table 1). The simplest example of reporter regulation in a Cre-dependent manner involves activation of reporter expression following recombination. For example, with the *Rosa26-tdTomato* reporter mouse line [36], expression of the fluorescent protein tdTOMATO is quenched with a STOP cassette flanked by *loxP* sites. In the presence of active Cre, the STOP

Table 1
Common Cre reporters

Mouse line	Reporter(s)	Repository	Stock No.
Rosa26–tdTomato	tdTomato	The Jackson Laboratory	007909
Rosa26–lacZ	β-galactosidase/lacZ	The Jackson Laboratory	003474
Rosa26–EGFP (or ZsGreen)	eGFP (or ZsGreen)	The Jackson Laboratory	004077 (007906)
Rosa26–Confetti (Brainbow)	Brainbow	The Jackson Laboratory	013731 017492
Rosa26–Luc	Luciferase	The Jackson Laboratory	005125
mT/mT	Membrane-bound tdTomato (-Cre); membrane-bound eGFP (+Cre)	The Jackson Laboratory	007676
Z/EG	β-galactosidase/lacZ (-Cre); GFP (+Cre)	The Jackson Laboratory	004178

List of Cre reporter mouse lines commonly used for lineage tracing that are available from public repositories such as the Jackson Laboratory. Available stock numbers are included. Note that some lines express a reporter in the absence of Cre (-Cre) and switch expression to another cassette following Cre-mediated recombination (+Cre)

cassette is excised, and expression of *tdTomato* is constitutively driven (similar to Fig. 1). Thus, founder cells with active Cre recombinase and all progeny derived from the founders are labeled with fluorescent expression of tdTOMATO.

An alternative approach is the switching of reporter expression. For example, with the *Rosa26-ACTB-tdTomato-EGFP* reporter mouse line [37] (commonly referred to as mT/mG), a membrane-bound tdTOMATO protein is constitutively expressed in all cell types, and Cre-mediated recombination excises the *tdTomato* cassette allowing for expression of a membrane-targeted EGFP. Thus, the dynamic switch from tdTOMATO+ to EGFP+ can be used for comparisons of Cre active versus inactive populations.

One important limitation to consider when using a single fluorescent reporter is that labeling between adjacent clones can be misinterpreted as a single clone. The overlap between adjacent clones is particularly worth noting in the male germline as the number of SSCs and their distribution within the testis remains contentious. Clonal overlap can be overcome by determining the minimum concentration of tamoxifen/4-OHT to achieve consistent labeling of a single clone within the testis. Low-efficiency labeling can accomplish similar outcomes. Unfortunately, this minimum labeling approach may not be advantageous in cases where the entire SSC population is being evaluated; given potentially heterogeneity within the SSC population [1], a single clone may not accurately represent the entire population. As a result, considerable efforts have been put forth to expand the number of labels that can be used to mark individual clones, thereby reducing the probability that adjacent clones intersect. For example, with the Brainbow *Rosa26-Confetti* mouse line [38], Cre activation induces recombination at interspersed *loxP* sites that create random combinations of deletions and inversions within an expression cassette composed of different fluorescent reporters. The resulting recombination leads to the random and combinatorial expression of different fluorescent proteins unique to a single clone.

Unfortunately, spectral overlap limits the number of fluorescent proteins that can be used in combination to create a large number of potential identifiers for each clone. To overcome this limitation, researchers have turned to different methods of generating cell 'barcodes" or sequences within the genome that identify individual cells or clones. These barcodes can then be assessed in bulk with DNA sequencing technologies. The Polylox transgenic mouse line was among the first to adopt this approach. Similar to the Brainbow mouse, the Polylox line utilizes a cassette composed of nine DNA elements with interspersed *loxP* sites that recombine in random arrangements in the presence of Cre recombinase. Temporal activation of Cre using a CreER^{T2} and tamoxifen is necessary to achieve incomplete recombination within the genomic locus creating random recombination events among the numerous *loxP* sequences. A number of other technologies utilizing viral integration sites, transposable elements, CRISPR-Cas9 genome editing, and other genomic modification techniques are under development to create a large diversity of clonal barcodes for applications in developmental and stem cell biology [39–41].

2.4 Reagents

1. Tamoxifen (CAS Registry No. 10540-29-1) or 4-hydroxytamoxifen (4-OHT; CAS Registry No. 68047-06-3).
2. Sterile corn oil (*see* **Note 1**).
3. Ethanol (100%).
4. Insulin syringe with needle.
5. Collagenase IV.
6. 4% paraformaldehyde solution.
7. LacZ staining buffer: 200 mM sodium phosphate, 2 mM magnesium chloride, 0.02% Triton X-100, and 0.01%.
8. X-Gal staining solution: 20 mM potassium ferricyanide, 20 mM potassium ferrocyanide, 1 mg/mL 5-bromo-4-chloro-2-indolyl-β-D galactosidase [X-gal] in LacZ staining buffer.

3 Methods

3.1 Prepare Tamoxifen/4-OHT Solutions

Tamoxifen and 4-OHT are available as both powders and pre-made solutions from commercial vendors (*see* **Note 2**). Both compounds are insoluble in aqueous solutions. Thus, careful considerations must be taken in generating tamoxifen/4-OHT solutions. Tamoxifen and 4-OHT can be either dissolved directly into sterile carrier solutions such as corn, peanut, or sunflower seed oils or first dissolved in organic solvents such as alcohols, dimethyl formamide (DMF), or dimethyl sulfoxide (DMSO) before preparing a final dilution in carrier solutions.

Dissolving tamoxifen directly in carrier solutions requires a few critical considerations. Both tamoxifen and 4-OHT readily dissolve at concentrations of 10–20 mg/mL. Complete dissolution of tamoxifen in oil typically requires 2 h to overnight incubation at warmer temperatures (37 °C) protected from light. After incubation, sonicate the tamoxifen solution in a warm water bath until any residual crystals are broken and dissolved in solution.

Alternatively, tamoxifen powder is first dissolved in 100% ethanol at room temperature. Once in solution, dilute in carrier oil to a final concentration of 20 mg/mL and 10% ethanol. For example, 20 mg tamoxifen is first dissolved in 100 μL of 100% ethanol before adding 900 μL sterile corn oil to generate a 20 mg/mL solution.

Tamoxifen powder can be stored for long periods of time at −20 °C to 4 °C. Solutions of tamoxifen can be aliquoted and stored at −20 °C for weeks until ready for use. Prolonged storage of either tamoxifen or 4-OHT can decrease potency likely due to isomerization (discussed above). Tamoxifen is UV sensitive and should therefore be stored protected from light. Prior to injection, warm the tamoxifen solution to room temperature and ensure that no crystals are present. Repeat sonication and/or incubation as needed.

3.2 Tamoxifen/4-OHT Injection

1. Produce male mice with the genotype $CreERT^{+}$; $Reporter^{+/-}$ by standard crosses. Age the animal to the target timeframe appropriate for the experimental outcome.
2. Perform intraperitoneal injection(s) with tamoxifen solution according to the dosage, frequency, and duration appropriate for each experiment.
3. Age the injected males to the target experimental endpoint.
4. Euthanize males and recover testes for either fluorescent or colorimetric analysis.

3.3 Whole-Mount Immunofluorescence Analysis

One effective method to both visualize labeled clones and quantify the extent of labeling is through the use of whole-mount preparations of seminiferous tubules. The testis is composed of numerous convoluted seminiferous tubules that can be gently dissociated into a uniform spread across a microscope slide. This effectively reorganizes intact tubules onto a two-dimensional plane from which images can be taken and quantified. Incubating the testis in a solution of collagenase aids with gentle dissociation while also removing interstitial Leydig cells attached to the outer surface of seminiferous tubules which can generate autofluorescence that commonly overlaps with fluorescently labeled clones.

1. Prepare 3 mL of collagenase IV solution (0.5 mg/mL dissolved in PBS) per testis. Prepare fresh before each tubule preparation.
2. Remove the tunica albuginea surrounding each testis and immerse each testis in collagenase solution.
3. Gently tease apart seminiferous tubules using fine-gauge needles under a dissection microscope at room temperature. Limit the time which tubules are incubated in collagenase to prevent breakdown of the tubule structure (<20 min; *see* **Note 3**).
4. Collect the tubules and collagenase solution into a 50 mL conical tube and place on ice.
5. Allow the tubules to settle to the bottom of the conical tube for 3 min.
6. Remove the collagenase solution without disturbing the tubules.
7. Wash the tubules with PBS and allow to settle for 3 min on ice.
8. Repeat the PBS wash for a total of two washes.
9. Carefully transfer dissociated tubules to a microscope slide and gently spread until the tubules do not overlap.
10. Mount with a coverslip (*see* **Note 4**) with media containing DAPI or other nuclear dyes.
11. Image all tubules from a single testis using a fluorescent microscope. "Tiling" or stitching together of multiple images is an effective means to image the large, two-dimensional area created from a tubule spread.

3.4 LacZ Staining for β-Galactosidase Reporters

For reporter constructs that drive β-galactosidase expression following recombination, staining with X-Gal is necessary to visualize labeled clones. Following X-Gal staining, labeled clones appear as dark blue regions along the length of a tubule. This reporter strategy is a viable alternative when advanced fluorescent microscopes are not available, particularly those microscopes that lack image tiling capabilities. Often, images can be taken at lower magnification and clonal area can be still quantified.

1. Remove the tunica albuginea from each testis and immerse in 4% paraformaldehyde solution for 1 h at 4 °C.

2. Wash the testes in LacZ staining buffer for >10 min at room temperature with rocking.
3. Repeat twice for a total of three washes.
4. Incubate testes in X-Gal staining solution overnight at 35 °C in the dark with rocking.
5. Wash the testes in LacZ staining buffer for >30 min at room temperature with rocking.
6. Repeat twice for a total of three washes.
7. Gently tease apart tubules using a fine-gauge needle under dissection microscope. Collagenase digest is not necessary as with fluorescence.
8. Wash tubules in PBS.
9. Mount with a coverslip and media.
10. Image.

3.5 Quantifying Whole-Mount Tubules

A useful metric from lineage tracing is the percent of tissue occupied by the labeled population(s). To quantify labeling, tiled images are imported into analysis software such as ImageJ. From there, threshold images are generated from each fluorescence channel and quantified to determine the area occupied within each two-dimensional image. The percentage of labeling is calculated as a ratio of the area detected for the fluorescent reporter over the area detected for the entire tubule labeled with DAPI (or other nuclear dye). It is important to note that this value does not directly measure the percentage of germ cells labeled because the nuclear stain signal corresponds to the entire tubule, composed of both germ and somatic cells. However, in adult testes, fluorescent signal from labeled clones occupies the majority of tubule area.

In addition to the total area occupied by labeled clones, the length of each clone may be relevant to the experimental outcomes of the study. For example, clone length can be used to quantitatively capture heterogeneity among clone sizes. ImageJ software can be used to measure the length of each labeled clone.

3.6 Calculating Labeling Efficiency

The efficiency of Cre-dependent recombination with inducible systems varies widely based on a myriad of factors, including the tamoxifen derivative (tamoxifen or 4-OHT), treatment dosage, treatment regime (singular or sequential), the level of Cre expression, the distance between *loxP* sites within the reporter (shorter distances typically recombine at higher frequencies [42]), developmental age, etc.

Labeling efficiency is calculated as the percentage of cells for which recombination occurs within all cells expressing the inducible Cre. Sophisticated inducible systems express both Cre and a different fluorescent reporter from the same cassette [1, 7], enabling the most direct quantification of labeling efficiency. Using this type of system, recombination efficiency is calculated as the ratio of cells labeled by the inducible reporter after tamoxifen administration compared to cells with expression from the combined Cre-reporter construct.

However, efficiency can be indirectly calculated using antibodies or transgene reporters to identify expression from a target gene that drives Cre expression. For example, recombination efficiency of a tamoxifen-inducible Cre driven from *Nanos2* regulatory elements was calculated as the proportion of lineage labeled clones to the number of clones expressing endogenous NANOS2 [3].

4 Notes

1. To sterilize oil, heat the appropriate volume at 90 °C for 20 min. Be sure not to boil. Alternatively, oil can be sterilized by autoclaving or filtering.
2. Tamoxifen and 4-OHT are hazardous chemicals. Review associated safety datasheets and documentation prior to handling and wear appropriate protective equipment.
3. Tubules incubated in a tissue culture dish during dissociation can be temporarily placed on ice (1–5 min) to slow the digestion process if incubation times exceed 20 min.
4. The diameter of seminiferous tubules may require slight modifications to effectively mount tubule spreads under a coverslip without compressing the tubule structure. This can be accomplished by either elevating coverslips with dots of nail polish under the corners of the cover glass or using coverslip spacers. Alternatively, slides with depression wells may be used with conventional coverglass.

References

1. La HM, Mäkelä J-A, Chan A-L et al (2018) Identification of dynamic undifferentiated cell states within the male germline. Nat Commun 9:2819. https://doi.org/10.1038/s41467-018-04827-z
2. Aloisio GM, Nakada Y, Saatcioglu HD et al (2014) PAX7 expression defines germline stem cells in the adult testis. J Clin Invest 124:3929–3944. https://doi.org/10.1172/JCI75943
3. Sada A, Suzuki A, Suzuki H, Saga Y (2009) The RNA-binding protein NANOS2 is required to maintain murine spermatogonial stem cells. Science 325:1394–1398. https://doi.org/10.1126/science.1172645
4. Sharma M, Srivastava A, Fairfield HE et al (2019) Identification of EOMES-expressing spermatogonial stem cells and their regulation by PLZF. eLife 8:e43352. https://doi.org/10.7554/eLife.43352
5. Tokue M, Ikami K, Mizuno S et al (2017) SHISA6 confers resistance to differentiation-promoting Wnt/β-catenin signaling in mouse spermatogenic stem cells. Stem Cell Reports 8:561–575. https://doi.org/10.1016/j.stemcr.2017.01.006
6. Chakraborty P, Buaas FW, Sharma M et al (2014) LIN28A marks the spermatogonial progenitor population and regulates its cyclic

expansion. Stem Cells 32:860–873. https://doi.org/10.1002/stem.1584

7. Sun F, Xu Q, Zhao D, Degui Chen C (2015) Id4 marks spermatogonial stem cells in the mouse testis. Sci Rep 5:17594. https://doi.org/10.1038/srep17594
8. Komai Y, Tanaka T, Tokuyama Y et al (2014) Bmi1 expression in long-term germ stem cells. Sci Rep 4:6175. https://doi.org/10.1038/srep06175
9. Hara K, Nakagawa T, Enomoto H et al (2014) Mouse spermatogenic stem cells continually interconvert between equipotent singly isolated and syncytial states. Cell Stem Cell 14:658–672. https://doi.org/10.1016/j.stem.2014.01.019
10. Takase HM, Nusse R (2016) Paracrine Wnt/β-catenin signaling mediates proliferation of undifferentiated spermatogonia in the adult mouse testis. Proc Natl Acad Sci U S A 113: E1489–E1497. https://doi.org/10.1073/pnas.1601461113
11. Law NC, Oatley JM (2020) Developmental underpinnings of spermatogonial stem cell establishment. Andrology 8:852–861. https://doi.org/10.1111/andr.12810
12. Nakagawa T, Sharma M, Nabeshima Y et al (2010) Functional hierarchy and reversibility within the murine spermatogenic stem cell compartment. Science 328:62–67. https://doi.org/10.1126/science.1182868
13. Nagano MC (2003) Homing efficiency and proliferation kinetics of male germ line stem cells following transplantation in mice. Biol Reprod 69:701–707. https://doi.org/10.1095/biolreprod.103.016352
14. Buchholz F, Ringrose L, Angrand PO et al (1996) Different thermostabilities of FLP and Cre recombinases: implications for applied site-specific recombination. Nucleic Acids Res 24: 4256–4262
15. Liu W, Hou SX (2008) Genetic tools used for cell lineage tracing and gene manipulation in drosophila germline stem cells. In: Hou SX, Singh SR (eds) Germline Stem Cells. Humana Press, Totowa, pp 61–70
16. Metzger D, Ali S, Bornert JM, Chambon P (1995) Characterization of the amino-terminal transcriptional activation function of the human estrogen receptor in animal and yeast cells. J Biol Chem 270:9535–9542. https://doi.org/10.1074/jbc.270.16.9535
17. Littlewood TD, Hancock DC, Danielian PS et al (1995) A modified oestrogen receptor ligand-binding domain as an improved switch for the regulation of heterologous proteins. Nucleic Acids Res 23:1686–1690. https://doi.org/10.1093/nar/23.10.1686
18. Feil R, Brocard J, Mascrez B et al (1996) Ligand-activated site-specific recombination in mice. Proc Natl Acad Sci U S A 93:10887–10890. https://doi.org/10.1073/pnas.93.20.10887
19. Feil R, Wagner J, Metzger D, Chambon P (1997) Regulation of Cre recombinase activity by mutated estrogen receptor ligand-binding domains. Biochem Biophys Res Commun 237:752–757. https://doi.org/10.1006/bbrc.1997.7124
20. Lindhorst A, Bechmann I, Gericke M (2020) Unspecific DNA recombination in AdipoqCre-ERT2 – mediated knockout approaches in transgenic mice is sex-, age- and genotype-dependent. Adipocyte 9:1–6. https://doi.org/10.1080/21623945.2019.1701394
21. Kristianto J, Johnson MG, Zastrow RK et al (2017) Spontaneous recombinase activity of Cre–ERT2 in vivo. Transgenic Res 26:411–417. https://doi.org/10.1007/s11248-017-0018-1
22. Hess RA, Cooke PS (2018) Estrogen in the male: a historical perspective. Biol Reprod 99:18
23. Free MJ, Jaffe RA (1979) Collection of rete testis fluid from rats without previous efferent duct ligation. Biol Reprod 20:269–278. https://doi.org/10.1095/biolreprod20.2.269
24. Kyrkanides S, Miller JH, Bowers WJ, Federoff HJ (2003) Transcriptional and posttranslational regulation of Cre recombinase by RU486 as the basis for an enhanced inducible expression system. Mol Ther J Am Soc Gene Ther 8:790–795. https://doi.org/10.1016/j.ymthe.2003.07.005
25. Chen M-R, Liu S-W, Wu T-C et al (2010) RU486-inducible recombination in the salivary glands of lactoferrin promoter-driven green fluorescent Cre transgenic mice. Genesis 48: 585–595. https://doi.org/10.1002/dvg.20666
26. Gossen M, Bujard H (1992) Tight control of gene expression in mammalian cells by tetracycline-responsive promoters. Proc Natl Acad Sci U S A 89:5547–5551. https://doi.org/10.1073/pnas.89.12.5547
27. Furth PA, St Onge L, Böger H et al (1994) Temporal control of gene expression in transgenic mice by a tetracycline-responsive promoter. Proc Natl Acad Sci U S A 91:9302–9306. https://doi.org/10.1073/pnas.91.20.9302

28. Crewe HK, Ellis SW, Lennard MS, Tucker GT (1997) Variable contribution of cytochromes P450 2D6, 2C9 and 3A4 to the 4-hydroxylation of tamoxifen by human liver microsomes. Biochem Pharmacol 53:171–178. https://doi.org/10.1016/s0006-2952(96)00650-8
29. Cronin-Fenton DP, Damkier P, Lash TL (2014) Metabolism and transport of tamoxifen in relation to its effectiveness: new perspectives on an ongoing controversy. Future Oncol Lond Engl 10:107–122. https://doi.org/10.2217/fon.13.168
30. Katzenellenbogen BS, Norman MJ, Eckert RL et al (1984) Bioactivities, estrogen receptor interactions, and plasminogen activator-inducing activities of tamoxifen and hydroxy-tamoxifen isomers in MCF-7 human breast cancer cells. Cancer Res 44:112–119
31. Wogan GN (1997) Review of the toxicology of tamoxifen. Semin Oncol 24:S1-87-S1-97
32. Kärki A, Mäntylä E, Hirsimäki Y et al (2000) Comparison of the effects of tamoxifen and toremifene on rat hepatocarcinogenesis. Arch Toxicol 74:249–256. https://doi.org/10.1007/s002040000116
33. Gill-Sharma MK, Balasinor N, Parte P et al (2001) Effects of tamoxifen metabolites on fertility of male rat. Contraception 63:103–109. https://doi.org/10.1016/S0010-7824(01)00178-0
34. Vrooman LA, Oatley JM, Griswold JE et al (2015) Estrogenic exposure alters the spermatogonial stem cells in the developing testis, permanently reducing crossover levels in the adult. PLoS Genet 11:e1004949. https://doi.org/10.1371/journal.pgen.1004949
35. Karmakar PC, Kang H-G, Kim Y-H et al (2017) Bisphenol a affects on the functional properties and proteome of testicular germ cells and spermatogonial stem cells in vitro culture model. Sci Rep 7:11858. https://doi.org/10.1038/s41598-017-12195-9
36. Madisen L, Zwingman TA, Sunkin SM et al (2010) A robust and high-throughput Cre reporting and characterization system for the whole mouse brain. Nat Neurosci 13:133–140. https://doi.org/10.1038/nn.2467
37. Muzumdar MD, Tasic B, Miyamichi K et al (2007) A global double-fluorescent Cre reporter mouse. Genesis 45:593–605. https://doi.org/10.1002/dvg.20335
38. Livet J, Weissman TA, Kang H et al (2007) Transgenic strategies for combinatorial expression of fluorescent proteins in the nervous system. Nature 450:56–62. https://doi.org/10.1038/nature06293
39. Gerrits A, Dykstra B, Kalmykowa OJ et al (2010) Cellular barcoding tool for clonal analysis in the hematopoietic system. Blood 115:2610–2618. https://doi.org/10.1182/blood-2009-06-229757
40. Sun J, Ramos A, Chapman B et al (2014) Clonal dynamics of native haematopoiesis. Nature 514:322–327. https://doi.org/10.1038/nature13824
41. McKenna A, Findlay GM, Gagnon JA et al (2016) Whole-organism lineage tracing by combinatorial and cumulative genome editing. Science 353:aaf7907. https://doi.org/10.1126/science.aaf7907
42. Zheng B, Sage M, Sheppeard EA et al (2000) Engineering mouse chromosomes with Cre-loxP: range, efficiency, and somatic applications. Mol Cell Biol 20:648–655

Chapter 17

Perspectives: Approaches for Studying Livestock Spermatogonia

Michela Ciccarelli and Jon M. Oatley

Abstract

At present, the knowledge base on characteristics and biology of spermatogonia in livestock is limited in comparison to rodents, yet the importance of studying these cells for comparative species analysis and enhancing reproductive capacity in food animals is high. Previous studies have established that although many core attributes of organ physiology and mechanisms governing essential cellular functions are conserved across eutherians, significant differences exist between mice and higher order mammals. In this chapter, we briefly discuss distinguishing aspects of testicular anatomy and the spermatogenic lineage in livestock and critical considerations for studying spermatogonial stem cell biology in these species.

Key words Pig, Cattle, Sheep, Goat, Conserved, Species-specific, Anatomy, Testis, Spermatogenesis

1 Testicular Anatomy and Spermatogenesis in Mammals

Testes of all mammals are formed by two compartments: the interstitial tissue (also referred to as the interstitium) and seminiferous tubules [1]. The interstitium is composed of multiple structures and cell types including lymphatic and blood vessels, nerves, mesenchymal cells, macrophages, and Leydig cells which serve as the main source of androgens [2]. The seminiferous tubules can be separated into two main components, the convoluted portion lined by the spermatogenetic epithelium (mature Sertoli cells and full complement of undifferentiated and differentiating germ cells) and the transition region (also referred to as intermediate region and Sertoli valve) that end as tubuli recti connection into the rete testis [3–5]. This transition region has been observed in a variety of mammals and consists of a modified form of Sertoli cells that remain immature and undifferentiated spermatogonia only [4, 6]. The rete testis is an elaborate network of channels that unify the seminiferous tubules to the efferent ducts that lead into tubules of the epididymis.

Jon M. Oatley and Brian P. Hermann (eds.), *Spermatogonial Stem Cells: Methods and Protocols*, Methods in Molecular Biology, vol. 2656, https://doi.org/10.1007/978-1-0716-3139-3_17,

Spermatogenesis produces terminally differentiated spermatozoa from undifferentiated spermatogonia. This process is divided into three main phases and during steady-state conditions will generate millions of spermatozoa daily in adult males. The proliferative phase includes mitotic divisions by diploid spermatogonia that coincide with steps of differentiation including the undifferentiated to differentiating transition and acquisition of responsiveness to retinoic acid signaling to eventually yield preleptotene spermatocytes which signifies start of the second or meiotic phase [7, 8]. After two meiotic divisions as primary and secondary spermatocytes, haploid spermatids are produced that embark on the third phase referred to as spermiogenesis during which morphological transformation occurs to yield elongate spermatids and eventually spermatozoa. Each of these phases of germ cell differentiation/maturation is supported by intimate association with Sertoli cells which are the only somatic cell type within seminiferous tubules [9]. Together, the germ cells and Sertoli cells make up the seminiferous epithelium. In all mammals studied to date, spermatogenesis proceeds in cycles or waves. For mice, a cycle of the seminiferous epithelium has been calculated to occur every 8.6 days and can be broken into 12 stages; a complete round of spermatogenesis requires ~4 cycles or 35 days [7, 10]. Although the basic stages of the seminiferous cycle are similar between rodents and livestock, the duration of a cycle and number of days for spermatogenesis is extended in livestock [11–15]. A comparative species perspective between mouse and livestock is provided in Table 1. For example, in bulls one cycle of the seminiferous epithelium lasts 13.5 days, and the entirety of spermatogenesis (from spermatogonial stem cells through spermiation) occurs over a 61-day period; thus 4.5 cycles are required for spermatozoa to be produced from spermatogonia [11, 15, 16].

Testicular tissue of all mammalian species contains a spermatogonial population which can be divided into undifferentiated and differentiating subpopulations [7, 17]. Actions of the

Table 1
List of stages, cycle length, and total duration of spermatogenesis for various livestock, mice, and humans

Species	Stages	One cycle	Spermatogenesis	References
Mouse	12	8.6	35	[10]
Human	6	16	64	[94]
Pig	8	8.3	39	[95]
Cattle	8	13.5	61	[15]
Sheep/goat	8	10.4	47	[13]

undifferentiated spermatogonial population provide the foundation for continual and robust spermatogenesis. Throughout a cycle of the seminiferous epithelium, undifferentiated spermatogonia undergo a series of mitotic divisions to build population numbers before a pulse of retinoic acid is generated that drives the progenitors to transition to a differentiating state which irreversibly commits them to terminal differentiation as spermatozoa [7, 18–20]. Thus, the transition induced by retinoic acid signaling is the initiation of a round of spermatogenesis. While nearly all the undifferentiated spermatogonia respond to retinoic acid and transition to a differentiating state, a small subset remains in the undifferentiated state and is tasked with rebuilding the population numbers in preparation for the next pulse [21]. Following the undifferentiated-to-differentiating transition, spermatogonia undergo another series of mitotic divisions while also gaining competence for meiotic initiation upon signaling by a second pulse of retinoic acid [8, 19, 22]. Because the retinoic acid induced transitions occur precisely at a single point during each cycle of the seminiferous epithelium, the generation of retinoic acid pulsation can be considered the rate limiter for establishing the timing of spermatogenic transitions.

Functional studies with rodent models have demonstrated that the portion of the undifferentiated spermatogonial population that functions as spermatogonial stem cells (SSCs) is a minor subset. During each cycle of the seminiferous epithelium, most of the undifferentiated spermatogonia undergo the differentiating transition and are irreversibly committed to terminal differentiation as sperm. In addition, outcomes of transplantation studies have demonstrated that only a small percentage of the undifferentiated spermatogonial population has the functional capacity to engraft within seminiferous tubules and restore continual spermatogenesis; the cells capable of accomplishing this feat are considered SSCs. Based on the number of colonies of continual spermatogenesis that are generated following transplantation of the isolated undifferentiated spermatogonial population from mouse testes, the SSC content can be calculated to be <10%. For non-rodent mammalian species, the percentages of undifferentiated spermatogonia that do or do not undergo the differentiating transition at each cycle of the seminiferous epithelium and are capable of regenerating continual spermatogenesis after transplantation are largely unknown. Thus, the size and complexity of the SSC pool in testes of livestock remains to be defined. Although rodents are valuable animal models for initial discoveries, the translation of findings to higher order mammals is not always straightforward. For example, spermatogonia in mice and rats can be classified based on intercellular bridge connections that create syncytium but empirical evidence that spermatogonia are interconnected in other mammals is currently lacking. In addition, culture conditions including growth factors that support in vitro propagation of rodent SSCs do not seem to support SSCs of other mammalian species. Caution should be exercised to not make assumptions about the biology of SSCs in livestock based on findings with rodents, and generation of independent empirical data should be prioritized.

Because translation of findings about SSC biology from rodent models to higher order animals has proven to not be direct in many instances, studying the cell population in large animals like livestock can advance knowledge about conserved and unique features across mammalian species. In addition, the remarkable ability of SSCs to regenerate sperm production following isolation from donor testicular tissue and transplantation into seminiferous tubules of a recipient male has enormous potential as an applied reproductive tool in livestock. Furthermore, SSC transplantation (SSCT) methodology, first reported for rodents over 25 years ago [23, 24], has potential utility as a fertility preservation strategy in men and endangered species, and developing the methodology in livestock could serve as a bridging model. In recent studies, donor-derived sperm production following SSCT was demonstrated in recipient male mice, pigs, and goats that were genetically ablated of endogenous germline [25]. This achievement has advanced SSCT as a tool for large-scale and widespread dissemination of desirable genetics in food animal production to the brink of application and provides higher order mammalian models to refine methodology with application to a variety of mammalian species.

2 Molecular Biomarkers of Livestock Spermatogonia

The study of spermatogonia in any mammalian species is enabled by a toolbox of valid morphological and/or molecular markers to clearly identify the heterogeneous population and discriminate the SSC containing undifferentiated subset. All spermatogonial subsets reside along the seminiferous tubule basement membrane and possess similar morphological characteristics. Although nuclear chromatin staining varies between undifferentiated and differentiating spermatogonia, clear identification requires appropriate histological staining. Thus, while most researchers with basic understanding of testicular biology can generally identify spermatogonia in vivo based on positioning within the seminiferous epithelium, distinguishing undifferentiated from differentiating spermatogonia is challenging without the use of molecular markers. In addition, preleptotene and leptotene spermatocytes which are the first step in meiotic initiation during spermatogenesis also reside along the basement membrane and many investigators mistakenly consider these cells as spermatogonia. For this chapter, we define a molecular biomarker as a gene product that can be detected at the protein level and used for labeling cells in situ or in vitro and/or for isolation with cell sorting techniques. Table 2 displays

Table 2
Expression of select biomarkers for germ cells and spermatogonia in various livestock and mice

Molecular marker	Spermatogonial expression (protein localization)				Germ cell specific?	Linked to SSC function?
	Mouse	Pig	Cattle	Sheep/goat		
DDX4	Yes	Yes	Yes	Yes	Yes	No
DAZL	Yes	Yes	Unknown	Unknown	Yes	No
NANOS2	Yes	Unknown	Unknown	Unknown	Yes	Yes
DND1	Yes	Unknown	Unknown	Unknown	Yes	Yes
ZBTB16	Yes	Yes	Yes	Yes	No	Yes
UCHL1	Yes	Yes	Yes	Yes	No	No
ID4	Yes	Unknown	Yes	Unknown	No	Yes
GFRA1	Yes	Unknown	Unknown	Unknown	No	No
THY1	Yes	Yes	Yes	Yes	No	Yes
ITGA6	Yes	Unknown	Unknown	Unknown	No	Yes
ITGB1	Yes	Unknown	Unknown	Unknown	No	Yes
CD9	Yes	Unknown	Unknown	Unknown	No	Yes

the current knowledge base of validated biomarker expression in testicular germ cells including spermatogonia of various livestock.

A first step in clearly identifying spermatogonia including the SSC containing undifferentiated population either *in situ or* in vitro should be labeling with a molecular biomarker that is specific for germ cells. Many molecules identified by previous studies to be present in spermatogonia are also expressed by various somatic cell populations. However, several RNA-binding proteins including DDX4, DAZL, DND1, and NANOS2 have been identified as having evolutionarily conserved germ cell-specific expression. Neither transcript from genes encoding these molecular nor translated protein are detectable in testicular somatic cells. Although DDX4 and DAZL are broadly expressed in the spermatogenic lineage, expression of NANOS2 and DND1 appears to be restricted to undifferentiated spermatogonia in mice, but whether that is the case in livestock is undefined. Regardless, by localizing expression of DDX4, DAZL, DND1, or NANOS2 to cells within testicular cross sections or cell populations isolated from testicular tissue of any mammalian species, one can draw valid conclusions about a germ cell nature.

Beyond the evolutionarily conserved set of RNA-binding proteins, knowledge of molecular markers for undifferentiated spermatogonia including SSCs in livestock is limited. Studies with mice were the first to identify spermatogonial expression of the transcriptional regulator ZBTB16 (aka PLZF) and de-ubiquitinase enzyme UCHL1 (aka PGP9.5) [26–28]. Subsequently, expression of both molecules was observed in testes of cattle [29, 30], goats [31], and sheep [32] and found to be localized generally to spermatogonia. In pigs, although expression of UCHL1 has been observed in cross sections of testicular tissue and localized to spermatogonia [33], similar detection of ZBTB16 expression has not been reported and can only be inferred from detection of transcript in isolated cell populations. Regardless, the discovery of ZBTB16 and UCHL1 expression in spermatogonia of multiple species suggest that these molecules may be useful for studying the population in most, if not all, mammals. However, to fully vet them as biomarkers, assessment of whether expression also occurs in testicular somatic cells is needed. In addition, evidence from mouse studies clearly indicate that expression of both molecules occurs in all undifferentiated and early differentiating spermatogonia [28, 34]. Thus, caution should be exercised to not overinterpret the use of ZBTB16 or UCHL1 as being specific to any spermatogonial subpopulation, especially SSCs.

In addition to intracellular localized molecules, several proteins localized to the cell surface including THY1 (aka CD90) [35], ITGA6 [36], ITGB1 [36], and CD9 [37] have also been found to be expressed by mouse SSCs and used as presumptive markers of SSCs in livestock [38–44]. Although transplantation analyses have demonstrated that mouse SSCs express these molecules, none are germ cell specific, and expression by various testicular somatic cell populations including macrophages [45], Leydig cells [46], mesenchymal cells [47], and Sertoli cells [37, 47] has clearly been shown. Thus, its use as standalone biomarkers for SSCs of any mammalian species is not scientifically sound and must be accompanied with co-expression of a germ cell-specific marker such as DDX4.

Due to their essential role in continuity of the spermatogenic lineage and capacity to regenerate spermatogenesis following transplantation, SSCs have garnered significant attention from scientists investigating male reproduction biology. However, only studies using definitive tools and criteria can make valid claims about SSCs and their biology. This consideration is especially important for studying SSCs in livestock. The challenges of experimentation with large animals that have long generation intervals and for which a deep toolbox of efficient functional and molecular approaches does not exist necessitate that specific criteria are applied to assessing SSCs in order to generate rigorous and trustworthy findings. Over the last 10 years, a multitude of studies have aimed to identify expression of molecular markers that distinguish SSCs from other spermatogonial subsets in mammalian testes [34, 48–52]. One of the earliest identified molecules to be selectively expressed by the SSC population in mammals was the transcriptional regulator ID4 [49, 50]. First reported in mice to distinguish SSC from progenitor spermatogonia based on cell identity in vivo and functional transplantation analyses, expression of ID4 has subsequently been found to associate with SSC identity in several higher order mammals [29, 53–55]. In addition to ID4, expression of the growth factor co-receptor GFRA1 has been identified as a potential maker that defines SSCs in mammalian testes [52, 56]. Immunostaining analyses for GFRA1+ spermatogonia in mice suggested that expression is restricted to a small subset of the spermatogonial population [34, 52, 56]. Also, lineage tracing analyses demonstrated that spermatogenesis originates from spermatogonia that are actively transcribing the Gfra1 gene, implying that GFRA1+ spermatogonia have the capacity to function as SSCs [56]. Somewhat paradoxically, transplantation analyses with spermatogonia isolated based on presence of GFRA1 protein has not clearly distinguished the population as being enriched for SSC capacity [57, 58]. Thus, what have become known as distinguishing markers of SSCs in mice and applied to other species may be misleading in that only a portion of the population expressing the marker(s) are actual SSCs and most lack the functional capacity to regenerate spermatogenesis. In livestock, clear determination of the testicular cell types that express either ID4 or GFRA1 has not been reported. This gap in knowledge should be filled before investigators use biomarkers identified in rodents to study SSCs in livestock and appreciate that functional analyses are the only manner, at present, to truly identify and assess SSCs for any species.

Beyond PLZF, UCHL1, ID4, and GFRA1 as molecular markers that have been identified for spermatogonia of several mammalian species including livestock, a host of other molecules are known to be expressed by the SSC containing undifferentiated spermatogonial population in mice and humans but are yet to be confirmed in livestock. Examples include LIN28 [59], c-KIT [60, 61], EOMES [62], PAX7 [51], and PDX1 [34]. While solid evidence exists in the peer-reviewed literature that LIN28 expression is restricted to the germline in testicular tissue, c-KIT is well known to also be expressed in subsets of the testicular somatic cell population, and whether EOMES, PAX7, or PDX1 are expressed by germ cells only or testicular somatic cells as well has not been clearly defined. Thus, at present, biomarker analysis of undifferentiated spermatogonia in livestock should include use of PLZF or UCH-L1 as a spermatogonial marker in conjunction with a germ cell-specific molecule such as DDX4, DAZL, DND1, or NANOS2. In addition, combining these markers with ID4 or GFRA1 expression, inferences about the SSC nature of isolated cell populations from livestock can be made.

3 Primary Cultures of Livestock Spermatogonia

The ability to maintain and grow primary populations of cells in vitro provides several advantages and opportunities. For example, rarity of stem cells in many tissues presents a challenge for in-depth study of molecular mechanisms underlying their biology which can be overcome by experimental platforms that support maintenance in vitro. In addition, the ability of SSCs to engraft in recipient seminiferous tubules and regenerate spermatogenesis following transplantation makes them an attractive cell population for widespread dissemination of elite genetics, direct germline genome editing, and preservation of desirable and ancestral germplasm. To achieve these feats, maintenance and expansion of a primary population in vitro is essential.

Seminal studies published between 2003 and 2004 established conditions that support long-term culture of primary mouse spermatogonial populations that contain an SSC pool [63, 64]. In these initial studies, mitotically inactivated feeder cell monolayers (STO and MEF), serum-free and chemically defined media, and growth factor cocktails containing GDNF and FGF2 were reported to be essential for maintenance of clumps of germ cells that expressed molecular markers of undifferentiated spermatogonia. Importantly, after several passages, generation of single cell suspensions from the clump-forming cells and subsequent transplantation into testes of recipient mice revealed that a portion (~1–20%) of the long-term cultured population could function as SSCs by regenerating spermatogenesis. Although some investigators have referred to the cultures as male germline stem (mGS) cells [63, 65–69], only a portion of the population has actual SSC capacity, and the percentage varies widely among individual cultures and over time of in vitro maintenance [64, 70, 71]. Subsequent studies showed that the SSC integrity to regenerate spermatogenesis declines with time of in vitro maintenance, but the timeframe for onset of this aging-related decline in function can be significantly extended by altering the atmospheric oxygen tension to favor glycolysis [71]. Despite these nuances, the conditions that have been reported to support primary cultures of mouse undifferentiated spermatogonia have been widely used as an experimental platform to advance knowledge of SSC biology and set the stage for devising conditions that support SSCs of livestock.

In many respects, the mouse is an outstanding model for discovering basic characteristics of mammalian cells and the mechanism governing their biological function. Thus, culture conditions defined for mouse undifferentiated spermatogonia have been used as a starting point to devise approaches for culturing livestock spermatogonia. Unfortunately, translation has not been straightforward. Over the last 10 years, a multitude of studies have been published with claims of devising culture conditions for SSCs from cattle [29, 43, 72, 73], pigs [38, 44, 74–76], goats [77–79], and sheep [80, 81]. Many of these have relied on observation of cells growing in clumps or expression or molecular markers that are unvalidated for livestock or are not germ cell specific as evidence for SSC maintenance. Considering that somatic cells are also capable of forming clumps in a culture dish and many supposed markers of SSCs are not germ cell specific, caution should be exercised when using these criteria to assess cell types growing in vitro. Importantly, claims of SSC presence in a primary culture should not be made unless the population has been functionally shown to have capacity for producing colonies of continual spermatogenesis following transplantation into testes of a recipient male. Without functional validation, claims of an SSC nature for primary cultures, even if they are clearly shown to be spermatogonia, are not supported by empirical evidence and can be misleading to the research community. Although some studies have demonstrated the maintenance of livestock spermatogonia in vitro via appropriate molecular marker analysis [29], no reports to date have unequivocally shown the presence of SSCs via transplantation. We propose that investigators in the field of livestock spermatogenesis should adopt a standard criterion for assessing spermatogonial identity in a primary culture established from testis cell populations. This should include determination of whether the cells express a panel of biomarker proteins known to be specific for germ cells (i.e., DDX4), validated to be conserved in spermatogonia (i.e., PLZF or UCHL1), and have been functionally linked to being expressed by SSCs in other species (i.e., ID4 or GFRA1). If the cells express the panel of biomarkers, claims of undifferentiated spermatogonial maintenance and growth can validly be made. However, claims of SSC capacity should not be made without assessment of whether the cells can generate colonies of continual spermatogenesis via transplantation into recipient testes or other functional strategies.

4 Transplantation of Spermatogonia in Livestock

The ability to transfer germline between males has utility for preserving fertility and unique germplasm or as a tool for achieving widespread and large-scale dissemination of desirable genetics. In livestock production, the latter could be useful in impacting genetic gain through a surrogate sire breeding concept to improve traits that influence resiliency, growth efficiency, and meat or milk production [25, 82]. In 1994, seminal studies published by Dr. Ralph Brinster and colleagues demonstrated that SSCs collected from testicular tissue of a donor male mouse could be transferred into the seminiferous tubules of a recipient male that had been depleted of endogenous germ cells [23, 24]. The transferred SSCs were shown to engraft in host tubules and serve as a source for continual spermatogenesis that yielded sperm containing the donor genome. The basic discovery in mice that germline could be transferred between males has enticed numerous investigators to pursue translation of the concept to higher order mammals including livestock.

A key aspect of achieving SSC transplantation (SSCT) in any animal is unequivocal demonstration of donor-derived sperm production in recipient testes. The initial discoveries in mice achieved this by using donor SSCs that possessed a *Rosa-LacZ* marker transgene (i.e., Gt(Rosa)26) and non-transgenic recipient males that lack endogenous spermatogenesis due to mutation in the KIT receptor tyrosine kinase [23]. Despite many attempts over a 20+ year period, unequivocal demonstration of donor-derived sperm production following germ cell transplantation in livestock was difficult to achieve due to lack of donor SSCs possessing unique transgenes and lack of naturally occurring genetic mutant recipients that are agametic [83–92]. Several studies used chemotoxic drugs or irradiation strategies to deplete endogenous germ cells in recipient males and followed with transplantation of cells isolated from testicular tissue of a donor male [83–90]. In other studies, recipient males were not subjected to endogenous germ cell depletion prior to transplantation of donor cells [83, 88, 91, 92]. Because endogenous germ cells persist in depleted or unprepared recipients, unless the donor cells are from an individual with a unique identifying trait, clear demonstration of donor sperm production is challenging. In addition, an ejaculate contains multiple different cell types including somatic cells; therefore, unless the sperm are specifically isolated and analyzed for donor genotype, DNA analysis of an unfractionated ejaculate cannot yield unequivocal assessment of donor-derived spermatogenesis. Through CRISPR-Cas9 editing of the *NANOS2* gene, male pigs, goats, and cattle with germline ablation were recently created and shown to be suitable hosts for generation of sperm possessing the donor genome following SSCT [25]. This advancement eliminates a major barrier to developing SSCT as a utilizable tool in essentially all animals for germplasm preservation and dissemination of desirable genetics within populations. For future studies on SSCT in livestock, use of germline ablated recipients such as NANOS2 knockout males should be prioritized as this strategy removes a confounding variable to clear identification of donor-derived spermatogenesis.

The utility of SSCT in any animal is influenced directly by the level of donor SSC engraftment that is achieved in recipient seminiferous tubules. Even with the use of endogenous germline ablated recipients, efficient dispersion of donor SSCs throughout the testicular parenchyma is critical. In early studies with mice, donor cells were microinjected directly into individual seminiferous

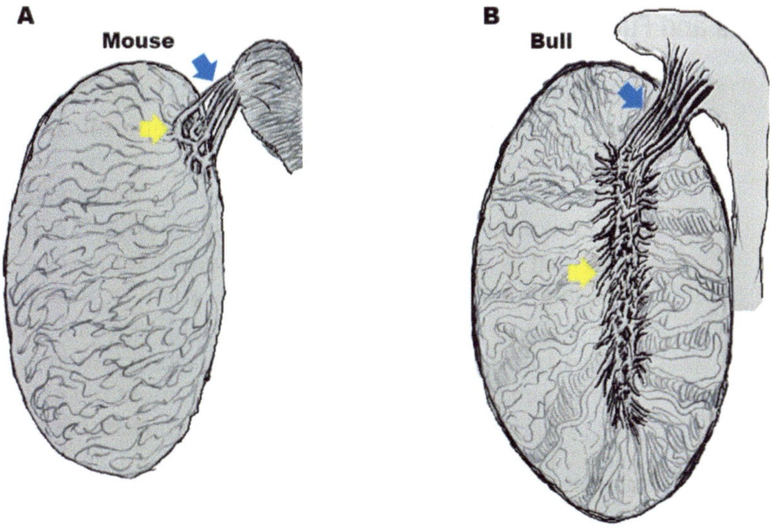

Fig. 1 Diagrammatic representation of gross testicular anatomy for the mouse (**a**) and livestock (**b**). Of note, the rete testis (denoted by yellow arrows) is positioned in the mediastinum of livestock testes whereas in mice and other rodents the positioning is at the periphery of the testis. In addition, the efferent ducts (denoted by blue arrows) that connect the rete testis tubules to the epididymal tubules are readily accessible in mice but covered tightly by the caput portion of the epididymis in livestock

tubules of recipient testes [23, 24, 93]. This approach was technically challenging and had low efficiency, thus necessitating a search for alternate routes of donor cell infusion. Follow-up studies with mice led to an approach in which the microinjection needle was inserted through the efferent ductile into the rete testis which became the standard for achieving mouse SSCT [93]. The rete testis is in the so-called mediastinum testis, a correct anatomical descriptor for higher order mammals but not in rodents owing to the location being at the periphery of the gonad rather than the middle (Fig. 1). During normal spermatogenesis, spermatozoa are released from the seminiferous epithelium, transit from the tubule proper through the intermediate region into the rete testis and finally the efferent ductile to reside in the epididymal tubules for final maturation before ejaculation. Thus, by infusing into the rete testis, a donor cell suspension can theoretically be dispersed throughout all seminiferous tubules. In livestock, access to the rete testis with an injection needle requires ultrasound-guided position because of the mediastinal location. Although this approach has been successful in achieving low levels of donor sperm production following SSCT in pigs and goats [25], technical refinements are needed to achieve greater coverage of the tissue mass and thereby yield spermatogenic regeneration that is like a normal condition.

5 Conclusions and Future

Numerous utilities exist for studying SSCs in livestock and developing technologies based on their unique functional attributes. Information produced from investigating characteristics of the cell type in pigs, cattle, sheep, and goats can advance knowledge about conserved features and mechanisms guiding biological function that is applicable across all mammalian species. In addition, devising methodologies with SSCs such as transplantation and in vitro amplification can not only yield tools for improving food animal production efficiency via impacting dissemination of trait diving genetics but also inform development of similar approaches for preserving fertility in men and germplasm of endangered species. Validity of findings from studying, culturing, and transplanting SSCs is impacted greatly by the criteria used to define them. Use of biomarkers should include a combination of those that are known to be expressed in germ cells only, spermatogonia in testicular tissue, and have been linked to SSC function in a mammalian model. Considering that stem cells are ultimately defined by functional capacity, claims of an SSC nature for isolated and cultured testis cell populations should be reserved for studies that have included functional analysis of a capacity to produce colonies of continual spermatogenesis such as SSCT. Lastly, developing and refining SSCT in livestock should include strategies that can unequivocally detect donor-derived spermatogenesis. In this regard, use of recipient males that are ablated of germline is a critical resource.

References

1. Russell LD, Ettlin RA, Hikim APS, Cless ED (1990) Histological and histopathological evaluation of the testis, 1st edn. Cache River Press, St. Louis
2. Bustos-Obregon E (1976) Ultrastructure and function of the lamina propria of mammalian seminiferous tubules. Andrologia 8(3): 179–185
3. Figueiredo AFA et al (2021) Insights into differentiation and function of the transition region between the seminiferous tubule and rete testis. Differentiation 120:36–47
4. Dym M (1974) The fine structure of monkey Sertoli cells in the transitional zone at the junction of the seminiferous tubules with the tubuli recti. Am J Anat 140(1):1–25
5. Hermo L, Dworkin J (1988) Transitional cells at the junction of seminiferous tubules with the rete testis of the rat: their fine structure, endocytic activity, and basement membrane. Am J Anat 181(2):111–131
6. Aiyama Y et al (2015) A niche for GFRalpha1-positive spermatogonia in the terminal segments of the seminiferous tubules in hamster testes. Stem Cells 33(9):2811–2824
7. de Rooij DG, Russell LD (2000) All you wanted to know about spermatogonia but were afraid to ask. J Androl 21(6):776–798
8. Griswold MD (2016) Spermatogenesis: the commitment to meiosis. Physiol Rev 96(1): 1–17
9. Griswold MD (2018) 50 years of spermatogenesis: Sertoli cells and their interactions with germ cells. Biol Reprod 99:87
10. Clermont Y (1972) Kinetics of spermatogenesis in mammals: seminiferous epithelium cycle and spermatogonial renewal. Physiol Rev 52(1):198–236

11. Staub C, Johnson L (2018) Review: spermatogenesis in the bull. Animal 12(s1):s27–s35
12. Franca LR, Becker-Silva SC, Chiarini-Garcia H (1999) The length of the cycle of seminiferous epithelium in goats (Capra hircus). Tissue Cell 31(3):274–280
13. Frankenhuis MT, Kramer MF, de Rooij DG (1982) Spermatogenesis in the boar. Vet Q 4(2):57–61
14. Courot M, Ortavant R (1981) Endocrine control of spermatogenesis in the ram. J Reprod Fertil Suppl 30:47–60
15. Amann RP (1962) Reproductive capacity of dairy bulls. IV. Spermatogenesis and testicular germ cell degeneration. Am J Anat 110:69–78
16. Johnson L, Petty CS, Neaves WB (1980) A comparative study of daily sperm production and testicular composition in humans and rats. Biol Reprod 22(5):1233–1243
17. Oatley JM, Brinster RL (2008) Regulation of spermatogonial stem cell self-renewal in mammals. Annu Rev Cell Dev Biol 24:263–286
18. de Rooij DG (1973) Spermatogonial stem cell renewal in the mouse. I Normal situation. Cell Tissue Kinet 6(3):281–287
19. Endo T et al (2015) Periodic retinoic acid-STRA8 signaling intersects with periodic germ-cell competencies to regulate spermatogenesis. Proc Natl Acad Sci U S A 112(18): E2347–E2356
20. Hogarth CA et al (2015) Processive pulses of retinoic acid propel asynchronous and continuous murine sperm production. Biol Reprod 92(2):37
21. Tegelenbosch RA, de Rooij DG (1993) A quantitative study of spermatogonial multiplication and stem cell renewal in the C3H/101 F1 hybrid mouse. Mutat Res 290(2):193–200
22. Koubova J et al (2006) Retinoic acid regulates sex-specific timing of meiotic initiation in mice. Proc Natl Acad Sci U S A 103(8):2474–2479
23. Brinster RL, Zimmermann JW (1994) Spermatogenesis following male germ-cell transplantation. Proc Natl Acad Sci U S A 91(24): 11298–11302
24. Brinster RL, Avarbock MR (1994) Germline transmission of donor haplotype following spermatogonial transplantation. Proc Natl Acad Sci U S A 91(24):11303–11307
25. Ciccarelli M et al (2020) Donor-derived spermatogenesis following stem cell transplantation in sterile NANOS2 knockout males. Proc Natl Acad Sci U S A 117(39):24195–24204
26. Buaas FW et al (2004) Plzf is required in adult male germ cells for stem cell self-renewal. Nat Genet 36(6):647–652
27. Costoya JA et al (2004) Essential role of Plzf in maintenance of spermatogonial stem cells. Nat Genet 36(6):653–659
28. Kwon J et al (2004) Developmental regulation of ubiquitin C-terminal hydrolase isozyme expression during spermatogenesis in mice. Biol Reprod 71(2):515–521
29. Oatley MJ et al (2016) Conditions for long-term culture of cattle undifferentiated spermatogonia. Biol Reprod 95(1):14
30. Herrid M, Davey RJ, Hill JR (2007) Characterization of germ cells from pre-pubertal bull calves in preparation for germ cell transplantation. Cell Tissue Res 330(2):321–329
31. Clotaire DZJ et al (2019) Functions of promyelocytic leukaemia zinc finger (Plzf) in male germline stem cell development and differentiation. Reprod Fertil Dev 31:1315
32. Borjigin U et al (2010) Expression of promyelocytic leukaemia zinc-finger in ovine testis and its application in evaluating the enrichment efficiency of differential plating. Reprod Fertil Dev 22(5):733–742
33. Luo J et al (2006) Protein gene product 9.5 is a spermatogonia-specific marker in the pig testis: application to enrichment and culture of porcine spermatogonia. Mol Reprod Dev 73(12): 1531–1540
34. La HM et al (2018) Identification of dynamic undifferentiated cell states within the male germline. Nat Commun 9(1):2819
35. Kubota H, Avarbock MR, Brinster RL (2003) Spermatogonial stem cells share some, but not all, phenotypic and functional characteristics with other stem cells. Proc Natl Acad Sci U S A 100(11):6487–6492
36. Shinohara T, Avarbock MR, Brinster RL (1999) beta1- and alpha6-integrin are surface markers on mouse spermatogonial stem cells. Proc Natl Acad Sci U S A 96(10):5504–5509
37. Kanatsu-Shinohara M, Toyokuni S, Shinohara T (2004) CD9 is a surface marker on mouse and rat male germline stem cells. Biol Reprod 70(1):70–75
38. Lee WY et al (2013) Establishment and in vitro culture of porcine spermatogonial germ cells in low temperature culture conditions. Stem Cell Res 11(3):1234–1249
39. Reding SC et al (2010) THY1 is a conserved marker of undifferentiated spermatogonia in the pre-pubertal bull testis. Reproduction 139(5):893–903
40. Abbasi H et al (2013) THY1 as a reliable marker for enrichment of undifferentiated spermatogonia in the goat. Theriogenology 80(8):923–932

41. Cai H et al (2016) Enrichment and culture of spermatogonia from cryopreserved adult bovine testis tissue. Anim Reprod Sci 166: 109–115
42. Kim YH et al (2013) Stage-specific embryonic antigen-1 expression by undifferentiated spermatogonia in the prepubertal boar testis. J Anim Sci 91(7):3143–3154
43. Suyatno et al (2018) Long-term culture of undifferentiated spermatogonia isolated from immature and adult bovine testes. Mol Reprod Dev 85(3):236–249
44. Goel S et al (2007) Identification, isolation, and in vitro culture of porcine gonocytes. Biol Reprod 77(1):127–137
45. Oatley JM et al (2009) Colony stimulating factor 1 is an extrinsic stimulator of mouse spermatogonial stem cell self-renewal. Development 136(7):1191–1199
46. Chen H et al (2017) Leydig cell stem cells: identification, proliferation and differentiation. Mol Cell Endocrinol 445:65–73
47. Tepekoy F et al (2015) CD90 and CD105 expression in the mouse ovary and testis at different stages of postnatal development. Reprod Biol 15(4):195–204
48. Helsel AR et al (2017) ID4 levels dictate the stem cell state in mouse spermatogonia. Development 144(4):624–634
49. Chan F et al (2014) Functional and molecular features of the Id4+ germline stem cell population in mouse testes. Genes Dev 28(12): 1351–1362
50. Oatley MJ et al (2011) Inhibitor of DNA binding 4 is expressed selectively by single spermatogonia in the male germline and regulates the self-renewal of spermatogonial stem cells in mice. Biol Reprod 85(2):347–356
51. Aloisio GM et al (2014) PAX7 expression defines germline stem cells in the adult testis. J Clin Invest 124(9):3929–3944
52. Hara K et al (2014) Mouse spermatogenic stem cells continually interconvert between equipotent singly isolated and syncytial states. Cell Stem Cell 14(5):658–672
53. Sachs C et al (2014) Evaluation of candidate spermatogonial markers ID4 and GPR125 in testes of adult human cadaveric organ donors. Andrology 2(4):607–614
54. Hermann BP et al (2018) The mammalian spermatogenesis single-cell transcriptome, from spermatogonial stem cells to spermatids. Cell Rep 25(6):1650–1667. e8
55. Shami AN et al (2020) Single-cell RNA sequencing of human, macaque, and mouse testes uncovers conserved and divergent features of mammalian spermatogenesis. Dev Cell 54(4):529–547. e12
56. Nakamura Y et al (2021) Transient suppression of transplanted spermatogonial stem cell differentiation restores fertility in mice. Cell Stem Cell 28(8):1443–1456. e7
57. Ebata KT, Zhang X, Nagano MC (2005) Expression patterns of cell-surface molecules on male germ line stem cells during postnatal mouse development. Mol Reprod Dev 72(2): 171–181
58. Garbuzov A et al (2018) Purification of GFRalpha1+ and GFRalpha1- spermatogonial stem cells reveals a niche-dependent mechanism for fate determination. Stem Cell Rep 10(2):553–567
59. Zheng K et al (2009) The pluripotency factor LIN28 marks undifferentiated spermatogonia in mouse. BMC Dev Biol 9:38
60. Manova K et al (1990) Gonadal expression of c-kit encoded at the W locus of the mouse. Development 110(4):1057–1069
61. Yoshinaga K et al (1991) Role of c-kit in mouse spermatogenesis: identification of spermatogonia as a specific site of c-kit expression and function. Development 113(2):689–699
62. Sharma M et al (2019) Identification of EOMES-expressing spermatogonial stem cells and their regulation by PLZF. elife 8:e43352
63. Kanatsu-Shinohara M et al (2003) Long-term proliferation in culture and germline transmission of mouse male germline stem cells. Biol Reprod 69(2):612–616
64. Kubota H, Avarbock MR, Brinster RL (2004) Growth factors essential for self-renewal and expansion of mouse spermatogonial stem cells. Proc Natl Acad Sci U S A 101(47): 16489–16494
65. Kanatsu-Shinohara M et al (2014) Improved serum- and feeder-free culture of mouse germline stem cells. Biol Reprod 91(4):88
66. Kanatsu-Shinohara M, Shinohara T (2007) Culture and genetic modification of mouse germline stem cells. Ann N Y Acad Sci 1120: 59–71
67. Shiromoto Y et al (2019) GPAT2 is required for piRNA biogenesis, transposon silencing, and maintenance of spermatogonia in micedagger. Biol Reprod 101(1):248–256
68. Mulder CL et al (2018) Long-term health in recipients of transplanted in vitro propagated spermatogonial stem cells. Hum Reprod 33(1): 81–90
69. Sato T et al (2013) In vitro sperm production from mouse spermatogonial stem cell lines using an organ culture method. Nat Protoc 8(11):2098–2104

70. Kanatsu-Shinohara M et al (2005) Genetic and epigenetic properties of mouse male germline stem cells during long-term culture. Development 132(18):4155–4163
71. Helsel AR, Oatley MJ, Oatley JM (2017) Glycolysis-optimized conditions enhance maintenance of regenerative integrity in mouse spermatogonial stem cells during long-term culture. Stem Cell Rep 8:1430
72. Aponte PM et al (2006) Basic features of bovine spermatogonial culture and effects of glial cell line-derived neurotrophic factor. Theriogenology 65(9):1828–1847
73. Izadyar F et al (2003) Proliferation and differentiation of bovine type A spermatogonia during long-term culture. Biol Reprod 68(1): 272–281
74. Zhao X et al (2021) Isolation and in vitro expansion of porcine spermatogonial stem cells. Reprod Domest Anim 57(2):210–220
75. Zhang P et al (2017) Long-term propagation of porcine undifferentiated spermatogonia. Stem Cells Dev 26(15):1121–1131
76. Zhao HM et al (2016) Isolation, proliferation, and induction of Bama mini-pig spermatogonial stem cells in vitro. Genet Mol Res 15(3). https://doi.org/10.4238/gmr.15038602
77. Heidari B et al (2012) Isolation, identification, and culture of goat spermatogonial stem cells using c-kit and PGP9.5 markers. J Assist Reprod Genet 29(10):1029–1038
78. Pramod RK, Mitra A (2014) In vitro culture and characterization of spermatogonial stem cells on Sertoli cell feeder layer in goat (Capra hircus). J Assist Reprod Genet 31(8): 993–1001
79. Zhu H et al (2014) Characterization of immortalized dairy goat male germline stem cells (mGSCs). J Cell Biochem 115(9):1549–1560
80. Qasemi-Panahi B et al (2018) Isolation and proliferation of spermatogonial cells from Ghezel sheep. Avicenna J Med Biotechnol 10(2): 93–97
81. Binsila KB et al (2018) Isolation and enrichment of putative spermatogonial stem cells from ram (Ovis aries) testis. Anim Reprod Sci 196:9–18
82. Giassetti MI, Ciccarelli M, Oatley JM (2019) Spermatogonial stem cell transplantation: insights and outlook for domestic animals. Annu Rev Anim Biosci 7:385–401
83. Herrid M et al (2006) Successful transplantation of bovine testicular cells to heterologous recipients. Reproduction 132(4):617–624
84. Honaramooz A et al (2003) Fertility and germline transmission of donor haplotype following germ cell transplantation in immunocompetent goats. Biol Reprod 69(4):1260–1264
85. Honaramooz A et al (2003) Germ cell transplantation in goats. Mol Reprod Dev 64(4): 422–428
86. Honaramooz A, Megee SO, Dobrinski I (2002) Germ cell transplantation in pigs. Biol Reprod 66(1):21–28
87. Stockwell S et al (2013) Transplanted germ cells persist long-term in irradiated ram testes. Anim Reprod Sci 142(3–4):137–140
88. Stockwell S et al (2009) Microsatellite detection of donor-derived sperm DNA following germ cell transplantation in cattle. Reprod Fertil Dev 21(3):462–468
89. Herrid M et al (2009) Irradiation enhances the efficiency of testicular germ cell transplantation in sheep. Biol Reprod 81(5):898–905
90. Izadyar F et al (2003) Autologous and homologous transplantation of bovine spermatogonial stem cells. Reproduction 126(6):765–774
91. Rodriguez-Sosa JR, Dobson H, Hahnel A (2006) Isolation and transplantation of spermatogonia in sheep. Theriogenology 66(9): 2091–2103
92. Rodriguez-Sosa JR et al (2009) Transduction and transplantation of spermatogonia into the testis of ram lambs through the extra-testicular rete. Reprod Domest Anim 44(4):612–620
93. Ogawa T et al (1997) Transplantation of testis germinal cells into mouse seminiferous tubules. Int J Dev Biol 41(1):111–122
94. Heller CH, Clermont Y (1964) Kinetics of the germinal epithelium in man. Recent Prog Horm Res 20:545–575
95. Swierstra EE (1968) Cytology and duration of the cycle of the seminiferous epithelium of the boar; duration of spermatozoan transit through the epididymis. Anat Rec 161(2): 171–185

Chapter 18

Perspectives: Methods for Evaluating Primate Spermatogonial Stem Cells

Sarah K. Munyoki and Kyle E. Orwig

Abstract

Mammalian spermatogenesis is a complex, highly productive process generating millions of sperm per day. Spermatogonial stem cells (SSCs) are at the foundation of spermatogenesis and can either self-renew, producing more SSCs, or differentiate to initiate spermatogenesis and produce sperm. The biological potential of SSCs to produce and maintain spermatogenesis makes them a promising tool for the treatment of male infertility. However, translating knowledge from rodents to higher primates (monkeys and humans) is challenged by different vocabularies that are used to describe stem cells and spermatogenic lineage development in those species. Furthermore, while rodent SSCs are defined by their biological potential to produce and maintain spermatogenesis in a transplant assay, there is no equivalent routine and accessible bioassay to test monkey and human SSCs or replicate their functions in vitro. This chapter describes progress characterizing, isolating, culturing, and transplanting SSCs in higher primates.

Key words Nonhuman primates, Humans, Spermatogonia, Transplantation, Grafting, Spermatogenesis, Adark, Apale

1 Introduction

1.1 SSCs and the SSC Niche

SSCs are the germline stem cells of the testis and are essential for male fertility. SSCs can either (1) self-renew producing more SSCs or (2) differentiate into daughter cells that undergo several rounds of transit-amplifying mitotic divisions, followed by two rounds of meiotic divisions to produce haploid spermatids that give rise to terminally differentiated sperm through a process called spermiogenesis. The delicate equilibrium between SSC self-renewal and differentiation is tightly regulated to maintain testis homeostasis and ensure continuous sperm production throughout an adult male's life [1]. SSCs reside on the basement membrane of the seminiferous tubules in specialized microenvironments known as the SSC niche, which provides factors that support SSC self-renewal and differentiation [2–5]. The location of the stem cell niche in the testes of fruit flies is well documented and confined to a

Jon M. Oatley and Brian P. Hermann (eds.), *Spermatogonial Stem Cells: Methods and Protocols*, Methods in Molecular Biology, vol. 2656, https://doi.org/10.1007/978-1-0716-3139-3_18,

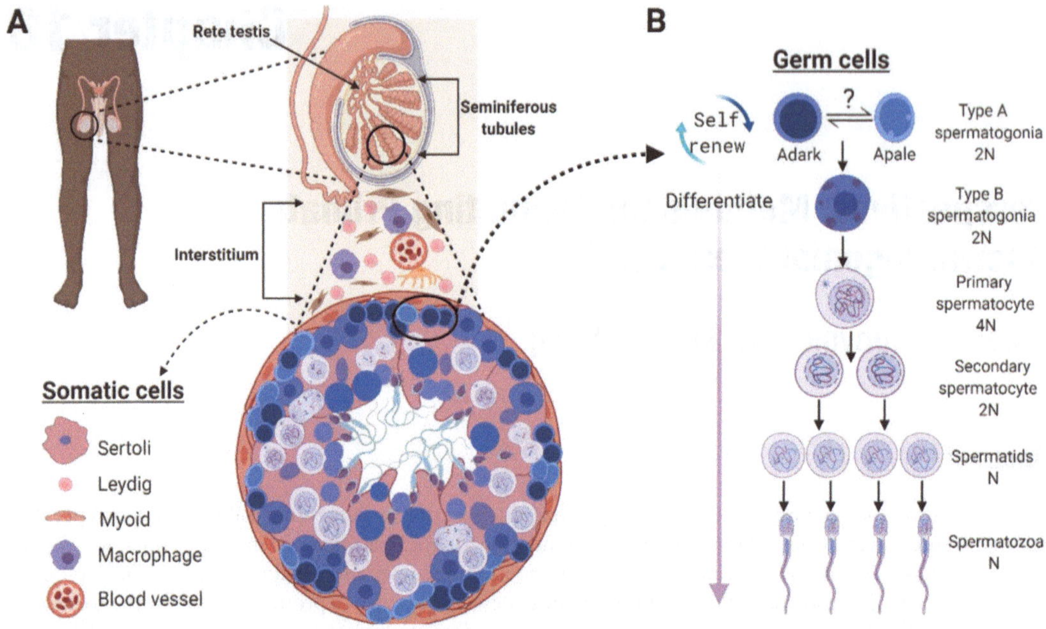

Fig. 1 Schematic representation of the human testis and spermatogenesis. (**a**) The testis is a complex tissue composed of germ and somatic cell types. The germ cells reside inside the seminiferous tubule in close contact with Sertoli cells. Peritubular myoid cells surround the tubule, while the remaining somatic cells are located in the interstitial space between tubules. (**b**) Stem cells in the human testis are thought to reside in the populations of Adark and Apale spermatogonia. The relative stem cell potential of Adark versus Apale spermatogonia is not known and it is not known whether these two cell types interconvert. Depending on the molecular signals they receive from their niche, the diploid SSCs (Adark and Apale) will either self-renew or differentiate undergoing mitosis and two rounds of meiosis to produce haploid spermatozoa

specific location in the blind-end apical tip of the testes. In that system, stem cells can easily be identified because they are the cells that are in direct contact with the niche (hub cells). When a stem cell divides, only one daughter cell can occupy the stem cell niche. The other daughter cell moves away from the niche and differentiates. This is known as asymmetric division [6]. In contrast, the location and precise characteristics of the stem niche in mammalian testes is not so well defined. Stem cells are not confined to the apical tip of the mammalian testis. They are distributed along the length of the seminiferous tubules. It is not known whether SSCs in mammalian testes undergo symmetrical divisions, asymmetric divisions, or a combination of both, depending on cues from the local microenvironment. A schematic of the human testis, seminiferous tubule cross section, and spermatogenic lineage development is shown in Fig. 1. Sertoli cells are the only somatic cells located within the seminiferous tubules and are intimately associated with the developing germ cells (Fig. 1). Murine Sertoli cells produce GDNF and FGF2 that are essential for maintaining and expanding mouse SSCs (mSSCs) in vivo and in vitro [7–10]. Peritubular myoid cells form layers around the seminiferous tubules near the SSCs, providing structural support to the tubules and secreting extracellular matrix components and growth factors, including GDNF and CSF1, that are known to affect SSC function [11, 12]. Sertoli cells also produce kit ligand that is required for spermatogonial differentiation [13]. Leydig cells in the interstitial space between seminiferous tubules secrete testosterone, which is required for spermatogenesis and CSF-1 that has been shown to stimulate mouse SSC self-renewal [12]. Although the somatic cells located outside of the seminiferous tubules do not have direct contact with spermatogonia, they secrete paracrine factors that have direct effects on spermatogonia or effects that are mediated by Sertoli cells that do have direct contact with germ cells [14–18]. An in depth understanding of mouse SSCs and experimental tools used to study those cells provides a foundation for understanding and studying SSCs in higher primates (monkeys and humans).

1.2 Spermatogenic Lineage Development and Clonal Arrangement

1.2.1 Rodents

Much of the seminal work on stem cells and spermatogenic lineage development in rodents, monkeys, and humans was performed in the laboratory of Yves Clermont at McGill University [19–23]. However, the prevailing A_{single} (A_s) model of SSCs and spermatogenic lineage development was described in a series of papers in 1971 by Claire Huckins (a trainee of Professor Clermont) in rats [24–26] and Oakberg in mice [27] with important contributions from Dirk de Rooij and others [28, 29]. In this model, A_{single} (A_s) spermatogonia are the SSCs [26, 27, 30–32]. The isolated A_s spermatogonia undergo mitosis to produce a pair of spermatogonia (A_{paired} (A_{pr})), which will complete cytokinesis to produce two new A_s spermatogonia (self-renewal) or remain attached by an intracytoplasmic bridge and divide again to produce a chain of 4 $A_{aligned}$ (A_{al}) spermatogonia. The 4 A_{al} cells further divide, forming chains of 8, 16, and sometimes 32 A_{al} (Fig. 2). A_{pr} that do not complete cytokinesis and all larger chains of A_{al} spermatogonia are considered to be committed progenitors. The A_s are rare, comprising only 0.03% of all germ cells in the murine testis [33]. The A_s, A_{pr}, and A_{al} collectively constitute the pool of undifferentiated stem and progenitor spermatogonia and comprise 0.3% of germ cells in the rodent testis [33]. Other models describing SSC self-renewal and differentiation dynamics have been described and elements of all models may be relevant under different circumstances [5, 34].

Yoshida and colleagues performed lineage tracing and live imaging experiments using a transgenic GFP reporter mouse and proposed a "clone fragmentation model" of spermatogonial stem cell maintenance [35, 36]. This model challenges the assumption that the A_s spermatogonia represent the entire stem cell pool and contends that the A_{pr} and A_{al} spermatogonia can sometimes break apart and produce new A_s that may contribute to the functional

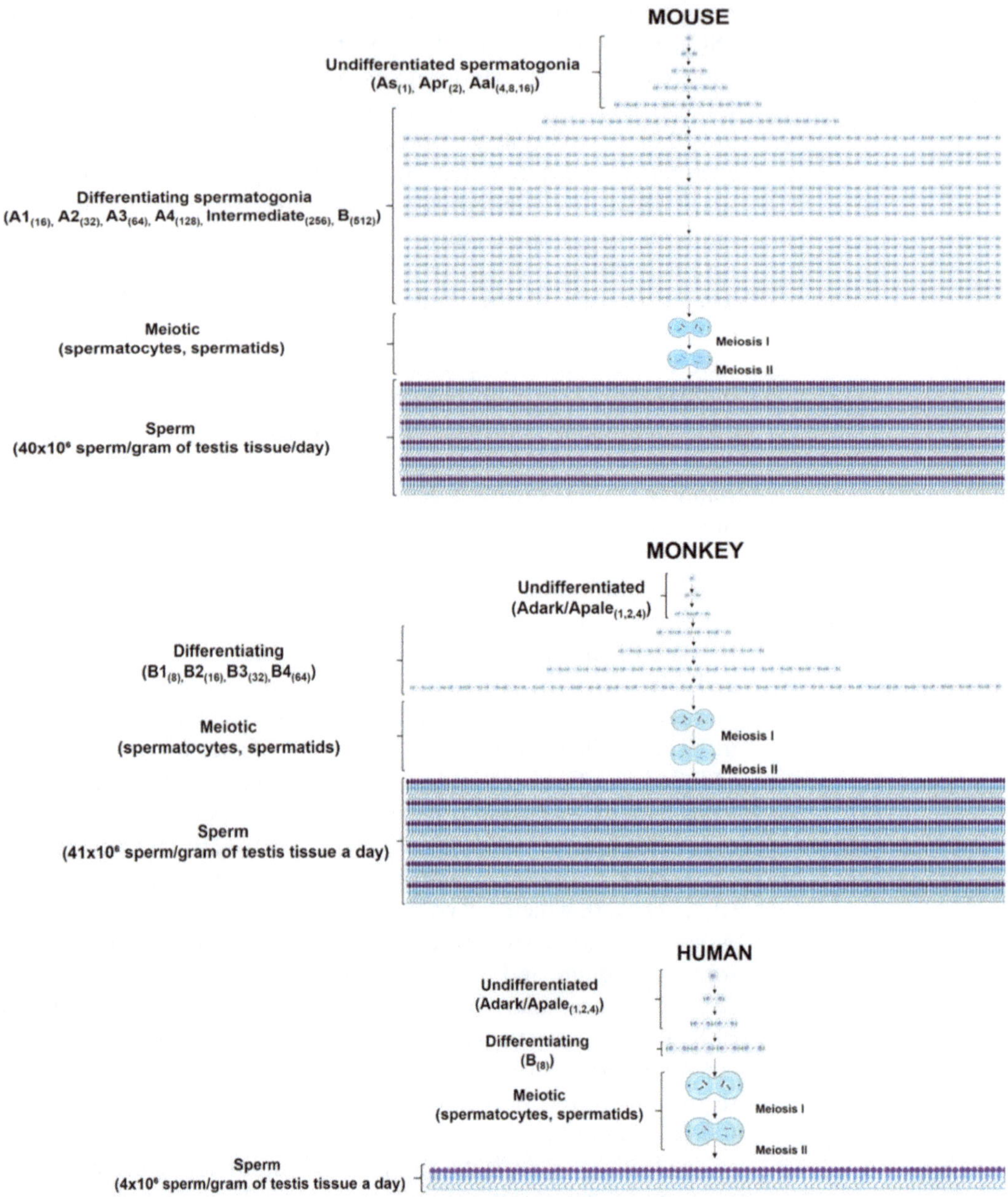

Fig. 2 SSC pool, clonal development, and sperm output in mouse, monkey, and human

stem cell pool. Hara and colleagues provided in vivo video evidence of clone fragmentation, but the fate of the resulting fragments could not be followed [36], and this requires further investigation. Also, the clone fragmentation model is based on observations of GFRa1-positive cells and does not account for the contribution of GFRa1-negative cells that are known to exist in the pool of A_s spermatogonia [37, 38] and the pool of transplantable stem cells

[39, 40]. Nonetheless, it seems reasonable that some clones identified as A_{pr} will complete their cytokinesis to produce two new A_s spermatogonia, a phenomenon described as false pairs by de Rooij and Griswold [41]. Odd size clones that might results from clone fragmentation (e.g., $A_{al(3, 5, 6, 7, 9, etc)}$) are occasionally seen in whole mount preparations of seminiferous tubules but they are rare [36–38, 42]. A revised A_s model known as the "hierarchical A_s model" [5] has been proposed stemming from SSC transplantation and lineage tracing studies by Oatley and others [43–46]. They proposed that SSC activity is contained within a sub-fraction of A_s spermatogonia that express the transcription factor ID4 coined the "ultimate SSC" while the other A_s spermatogonia exhibiting low or no ID4 expression have a limited stem cell capacity. The A_{pr}, A_{al}, and perhaps some A_s spermatogonia comprise the progenitor population that is committed to differentiation [44–46].

Undifferentiated A_{al} spermatogonia typically reach a chain size of 16 (4 transit-amplifying divisions) before differentiating to produce A1 spermatogonia. Six additional transit-amplifying mitotic divisions produce A2, A3, A4, intermediate, and type B differentiating spermatogonia and finally primary spermatocytes. Primary spermatocytes undergo a first meiotic division to produce secondary spermatocytes, which undergo a second meiotic division to produce haploid round spermatids (Fig. 2). The round spermatids undergo morphological transformation through a process known as spermiogenesis to form mature elongated spermatids that undergo a final remodeling on the luminal edge of the seminiferous tubules, becoming spermatozoa that are released into the lumen and transported to the epididymis [47, 48]. Through this complex series of ~10 transit-amplifying mitotic divisions and two meiotic divisions, the rodent stem cell pool produces ~40 million sperm per gram of testis tissue each day [49] (Fig. 2).

1.2.2 Higher Primates

In humans and monkeys, two morphologically distinct type A spermatogonia have been described, A_{dark} (A_d) and A_{pale} (A_p), and are thought to represent the pool of stem and progenitor spermatogonia [50–52]. They are not typically described by their clone size (e.g., A_s, A_{pr}, A_{al}) but by their position on the basement membrane of the seminiferous tubule, nuclear morphology, and intensity of staining with hematoxylin [20, 21, 53, 54]. Clermont and Antar described A_d spermatogonia as having a spherical nucleus with fine granular densely packed (dust-like) chromatin that is darkly and uniformly stained with hematoxylin. Some A_d spermatogonia contain 1–3 intranuclear chromatin rarefaction zones (vacuoles) [55]. These vacuolar spaces may include a nucleolus attached to the nuclear membrane [21, 53, 54]. A_p spermatogonia have an ovoid nucleus containing coarse granular chromatin that stains less intensely with hematoxylin and have a ground-glass texture [20, 21, 55]. The A_p nucleus contains 1–2 faintly stained

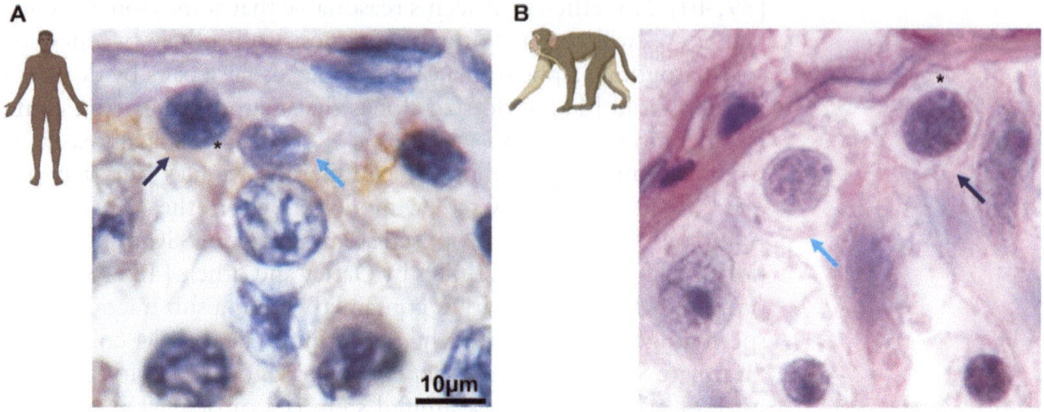

Fig. 3 Histological examination of A_d and A_p spermatogonia in higher primates. Hematoxylin (PAS-H) staining of adult testis cross sections showing the A_d (navy arrow) and A_p (light blue arrow) spermatogonia on the basement membrane of the seminiferous epithelium. Stars* indicate vacuolar spaces that contain a nucleolus attached to the nuclear membrane. Scale bar 10 μm. (**a**) Human (**b**) Monkey

nucleoli attached to the nuclear membrane but with no vacuoles (Fig. 3a-b). A_d and A_p spermatogonia appear in almost equal proportion in the primate testis [55–57]. Ramaswamy and colleagues showed that the A_d and A_p comprise 1.6% and 2% of all germ cells in the monkey testis, respectively [58], suggesting that the pool of stem and progenitor spermatogonia in nonhuman primates is much larger than rodents where undifferentiated A_s, A_{pr}, and A_{al} spermatogonia comprise 0.3% of all germ cells. Like nonhuman primates, human A_d and A_p spermatogonia are present in equal numbers [54, 59, 60] and make up 22% of germ cells in the testis [60, 61].

Our knowledge of the clonal arrangement of A_d and A_p spermatogonia is limited. Clermont reported that A_d and A_p spermatogonia are organized in clones of 1, 2, or 4 cells using camera lucida drawings and 3D reconstruction mapping of serial cross sections of the monkey testis. Similar observations were made by Ehmcke and colleagues in rhesus macaques. Apale spermatogonia were observed as clones of two or four cells, while differentiated B1 spermatognia were observed as clones of four or eight cells [50]. Clonal analysis using morphological and topographical criteria from rodent studies identified A_d and A_p clones of 1, 2, 4, 8, and ≥16 on rhesus macaques treated with low dose radiation, which may not equate to a normal clonal arrangement [62, 63]. Like rodents, Valli and colleagues demonstrated that smaller clones are more likely to have an undifferentiated phenotype (cKIT negative) and larger clones are more likely to have a differentiated phenotype (cKIT positive in nonhuman primates and humans [64]), although clone sizes in monkeys and humans are smaller than in rodents (Fig. 2, *see* next section).

Primate undifferentiated spermatogonia have fewer transit-amplifying divisions before differentiation than rodents (Fig. 2). Rodent A_{al} spermatogonia differentiate sequentially into A1, A2, A3, A4, intermediate, and B spermatogonia, while primate type A spermatogonia differentiate directly to type B spermatogonia (Fig. 2). Monkeys have four generations of B spermatogonia (B1, B2, B3, and B4) that display similar morphology, while humans have one generation of B spermatogonia. The primate type B nucleus is large and ovoid with granulated pale stained chromatin, a non-uniform amount of darkly stained chromatin flakes, and granules with a large polymorphous nucleolus in the center with no connection to the nuclear membrane. The chromatin flakes and granules may be associated with the nucleolus and nuclear membrane and became coarser and more abundant from B1 to B4 spermatogonia [43–47]. Like rodents, primate B spermatogonia subsequently produce primary spermatocytes, secondary spermatocytes, spermatids, and terminally differentiated sperm. In total, there are eight transit-amplifying divisions in monkeys and five in humans from a single A_d/A_p spermatogonia to mature sperm compared to the 12 divisions in rodents from a single A_s spermatogonia [61] (Fig. 2).

Rodents should yield 4096 ($2^{12} = 4096$) sperm for each A_s spermatogonia that commits to differentiation. But this is not the case; ~50% of the A1–A4 spermatogonia are lost to apoptosis during spermatogenesis reducing the yield [28, 30]. One might assume that an isolated A_d/A_p spermatogonia would produce 256 ($2^8 = 256$) monkey and 32 ($2^5 = 32$) human sperm, which compared to mice would be a reduced yield of 16- and 128-fold, respectively. However, monkey sperm output is equivalent to the rodent at ~40 million sperm/per gram of testis tissue/day. Human sperm output is only reduced by tenfold at ~4.4 million sperm/gram of testis tissue/day [61]. Although the impact of apoptosis in primates is unknown, differences in the size of the stem/progenitor cell pool across species (0.3%, 4%, and 22% of germ cells in the testis, respectively) may contribute to the differences in sperm output. Larger stem/progenitor pools in primates may partially compensate for the reduced number of transit-amplifying divisions (Fig. 2).

1.2.2.1 Primate SSC Proliferation Dynamics

In 1959, Clermont and Leblond proposed that A1 (later known as A_{dark}) spermatogonia are the stem cells that divide and either self-renew to maintain the pool of A1 spermatogonial stem cells or produce A2 (later known as A_{pale}) progenitor spermatogonia that may undergo one or more transit amplifying divisions before differentiating to produce B1 spermatogonia. Clermont revised that model 10 years later based on in vivo labeling with 3H-thymidine in vervet monkeys, which indicated that A_p spermatogonia, but not

A_d spermatogonia, incorporated 3H-thymidine. Based on those results, Clermont proposed that the A_d are the "reserve" stem cells that only divide after an insult to the testis to regenerate spermatogenesis, while the A_p are the "active" stem cells that continually proliferate and maintain steady-state spermatogenesis under normal conditions [15]. Additional studies using 3H-thymidine or bromodeoxyuridine (BrdU) have reported the labeling index of primate type A spermatogonia. Some studies observed no labeling of A_d spermatogonia [50, 65] replicating Clermont's findings, while others observed labeling of A_d spermatogonia [51, 52, 66, 67]. These studies administered a single bolus of 3H-thymidine or BrdU which may not efficiently label slow-cycling stem cells. Long-term labeling experiments are needed to effectively determine if A_d spermatogonia are quiescent or slow-cycling cells.

Nevertheless, this model has been supported by experimental evidence showing X-irradiation of the monkey testis caused significant ablation of spermatogenesis, including the entire population of A_p spermatogonia, which were replenished over time from the surviving pool of A_d spermatogonia that were more resistant to the radiation insult [68, 69]. These observations have led some to claim that the low mitotic index of the A_d spermatogonia is an attribute of a "true stem cell" while the regularly proliferating A_p represent the "renewing progenitors" [70–72]. Still, others have offered a differing opinion based on studies in monkeys showing that A_d and A_p share a common molecular phenotype. Hermann and colleagues proposed that A_d and A_p are the same cell type and that their nuclear phenotypes correlate with different cell cycle stages (i.e., A_d: G0 versus A_p: G1/S/G2/M) [73, 74]. Just like the rodent system, there are differences of opinion about the identity of stem cells in monkey and human testes. It may be that the stem cell pool in mammalian testes is heterogeneous and cannot be completely defined by a particular phenotype.

2 Molecular Characteristics of Primate SSCs

To date, no protein marker, in any species, has been identified that is restricted to the stem cell pool and that marks the entire population of stem cells [61]. The only method to unequivocally identify an SSC is to retrospectively evaluate its capacity to regenerate spermatogenesis when transplanted into an infertile testis. Although no SSC-specific marker exists, several conserved markers of undifferentiated spermatogonia have been identified by assessing their localization on the basement membrane of seminiferous tubules and co-expression with established spermatogonial markers. GFRA1, UTF1, PLZF, SALL4, and LIN28 are conserved markers of undifferentiated spermatogonia that are conserved from rodents to higher primates [38, 39, 64, 73, 75–87]. Undifferentiated spermatogonia of rodents [44] and humans [86] express ID4, while nonhuman primate has yet to be reported. Currently, there are no markers that can distinguish the entire A_d or A_p pool, but it has been demonstrated that expression of EXOSC10, FGFR3, and OCT2 (POU2F2) are restricted to the subpopulation of A_d spermatogonia containing a nuclear vacuole [88–90]. Recently, Caldeira-Brant and colleagues identified a subtype of human vacuolated A_d spermatogonia that they classified as the "reserve SSCs" because they were entirely quiescent with a UTF1+/GFRA1-/MCM7- protein expression signature [91]. SSX2-4 expression is restricted to a subpopulation of A_p spermatogonia [92], while a subpopulation of A_p express cKIT and NGN3 [73]. There is broad acceptance that the shift from undifferentiated to differentiating spermatogonia is associated with the gain of cKIT expression [61, 93]. cKIT is not expressed by A_d spermatogonia but is expressed by about ~25% of A_p spermatogonia in rhesus macaques [73]. Furthermore, Izadyar reported that human spermatogonia exhibiting SSC colonization activity on the human to nude mouse xenotransplant assay are in the cKIT low/negative fraction of human testis cells [94].

2.1 Cell Sorting Techniques for Isolating Primate SSCs

Flow cytometry has become a valuable tool for analyzing the heterogeneity of state-specific spermatogonia populations. Fluorescence-activated cell sorting (FACS) allows for gating and isolation/purification of specific cells in a heterogeneous cell population based on several parameters such as viability, cell size, cell complexity, and expression of specific molecular markers. FACS coupled with xenotransplantation (Fig. 4) have identified SSEA4, ITGA6, EPCAM, THY1, and CD9 as cell surface markers that can be used to isolate transplantable human or monkey undifferentiated spermatogonia [93–102]. Using these markers, FACS can segregate heterogenous testicular cell suspensions into subpopulations that can be tested for colonization potential. However, many of these markers are broadly expressed by testicular germ cells and somatic cells, so do not produce a pure population of spermatogonia [61, 64, 103]. Other cell surface markers of monkey and/or human spermatogonia are GFRA1 and GPR125 [104].

Magnetic activated cell sorting (MACS) can also be used to achieve significant enrichment of undifferentiated spermatogonia. ITGA6 and CD9 can effectively isolate and enrich undifferentiated spermatogonia via MACS [64, 100, 104]. Both FACS and MACS experiments have demonstrated that a cKIT negative phenotype is associated with transplantable primate spermatogonia [64, 73, 105, 106]. Although MACS has a lower sorting resolution, it is high throughput, technically accessible, and affordable. In comparison, FACS allows higher resolution and more precise gating but more

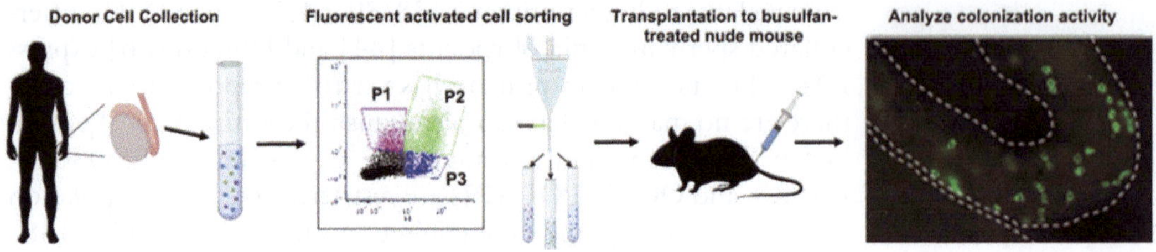

Fig. 4 Workflow of the human to nude mouse xenotransplantation assay. FACS coupled with xenotransplantation to identify transplantable human spermatogonia. FACS allows for gating and isolating spermatogonia cell populations expressing specific cell surface markers from a heterogeneous cell suspension of human testicular cells. The FACS purified cell populations can then be microinjected into the testes of busulfan-treated nude mice. Testes are recovered 8 weeks after xenotransplantation for whole-mount immunofluorescence. Transplantable human spermatogonia are identified by their ability to migrate and colonize the basal lamina of the mouse seminiferous tubules, proliferate to produce chains of human spermatogonia, and survive in the mouse testes for months

limited throughput and requires expensive, specialized equipment and a skilled operator [64].

3 Functional Evaluation of Primate SSCs

3.1 SSC Transplantation

Over 25 years ago, Brinster and colleagues developed a transplantation method that established a functional assay for identifying and quantifying SSCs [107]. mSSCs were microinjected into the testes of infertile recipient mice, where they migrated to the basement membrane of seminiferous tubules and regenerated spermatogenesis producing sperm [107, 108] and offspring [108]. Transplant experiments can be carefully designed so that each colony of donor-derived spermatogenesis in recipient testes arises from a single transplanted stem cell [109, 110], making transplantation a quantitative assay for SSCs. This technique opened the gateway to address fundamental questions on SSC biology and remains the only method to unequivocally identify functional SSCs.

SSC transplantation is a robust technique that has been successfully replicated in numerous animal models. Allogeneic transplantation of mouse, goat, chicken, sheep, and rat SSCs result in restoration of spermatogenesis and production of offspring [108, 111–117]. Autologous or allogeneic SSC transplantation into infertile macaques restored spermatogenesis [118–120] and could produce fertilization competent sperm [119]. Furthermore, rat SSCs xenotransplanted into the testes of immune-deficient mice generated colonies of donor-derived spermatogenesis [121, 122] and fertilization-competent sperm [123]. Hamster SSCs can also be xenotransplanted to recipient mouse testis and regenerate spermatogenesis, but spermatid and sperm defects were noted [124]. Xenotransplanted rabbit, dog, pig, bull, cat, horse, monkey, and human SSCs colonize the basement membrane of the mouse seminiferous tubule and begin to form chains and patches of spermatogonia that can survive in the recipient testis for months, but do not produce complete spermatogenesis [73, 91, 94, 98–100, 125–137]. Although spermatogonia from primates and other large animals do not produce complete spermatogenesis in mouse recipients, they do exhibit several SSC-like characteristics. They (1) migrate and engraft the basement membrane of mouse seminiferous tubules; (2) proliferate to produce characteristic chains and clusters of spermatogonia, frequently connected by intercellular cytoplasmic bridges; and (3) persist long-term in the mouse testes. Thus, xenotransplantation to mouse testes has emerged as an experimentally tractable bioassay for quantifying spermatogonial colonization in primates and other large animal species where autologous or allogeneic transplantation is not feasible [61, 107] (Fig. 4).

3.2 In Vitro Culture of Primate SSCs

Methods to maintain and expand mouse SSCs in long-term culture were first described in 2003 [9] and 2004 [10] and extended to rats in 2005 [138, 139]. SSC culture has become a powerful tool to dissect factors and pathways that are important for SSC renewal and differentiation [12, 140–146]. Keys to the initial successes of rodent SSC culture included development of methods to isolate and enrich SSCs (e.g., differential plating, Percoll gradients, MACS, and FACS) while reducing the number of testicular somatic cells that could rapidly overwhelm the cultures. In addition, development of serum free culture conditions facilitated identification of single growth factors that supported the maintenance and expansion of rodent SSCs in culture. While Kanatsu-Shinohara and colleagues initially used GDNF, bFGF, EGF, and LIF in mouse SSC cultures [9], others have found that GDNF and bFGF are necessary and sufficient to maintain and expand mouse SSCs in long-term culture [10]. Feeder cells are frequently used to provide a substrate for SSC culture and perhaps provide unknown factors that support SSC survival or proliferation [9, 10, 147], although feeder free conditions have also been described [148, 149].

The growth factors needed for long-term in vitro culture of primate SSCs have not yet been identified, but several research groups have reported on primate SSC culture [96, 102, 129, 150–171]. Based on experiences with mouse SSC culture, many of these studies employed FACS, MACS, and differential plating on either plastic, gelatin, or collagen to enrich SSCs and eliminate contaminating somatic cells that can overgrow the culture. Other studies supplemented differential plating with Percoll gradient selection and positive/negative selection for cell surface markers using FACS or MACS. Positive selection markers have included ITGA6, CD9, GPR125, SSEA4, and EPCAM, while negative selection markers include cKIT, CD45, and THY1 (reviewed in

[103]). It is noteworthy that THY1 is a positive selection marker for mouse spermatogonia but a negative selection marker for transplantable human spermatogonia [166, 172].

In 2009, Sadri-Ardekani and colleagues were the first to report on the long-term culture of adult hSSCs using rodent SSC culture conditions [130]. Germ cells were enriched by differential plating and cultured on human placental laminin in media containing a growth factor cocktail of GDNF, bFGF, LIF, and EGF [130, 131]. The authors reported that cells expressing GDNF, UCHL1, GPR125, and PLZF could be maintained in vitro for several months with and 18,000-fold expansion in numbers. One other group reported significant expansion of hSSC numbers under those conditions [161], while others reported a rapid decline in hSSC numbers using the same conditions; some reporting that these conditions support the expansion of mesenchymal stem cells [96, 155, 162, 167]. These discrepancies may be explained by differences in starting cell populations, culture conditions, and approaches that were employed to identify and quantify hSSCs in culture. Many research groups have now reported human SSC survival and/or expansion for periods ranging from 1 week to 6 months [96, 129–131, 150–156, 158, 161, 162, 164–167, 173–175]. However, there is no consensus "best method" for culturing human SSCs that has been consistently and independently replicated in different laboratories and no consensus on best methods to identify and quantify bona fide hSSCs in culture. While the results for hSSC culture are promising, they are challenged by the inability to test the full spermatogenic potential of these cells by xenotransplantation into mice; and autologous or homologous transplantation for experimental purposes in humans is not possible.

3.3 hSSC Transplantation as a Therapy for Male Infertility

Infertility is a devastating condition that impacts 9–14% of American men [70]. The inability to have biological children can deleteriously impact an individual's relationships, psychological well-being, finances, and life expectancy [71–76]. Additionally, an impaired reproductive system has been associated with an increased risk of several cancers, metabolic disease, diabetes, cardiovascular disease, and multiple sclerosis [77–82]. Each year in the United States, approximately 1800 boys will be rendered permanently infertile after radiation, chemotherapy, and surgical cancer treatment [83]. Unlike adult men, these prepubertal boys are not yet reproductively mature and cannot safeguard their future fertility by cryopreserving sperm before treatment.

Although prepubertal testes do not have sperm, they contain SSCs that can initiate spermatogenesis at puberty or regenerate spermatogenesis and restore fertility in infertile individuals [112, 119]. Therefore, cryopreservation of immature testicular tissues has emerged as an experimental fertility preservation option for these patients because it is predicted those tissues or SSCs can be matured in the future to produce sperm [130, 131, 176–194]. SSC or testicular tissue transplantation, in particular, have emerged as mature technologies that have been replicated in numerous mammalian species and may be ready for translation to the human clinic [111–114, 117, 118, 195–201].

There are important challenges on the path to translating stem cell-based therapies into clinical practice for treating male infertility. In some cases (e.g., testicular cancer and leukemia), transplantation of cells or tissues poses the risk of reintroducing malignant cells back into patient survivors. A rat model showed that transplantation of testicular cells from leukemic donors consistently transmitted leukemia to healthy recipients [202]. Furthermore, evaluation of testicular biopsies taken from boys with acute lymphocytic leukemia revealed that ~20% contained leukemia [203]. Reintroducing malignant cells into a survivor that has been cured of their primary disease is an unacceptable risk. Several groups have reported FACS-based sorting methods to eliminate malignant contamination from testicular cell suspensions [132, 204–206]. Others have reported that leukemic cells can be depleted over time in human SSC culture [207].

Transgender females may not want to go through male puberty that would be required to mature their testicular tissues or initiate spermatogenesis from their SSCs in the autologous transplantation paradigm. Therefore, it is necessary to develop and translate toward the clinic, technologies to mature testicular tissues or SSCs outside the body, such as human to animal xenografting [208–210] or testicular tissue organ culture [211–213]. To our knowledge, these approaches have been applied for maturation of immature animal testicular tissues, but sperm have not yet been produced from immature human tissues with these methods.

Small biopsies obtained from the testes of young cancer patients may contain only a small number of SSCs that may limit the extent of engraftment and spermatogenesis. Therefore, it may be necessary to isolate and purify hSSCs and expand them in vitro to ensure robust engraftment after transplantation. As indicated above, reproducible methods to maintain and expand human SSCs in long-term culture have not been described. Also, autologous cell-based therapies that involve extensive manipulation outside the body may require premarket review and approval by the FDA.

4 Conclusion

Over the past 70 years, different vocabularies and models have emerged to describe SSC self-renewal/differentiation dynamics and spermatogenic lineage development in rodents compared with higher primates. In the rodent A_s model of spermatogenic lineage development, undifferentiated spermatogonia are described by clone size (A_s, A_{pr}, A_{al}), while in higher primates, A_d and A_p undifferentiated spermatogonia are described by differences in nuclear morphology and staining with hematoxylin. These differences may give the impression that spermatogenesis in rodents is different than in monkeys and humans but in fact, the systems are very similar. In rodents and humans, SSCs are located on the basement membrane of seminiferous tubules. SSCs differentiate to produce progenitor spermatogonia that undergo a species-specific number of transit-amplifying divisions, producing clones of cells connected by intercellular cytoplasmic bridges. Many molecular markers of undifferentiated stem/progenitor spermatogonia are conserved from mice to monkeys to humans. In both rodents and higher primates, smaller clones are associated with a more undifferentiated molecular phenotype (ID4, GFRA1, UTF1, SALL4) while larger clones are associated with a more differentiated molecular phenotype (cKIT). There are important differences in the size of the SSC pool, SSC cell cycle length, number of transit amplification divisions, and sperm output between rodents and primates that could impact how SSC-based therapies can be most effectively deployed in the human clinic. Primate to nude mouse xenotransplantation has emerged as the "gold standard" functional assay for primate spermatogonia but those transplant results should be interpreted with caution because nonhuman primate and human cells do not recapitulate complete spermatogenesis in recipient mouse testes. SSC culture has been a powerful tool to dissect molecules and pathways that regulate rodent SSC proliferation and differentiation dynamics. More work is needed to establish a robust nonhuman primate or human SSC culture system. This would have important implication for fundamental investigations of primate SSCs or development of hSSC therapeutics. This chapter summarizes classical and contemporary understanding of primate SSC identity and behavior. We described the techniques and reagents used to investigate primate SSC biology and identified knowledge or technology gaps that are opportunities for future investigations.

Acknowledgments

SKM was supported by a *Eunice Kennedy Shriver* National Institute for Child Health and Human Development F31 training grant HD101323. KEO was supported by *Eunice Kennedy Shriver* National Institute for Child Health and Human Development grants P50 HD096723, R01 HD100197 as well as R01 OD028223.

References

1. Kubota H, Brinster RL (2018) Spermatogonial stem cells. Biol Reprod 99(1):52–74
2. de Rooij DG (2009) The spermatogonial stem cell niche. Microsc Res Tech 72(8): 580–585
3. Oatley JM, Brinster RL (2012) The germline stem cell niche unit in mammalian testes. Physiol Rev 92(2):577–595
4. Law NC, Oatley JM (2018) Spermatogonial stem cell and niche. In: Reference module in biomedical sciences. Elsevier
5. de Rooij DG (2017) The nature and dynamics of spermatogonial stem cells. Development 144(17):3022–3030
6. de Cuevas M, Matunis EL (2011) The stem cell niche: lessons from the drosophila testis. Development 138(14):2861–2869
7. Mullaney BP, Skinner MK (1992) Basic fibroblast growth factor (bFGF) gene expression and protein production during pubertal development of the seminiferous tubule: follicle-stimulating hormone-induced Sertoli cell bFGF expression. Endocrinology 131(6): 2928–2934
8. Tadokoro Y, Yomogida K, Ohta H, Tohda A, Nishimune Y (2002) Homeostatic regulation of germinal stem cell proliferation by the GDNF/FSH pathway. Mech Dev 113(1): 29–39
9. Kanatsu-Shinohara M, Ogonuki N, Inoue K, Miki H, Ogura A, Toyokuni S, Shinohara T (2003) Long-term proliferation in culture and germline transmission of mouse male germline stem cells. Biol Reprod 69(2): 612–616
10. Kubota H, Avarbock MR, Brinster RL (2004) Growth factors essential for self-renewal and expansion of mouse spermatogonial stem cells. Proc Natl Acad Sci U S A 101(47): 16489–16494
11. Spinnler K, Kohn FM, Schwarzer U, Mayerhofer A (2010) Glial cell line-derived neurotrophic factor is constitutively produced by human testicular peritubular cells and may contribute to the spermatogonial stem cell niche in man. Hum Reprod 25(9): 2181–2187
12. Oatley JM, Oatley MJ, Avarbock MR, Tobias JW, Brinster RL (2009) Colony stimulating factor 1 is an extrinsic stimulator of mouse spermatogonial stem cell self-renewal. Development 136(7):1191–1199
13. Flanagan JG, Chan DC, Leder P (1991) Transmembrane form of the kit ligand growth factor is determined by alternative splicing and is missing in the Sld mutant. Cell 64(5): 1025–1035
14. Bellve AR, Zheng W (1989) Growth factors as autocrine and paracrine modulators of male gonadal functions. J Reprod Fertil 85(2): 771–793
15. Clermont Y (1972) Kinetics of spermatogenesis in mammals: seminiferous epithelium cycle and spermatogonial renewal. Physiol Rev 52(1):198–236
16. Saez JM, Tabone E, Perrard-Sapori MH, Rivarola MA (1986) Paracrine role of Sertoli cells. Med Biol 63(5–6):225–236
17. Mullaney BP, Skinner MK (1991) Growth factors as mediators of testicular cell-cell interactions. Bailliere Clin Endocrinol Metab 5(4): 771–790
18. Skinner MK, Norton JN, Mullaney BP, Rosselli M, Whaley PD, Anthony CT (1991) Cell-cell interactions and the regulation of testis function. Ann N Y Acad Sci 637:354–363
19. Clermont Y, Leblond CP (1953) Renewal of spermatogonia in the rat. Am J Anat 93(3): 475–501
20. Clermont Y, Leblond CP (1959) Differentiation and renewal of spermatogonia in the monkey, Macacus rhesus. Am J Anat 104: 237–273
21. Clermont Y (1966) Spermatogenesis in man. A study of the spermatogonial population. Fertil Steril 17(6):705–721
22. Clermont Y (1969) Two classes of spermatogonial stem cells in the monkey (Cercopithecus aethiops). Am J Anat 126(1):57–71
23. Dym M, Clermont Y (1970) Role of spermatogonia in the repair of the seminiferous epithelium following x-irradiation of the rat testis. Am J Anat 128(3):265–282
24. Huckins C (1971) The spermatogonial stem cell population in adult rats. III. Evidence for a long-cycling population. Cell Tissue Kinet 4(4):335–349
25. Huckins C (1971) The spermatogonial stem cell population in adult rats. II. A radioautographic analysis of their cell cycle properties. Cell Tissue Kinet 4(4):313–334
26. Huckins C (1971) The spermatogonial stem cell population in adult rats. I. Their morphology, proliferation and maturation. Anat Rec 169(3):533–557
27. Oakberg EF (1971) Spermatogonial stem-cell renewal in the mouse. Anat Rec 169(3): 515–531

28. de Rooij DG (1973) Spermatogonial stem cell renewal in the mouse. I. Normal situation. Cell Tissue Kinet 6(3):281–287
29. De Rooij DG (1983) Proliferation and differentiation of undifferentiated spermatogonia in the mammalian restis (review). In: Potten CS (ed) Stem cells, their identification and characterization. Churchill Livingstone, Edinburgh, pp 89–117
30. Huckins C, Oakberg EF (1978) Morphological and quantitative analysis of spermatogonia in mouse testes using whole mounted seminiferous tubules, I. the normal testes. Anat Rec 192(4):519–528
31. Huckins C, Oakberg EF (1978) Morphological and quantitative analysis of spermatogonia in mouse testes using whole mounted seminiferous tubules. II. The irradiated testes. Anat Rec 192(4):529–542
32. Huckins C (1978) The morphology and kinetics of spermatogonial degeneration in normal adult rats: an analysis using a simplified classification of the germinal epithelium. Anat Rec 190(4):905–926
33. Tegelenbosch RA, de Rooij DG (1993) A quantitative study of spermatogonial multiplication and stem cell renewal in the C3H/101 F1 hybrid mouse. Mutat Res 290(2):193–200
34. Lord T, Oatley JM (2017) A revised a(single) model to explain stem cell dynamics in the mouse male germline. Reproduction 154(2): R55–r64
35. Nakagawa T, Sharma M, Nabeshima Y, Braun RE, Yoshida S (2010) Functional hierarchy and reversibility within the murine spermatogenic stem cell compartment. Science 328(5974):62–67
36. Hara K, Nakagawa T, Enomoto H, Suzuki M, Yamamoto M, Simons BD, Yoshida S (2014) Mouse Spermatogenic stem cells continually interconvert between equipotent singly isolated and syncytial states. Cell Stem Cell 14(5):658–672
37. Suzuki H, Sada A, Yoshida S, Saga Y (2009) The heterogeneity of spermatogonia is revealed by their topology and expression of marker proteins including the germ cell-specific proteins Nanos2 and Nanos3. Dev Biol 336(2):222–231
38. Gassei K, Orwig KE (2013) SALL4 expression in gonocytes and spermatogonial clones of postnatal mouse testes. PLoS One 8(1): e53976
39. Grisanti L, Falciatori I, Grasso M, Dovere L, Fera S, Muciaccia B, Fuso A, Berno V, Boitani C, Stefanini M et al (2009) Identification of spermatogonial stem cell subsets by morphological analysis and prospective isolation. Stem Cells 27(12):3043–3052
40. Garbuzov A, Pech MF, Hasegawa K, Sukhwani M, Zhang RJ, Orwig KE, Artandi SE (2018) Purification of GFRα1+ and GFRα1− spermatogonial stem cells reveals a niche-dependent mechanism for fate determination. Stem Cell Reports 10(2):553–567
41. de Rooij DG, Griswold MD (2012) Questions about spermatogonia posed and answered since 2000. J Androl 33(6): 1085–1095
42. Tokuda M, Kadokawa Y, Kurahashi H, Marunouchi T (2007) CDH1 is a specific marker for undifferentiated spermatogonia in mouse testes. Biol Reprod 76(1):130–141
43. Chan F, Oatley MJ, Kaucher AV, Yang Q-E, Bieberich CJ, Shashikant CS, Oatley JM (2014) Functional and molecular features of the Id4+ germline stem cell population in mouse testes. Genes Dev 28(12):1351–1362
44. Oatley MJ, Kaucher AV, Racicot KE, Oatley JM (2011) Inhibitor of DNA binding 4 is expressed selectively by single spermatogonia in the male germline and regulates the self-renewal of spermatogonial stem cells in mice. Biol Reprod 85(2):347–356
45. Helsel AR, Yang QE, Oatley MJ, Lord T, Sablitzky F, Oatley JM (2017) ID4 levels dictate the stem cell state in mouse spermatogonia. Development 144(4):624–634
46. Sun F, Xu Q, Zhao D, Degui Chen C (2015) Id4 marks spermatogonial stem cells in the mouse testis. Sci Rep 5:17594
47. O'Donnell L (2014) Mechanisms of spermiogenesis and spermiation and how they are disturbed. Spermatogenesis 4(2):e979623
48. Hess RA (1999) Spermatogenesis, Overview. In: Knobil E, Neill JD (eds) Encyclopedia of reproduction, vol 4. Academic Press, San Diego, pp 539–545
49. Thayer KA, Ruhlen RL, Howdeshell KL, Buchanan DL, Cooke PS, Preziosi D, Welshons WV, Haseman J, vom Saal FS (2001) Altered prostate growth and daily sperm production in male mice exposed prenatally to subclinical doses of 17a-ethinyl oestradiol. Hum Reprod 16(5):988–996
50. Ehmcke J, Luetjens CM, Schlatt S (2005) Clonal organization of proliferating spermatogonial stem cells in adult males of two species of non-human primates, Macaca mulatta and Callithrix jacchus. Biol Reprod 72(2):293–300
51. Ehmcke J, Simorangkir DR, Schlatt S (2005) Identification of the starting point for spermatogenesis and characterization of the

testicular stem cell in adult male rhesus monkeys. Hum Reprod 20(5):1185–1193
52. Simorangkir DR, Marshall GR, Ehmcke J, Schlatt S, Plant TM (2005) Prepubertal expansion of dark and pale type A spermatogonia in the rhesus monkey (Macaca mulatta) results from proliferation during infantile and juvenile development in a relatively gonadotropin independent manner. Biol Reprod 73(6):1109–1115
53. Clermont Y (1963) The cycle of the seminiferous epithelium in man. Am J Anat 112:35–51
54. Clermont Y (1966) Renewal of spermatogonia in man. Am J Anat 118(2):509–524
55. Nistal M, Codesal J, Paniagua R, Santamaria L (1987) Decrease in the number of human Ap and Ad spermatogonia and in the Ap-Ad ratio with advancing age - new data on the spermatogonial stem-cell. J Androl 8(2):64–68
56. Clermont Y, Antar M (1973) Duration of the cycle of the seminiferous epithelium and the spermatogonial renewal in the monkey Macaca arctoides. Am J Anat 136(2):153–165
57. Marshall GR, Plant TM (1996) Puberty occurring either spontaneously or induced precociously in rhesus monkey (Macaca mulatta) is associated with a marked proliferation of Sertoli cells. Biol Reprod 54(6): 1192–1199
58. Ramaswamy S, Marshall GR, McNeilly AS, Plant TM (2000) Dynamics of the follicle-stimulating hormone (FSH)-inhibin B feedback loop and its role in regulating spermatogenesis in the adult male rhesus monkey (Macaca mulatta) as revealed by unilateral orchidectomy. Endocrinology 141(1):18–27
59. Schulze W (1978) Light and electron microscope studies of the morphology of A spermatogonia in men with normal spermatogenesis and in patients treated with antiandrogens. Andrologia 10(4):307–320
60. Paniagua R, Codesal J, Nistal M, Rodriguez MC, Santamaria L (1987) Quantification of cell types throughout the cycle of the human seminiferous epithelium and their DNA content. A new approach to the spermatogonial stem cell in man. Anat Embryol 176(2): 225–230
61. Fayomi AP, Orwig KE (2018) Spermatogonial stem cells and spermatogenesis in mice, monkeys and men. Stem Cell Res 29:207–214
62. van Alphen MM, van de Kant HJ, de Rooij DG (1988) Repopulation of the seminiferous epithelium of the rhesus monkey after X irradiation. Radiat Res 113(3):487–500
63. Valli H, Phillips BT, Gassei K, Nagano MC, Orwig KE (2015) Spermatogonial stem cells and spermatogenesis. In: Plant TM, Zeleznik AJ (eds) Knobil and Neill's physiology of reproduction, vol 1, 4th edn. Elsevier, San Diego, pp 595–635
64. Valli H, Sukhwani M, Dovey SL, Peters KA, Donohue J, Castro CA, Chu T, Marshall GR, Orwig KE (2014) Fluorescence- and magnetic-activated cell sorting strategies to isolate and enrich human spermatogonial stem cells. Fertil Steril 102(2):566–580
65. Simorangkir DR, Marshall GR, Plant TM (2009) A re-examination of proliferation and differentiation of type A spermatogonia in the adult rhesus monkey (Macaca mulatta). Hum Reprod 24(7):1596–1604
66. Kluin PM, Kramer MF, de Rooij DG (1983) Testicular development in Macaca irus after birth. Int J Androl 6(1):25–43
67. Fouquet JP, Dadoune JP (1986) Renewal of spermatogonia in the monkey (Macaca fascicularis). Biol Reprod 35(1):199–207
68. van Alphen MM, van de Kant HJ, de Rooij DG (1988) Depletion of the spermatogonia from the seminiferous epithelium of the rhesus monkey after X irradiation. Radiat Res 113(3):473–486
69. van Alphen MM, de Rooij DG (1986) Depletion of the seminiferous epithelium of the rhesus monkey, Macaca mulatta, after X-irradiation. Br J Cancer Suppl 7:102–104
70. Dadoune JP (2007) New insights into male gametogenesis: what about the spermatogonial stem cell niche? Folia Histochem Cytobiol 45(3):141–147
71. Ehmcke J, Schlatt S (2006) A revised model for spermatogonial expansion in man: lessons from non-human primates. Reproduction 132(5):673–680
72. Ehmcke J, Wistuba J, Schlatt S (2006) Spermatogonial stem cells: questions, models and perspectives. Hum Reprod Update 12(3): 275–282
73. Hermann BP, Sukhwani M, Simorangkir DR, Chu T, Plant TM, Orwig KE (2009) Molecular dissection of the male germ cell lineage identifies putative spermatogonial stem cells in rhesus macaques. Hum Reprod 24(7): 1704–1716
74. Hermann BP, Sukhwani M, Hansel MC, Orwig KE (2010) Spermatogonial stem cells in higher primates: are there differences from those in rodents? Reproduction 139(3): 479–493
75. van Bragt MP, Roepers-Gajadien HL, Korver CM, Bogerd J, Okuda A, Eggen BJ, de Rooij

DG, van Pelt AM (2008) Expression of the pluripotency marker UTF1 is restricted to a subpopulation of early a spermatogonia in rat testis. Reproduction 136(1):33–40

76. Meng X, Lindahl M, Hyvonen ME, Parvinen M, de Rooij DG, Hess MW, Raatikainen-Ahokas A, Sainio K, Rauvala H, Lakso M et al (2000) Regulation of cell fate decision of undifferentiated spermatogonia by GDNF. Science 287(5457):1489–1493
77. Costoya JA, Hobbs RM, Barna M, Cattoretti G, Manova K, Sukhwani M, Orwig KE, Wolgemuth DJ, Pandolfi PP (2004) Essential role of Plzf in maintenance of spermatogonial stem cells. Nat Genet 36(6):653–659
78. Buaas FW, Kirsh AL, Sharma M, McLean DJ, Morris JL, Griswold MD, de Rooij DG, Braun RE (2004) Plzf is required in adult male germ cells for stem cell self-renewal. Nat Genet 36(6):647–652
79. Hobbs RM, Fagoonee S, Papa A, Webster K, Altruda F, Nishinakamura R, Chai L, Pandolfi PP (2012) Functional antagonism between Sall4 and Plzf defines germline progenitors. Cell Stem Cell 10(3):284–298
80. Eildermann K, Aeckerle N, Debowski K, Godmann M, Christiansen H, Heistermann M, Schweyer S, Bergmann M, Kliesch S, Gromoll J et al (2012) Developmental expression of the pluripotency factor Sal-like protein 4 in the monkey, human and mouse testis: restriction to Premeiotic germ cells. Cells Tissues Organs 196:206–220
81. Gassei K, Ehmcke J, Dhir R, Schlatt S (2010) Magnetic activated cell sorting allows isolation of spermatogonia from adult primate testes and reveals distinct GFRa1-positive subpopulations in men. J Med Primatol 39(2):83–91
82. Zheng K, Wu X, Kaestner K, Wang P (2009) The pluripotency factor LIN28 marks undifferentiated spermatogonia in mouse. BMC Dev Biol 9(1):38
83. Aeckerle N, Eildermann K, Drummer C, Ehmcke J, Schweyer S, Lerchl A, Bergmann M, Kliesch S, Gromoll J, Schlatt S et al (2012) The pluripotency factor LIN28 in monkey and human testes: a marker for spermatogonial stem cells? Mol Hum Reprod 18(10):477–488
84. Lin ZY, Hirano T, Shibata S, Seki NM, Kitajima R, Sedohara A, Siomi MC, Sasaki E, Siomi H, Imamura M et al (2015) Gene expression ontogeny of spermatogenesis in the marmoset uncovers primate characteristics during testicular development. Dev Biol 400(1):43–58
85. Di Persio S, Saracino R, Fera S, Muciaccia B, Esposito V, Boitani C, Berloco BP, Nudo F, Spadetta G, Stefanini M et al (2017) Spermatogonial kinetics in humans. Development 144(19):3430–3439
86. Sachs C, Robinson BD, Andres Martin L, Webster T, Gilbert M, Lo HY, Rafii S, Ng CK, Seandel M (2014) Evaluation of candidate spermatogonial markers ID4 and GPR125 in testes of adult human cadaveric organ donors. Andrology 2(4):607–614
87. Rodriguez-Sosa JR, Costa GM, Rathi R, Franca LR, Dobrinski I (2012) Endocrine modulation of the recipient environment affects development of bovine testis tissue ectopically grafted in mice. Reproduction 144(1):37–51
88. von Kopylow K, Staege H, Schulze W, Will H, Kirchhoff C (2012) Fibroblast growth factor receptor 3 is highly expressed in rarely dividing human type A spermatogonia. Histochem Cell Biol 138(5):759–772
89. von Kopylow K, Staege H, Spiess AN, Schulze W, Will H, Primig M, Kirchhoff C (2012) Differential marker protein expression specifies rarefaction zone-containing human Adark spermatogonia. Reproduction 143(1): 45–57
90. Lim J, Goriely A, Turner GD, Ewen KA, Jacobsen GK, Graem N, Wilkie AO, Rajpert-De Meyts E (2011) OCT2, SSX and SAGE1 reveal the phenotypic heterogeneity of spermatocytic seminoma reflecting distinct subpopulations of spermatogonia. J Pathol 224(4):473–483
91. Dobrinski I, Avarbock MR, Brinster RL (2000) Germ cell transplantation from large domestic animals into mouse testes. Mol Reprod Dev 57(3):270–279
92. Caldeira-Brant AL, Martinelli LM, Marques MM, Reis AB, Martello R, Almeida F, Chiarini-Garcia H (2020) A subpopulation of human Adark spermatogonia behaves as the reserve stem cell. Reproduction 159(4): 437–451
93. Eildermann K, Gromoll J, Behr R (2012) Misleading and reliable markers to differentiate between primate testis-derived multipotent stromal cells and spermatogonia in culture. Hum Reprod 27(6):1754–1767
94. Izadyar F, Wong J, Maki C, Pacchiarotti J, Ramos T, Howerton K, Yuen C, Greilach S, Zhao HH, Chow M et al (2011) Identification and characterization of repopulating spermatogonial stem cells from the adult human testis. Hum Reprod 26(6): 1296–1306

95. Muller T, Eildermann K, Dhir R, Schlatt S, Behr R (2008) Glycan stem-cell markers are specifically expressed by spermatogonia in the adult non-human primate testis. Hum Reprod 23(10):2292–2298
96. Zheng Y, Thomas A, Schmidt CM, Dann CT (2014) Quantitative detection of human spermatogonia for optimization of spermatogonial stem cell culture. Hum Reprod 29(11): 2497–2511
97. Altman E, Yango P, Moustafa R, Smith JF, Klatsky PC, Tran ND (2014) Characterization of human spermatogonial stem cell markers in fetal, pediatric, and adult testicular tissues. Reproduction 148(4):417–427
98. Shami AN, Zheng X, Munyoki SK, Ma Q, Manske GL, Green CD, Sukhwani M, Orwig KE, Li JZ, Hammoud SS (2020) Single-cell RNA sequencing of human, macaque, and mouse testes uncovers conserved and divergent features of mammalian spermatogenesis. Dev Cell 54(4):529–547. e512
99. Tan K, Song HW, Thompson M, Munyoki S, Sukhwani M, Hsieh TC, Orwig KE, Wilkinson MF (2020) Transcriptome profiling reveals signaling conditions dictating human spermatogonia fate in vitro. Proc Natl Acad Sci U S A 117(30):17832–17841
100. Zohni K, Zhang X, Tan SL, Chan P, Nagano M (2012) CD9 is expressed on human male germ cells that have a long-term repopulation potential after transplantation into mouse testes. Biol Reprod 87(2):27
101. Maki CB, Pacchiarotti J, Ramos T, Pascual M, Pham J, Kinjo J, Anorve S, Izadyar F (2009) Phenotypic and molecular characterization of spermatogonial stem cells in adult primate testes. Hum Reprod 24(6):1480–1491
102. Kim YH, Kang HG, Kim BJ, Jung SE, Karmakar PC, Kim SM, Hwang S, Ryu BY (2017) Enrichment and in vitro culture of spermatogonial stem cells from pre-pubertal monkey testes. Tissue Eng Regen Med 14(5): 557–566
103. David S, Orwig KE (2020) Spermatogonial stem cell culture in oncofertility. Urol Clin North Am 47(2):227–244
104. Nickkholgh B, Mizrak SC, Korver CM, van Daalen SK, Meissner A, Repping S, van Pelt AM (2014) Enrichment of spermatogonial stem cells from long-term cultured human testicular cells. Fertil Steril 102(2): 558–565. e555
105. von Schonfeldt V, Krishnamurthy H, Foppiani L, Schlatt S (1999) Magnetic cell sorting is a fast and effective method of enriching viable spermatogonia from Djungarian hamster, mouse, and marmoset monkey testes. Biol Reprod 61(3):582–589
106. Murdock MH, David S, Swinehart IT, Reing JE, Tran K, Gassei K, Orwig KE, Badylak SF (2018) Human testis extracellular matrix enhances human spermatogonial stem cell survival in vitro. Tissue Eng Part A 25(7-8): 663–676
107. Brinster RL, Zimmermann JW (1994) Spermatogenesis following male germ-cell transplantation. Proc Natl Acad Sci U S A 91(24): 11298–11302
108. Brinster RL, Avarbock MR (1994) Germline transmission of donor haplotype following spermatogonial transplantation. Proc Natl Acad Sci U S A 91(24):11303–11307
109. Zhang X, Ebata KT, Nagano MC (2003) Genetic analysis of the clonal origin of regenerating mouse spermatogenesis following transplantation. Biol Reprod 69(6): 1872–1878
110. Kanatsu-Shinohara M, Inoue K, Miki H, Ogonuki N, Takehashi M, Morimoto T, Ogura A, Shinohara T (2006) Clonal origin of germ cell colonies after spermatogonial transplantation in mice. Biol Reprod 75(1): 68–74
111. Ogawa T, Dobrinski I, Avarbock MR, Brinster RL (2000) Transplantation of male germ line stem cells restores fertility in infertile mice. Nat Med 6(1):29–34
112. Shinohara T, Orwig KE, Avarbock MR, Brinster RL (2001) Remodeling of the postnatal mouse testis is accompanied by dramatic changes in stem cell number and niche accessibility. Proc Natl Acad Sci U S A 98(11): 6186–6191
113. Kanatsu-Shinohara M, Ogonuki N, Inoue K, Ogura A, Toyokuni S, Honjo T, Shinohara T (2003) Allogeneic offspring produced by male germ line stem cell transplantation into infertile mouse testis. Biol Reprod 68(1): 167–173
114. Honaramooz A, Behboodi E, Megee SO, Overton SA, Galantino-Homer H, Echelard Y, Dobrinski I (2003) Fertility and germline transmission of donor haplotype following germ cell transplantation in immunocompetent goats. Biol Reprod 69(4): 1260–1264
115. Lee YM, Jung JG, Kim JN, Park TS, Kim TM, Shin SS, Kang DK, Lim JM, Han JY (2006) A testis-mediated germline chimera production based on transfer of chicken testicular cells directly into heterologous testes. Biol Reprod 75(3):380–386

116. Trefil P, Micakova A, Mucksova J, Hejnar J, Poplstein M, Bakst MR, Kalina J, Brillard JP (2006) Restoration of spermatogenesis and male fertility by transplantation of dispersed testicular cells in the chicken. Biol Reprod 75(4):575–581
117. Herrid M, Olejnik J, Jackson M, Suchowerska N, Stockwell S, Davey R, Hutton K, Hope S, Hill JR (2009) Irradiation enhances the efficiency of testicular germ cell transplantation in sheep. Biol Reprod 81(5): 898–905
118. Jahnukainen K, Ehmcke J, Quader MA, Saiful Huq M, Epperly MW, Hergenrother S, Nurmio M, Schlatt S (2011) Testicular recovery after irradiation differs in prepubertal and pubertal non-human primates, and can be enhanced by autologous germ cell transplantation. Hum Reprod 26(8):1945–1954
119. Hermann BP, Sukhwani M, Winkler F, Pascarella JN, Peters KA, Sheng Y, Valli H, Rodriguez M, Ezzelarab M, Dargo G et al (2012) Spermatogonial stem cell transplantation into rhesus testes regenerates spermatogenesis producing functional sperm. Cell Stem Cell 11(5):715–726
120. Shetty G, Uthamanthil RK, Zhou W, Shao SH, Weng CC, Tailor RC, Hermann BP, Orwig KE, Meistrich ML (2013) Hormone suppression with GnRH antagonist promotes spermatogenic recovery from transplanted spermatogonial stem cells in irradiated cynomolgus monkeys. Andrology 1(6):886–898
121. Clouthier DE, Avarbock MR, Maika SD, Hammer RE, Brinster RL (1996) Rat spermatogenesis in mouse testis. Nature 381(6581):418–421
122. Ogawa T, Dobrinski I, Avarbock MR, Brinster RL (1999) Xenogeneic spermatogenesis following transplantation of hamster germ cells to mouse testes. Biol Reprod 60(2): 515–521
123. Shinohara T, Kato M, Takehashi M, Lee J, Chuma S, Nakatsuji N, Kanatsu-Shinohara M, Hirabayashi M (2006) Rats produced by interspecies spermatogonial transplantation in mice and in vitro microinsemination. Proc Natl Acad Sci U S A 103(37):13624–13628
124. Ogawa T, Dobrinski I, Brinster RL (1999) Recipient preparation is critical for spermatogonial transplantation in the rat. Tissue Cell 31(5):461–472
125. Dobrinski I, Avarbock MR, Brinster RL (1999) Transplantation of germ cells from rabbits and dogs into mouse testes. Biol Reprod 61(5):1331–1339
126. Nagano M, McCarrey JR, Brinster RL (2001) Primate spermatogonial stem cells colonize mouse testes. Biol Reprod 64(5):1409–1416
127. Nagano M, Patrizio P, Brinster RL (2002) Long-term survival of human spermatogonial stem cells in mouse testes. Fertil Steril 78(6): 1225–1233
128. Hermann BP, Sukhwani M, Lin CC, Sheng Y, Tomko J, Rodriguez M, Shuttleworth JJ, McFarland D, Hobbs RM, Pandolfi PP et al (2007) Characterization, cryopreservation, and ablation of spermatogonial stem cells in adult rhesus macaques. Stem Cells 25(9): 2330–2338
129. Wu X, Schmidt JA, Avarbock MR, Tobias JW, Carlson CA, Kolon TF, Ginsberg JP, Brinster RL (2009) Prepubertal human spermatogonia and mouse gonocytes share conserved gene expression of germline stem cell regulatory molecules. Proc Natl Acad Sci U S A 106(51):21672–21677
130. Sadri-Ardekani H, Mizrak SC, van Daalen SK, Korver CM, Roepers-Gajadien HL, Koruji M, Hovingh S, de Reijke TM, de la Rosette JJ, van der Veen F et al (2009) Propagation of human spermatogonial stem cells in vitro. JAMA 302(19):2127–2134
131. Sadri-Ardekani H, Akhondi MA, van der Veen F, Repping S, van Pelt AM (2011) In vitro propagation of human prepubertal spermatogonial stem cells. JAMA 305(23): 2416–2418
132. Dovey SL, Valli H, Hermann BP, Sukhwani M, Donohue J, Castro CA, Chu T, Sanfilippo JS, Orwig KE (2013) Eliminating malignant contamination from therapeutic human spermatogonial stem cells. J Clin Invest 123(4):1833–1843
133. Clark AT, Gkountela S, Chen D, Liu W, Sosa E, Sukhwani M, Hennebold JD, Orwig KE (2017) Primate primordial germ cells acquire transplantation potential by Carnegie stage 23. Stem Cell Reports 9(1):329–341
134. Durruthy Durruthy J, Ramathal C, Sukhwani M, Fang F, Cui J, Orwig KE, Reijo Pera RA (2014) Fate of induced pluripotent stem cells following transplantation to murine seminiferous tubules. Hum Mol Genet 23(12):3071–3084
135. Ramathal C, Durruthy-Durruthy J, Sukhwani M, Arakaki JE, Turek PJ, Orwig KE, Reijo Pera RA (2014) Fate of iPSCs derived from azoospermic and fertile men following xenotransplantation to murine seminiferous tubules. Cell Rep 7(4): 1284–1297

136. Oatley JM, de Avila DM, Reeves JJ, McLean DJ (2004) Spermatogenesis and germ cell transgene expression in xenografted bovine testicular tissue. Biol Reprod 71(2):494–501
137. Kim Y, Selvaraj V, Dobrinski I, Lee H, Mcentee MC, Travis AJ (2006) Recipient preparation and mixed germ cell isolation for spermatogonial stem cell transplantation in domestic cats. J Androl 27(2):248–256
138. Hamra FK, Chapman KM, Nguyen DM, Williams-Stephens AA, Hammer RE, Garbers DL (2005) Self renewal, expansion, and transfection of rat spermatogonial stem cells in culture. Proc Natl Acad Sci U S A 102(48): 17430–17435
139. Ryu BY, Kubota H, Avarbock MR, Brinster RL (2005) Conservation of spermatogonial stem cell self-renewal signaling between mouse and rat. Proc Natl Acad Sci U S A 102(40):14302–14307
140. Oatley JM, Avarbock MR, Brinster RL (2007) Glial cell line-derived neurotrophic factor regulation of genes essential for self-renewal of mouse spermatogonial stem cells is dependent on Src family kinase signaling. J Biol Chem 282(35):25842–25851
141. Oatley JM, Brinster RL (2008) Regulation of spermatogonial stem cell self-renewal in mammals. Annu Rev Cell Dev Biol 24:263–286
142. Lee J, Kanatsu-Shinohara M, Morimoto H, Kazuki Y, Takashima S, Oshimura M, Toyokuni S, Shinohara T (2009) Genetic reconstruction of mouse spermatogonial stem cell self-renewal in vitro by Ras-cyclin D2 activation. Cell Stem Cell 5(1):76–86
143. Takashima S, Kanatsu-Shinohara M, Tanaka T, Takehashi M, Morimoto H, Shinohara T (2011) Rac mediates mouse spermatogonial stem cell homing to germline niches by regulating transmigration through the blood-testis barrier. Cell Stem Cell 9(5):463–475
144. Ishii K, Kanatsu-Shinohara M, Toyokuni S, Shinohara T (2012) FGF2 mediates mouse spermatogonial stem cell self-renewal via upregulation of Etv5 and Bcl6b through MAP2K1 activation. Development 139(10): 1734–1743
145. Ebata KT, Yeh JR, Zhang X, Nagano MC (2011) Soluble growth factors stimulate spermatogonial stem cell divisions that maintain a stem cell pool and produce progenitors in vitro. Exp Cell Res 317(10):1319–1329
146. Yeh JR, Zhang X, Nagano MC (2011) Wnt5a is a cell-extrinsic factor that supports self-renewal of mouse spermatogonial stem cells. J Cell Sci 124(Pt 14):2357–2366
147. Nagano M, Avarbock MR, Leonida EB, Brinster CJ, Brinster RL (1998) Culture of mouse spermatogonial stem cells. Tissue Cell 30(4): 389–397
148. Kanatsu-Shinohara M, Inoue K, Lee J, Miki H, Ogonuki N, Toyokuni S, Ogura A, Shinohara T (2006) Anchorage-independent growth of mouse male germline stem cells in vitro. Biol Reprod 74(3):522–529
149. Kanatsu-Shinohara M, Miki H, Inoue K, Ogonuki N, Toyokuni S, Ogura A, Shinohara T (2005) Long-term culture of mouse male germline stem cells under serum-or feeder-free conditions. Biol Reprod 72(4):985–991
150. Nowroozi MR, Ahmadi H, Rafiian S, Mirzapour T, Movahedin M (2011) In vitro colonization of human spermatogonia stem cells: effect of Patient's clinical characteristics and testicular histologic findings. Urology 78(5):1075–1081
151. Mirzapour T, Movahedin M, Ibrahim TAT, Koruji M, Haron AW, Nowroozi MR, Rafieian SH (2012) Effects of basic fibroblast growth factor and leukaemia inhibitory factor on proliferation and short-term culture of human spermatogonial stem cells. Andrologia 44:41–55
152. Goharbakhsh L, Mohazzab A, Salehkhou S, Heidari M, Zarnani AH, Parivar K, Akhondi MM (2013) Isolation and culture of human spermatogonial stem cells derived from testis biopsy. Avicenna J Med Biotechnol 5(1): 54–61
153. Akhondi MM, Mohazzab A, Jeddi-Tehrani M, Sadeghi MR, Eidi A, Khodadadi A, Piravar Z (2013) Propagation of human germ stem cells in long-term culture. Iran J Reprod Med 11(7):551–558
154. Piravar Z, Jeddi-Tehrani M, Sadeghi MR, Mohazzab A, Eidi A, Akhondi MM (2013) In vitro culture of human testicular stem cells on feeder-free condition. J Reprod Infertil 14(1):17–22
155. Baert Y, Braye A, Struijk RB, van Pelt AM, Goossens E (2015) Cryopreservation of testicular tissue before long-term testicular cell culture does not alter in vitro cell dynamics. Fertil Steril 104(5):1244–1252.e1–4
156. Gat I, Maghen L, Filice M, Wyse B, Zohni K, Jarvi K, Lo KC, Gauthier Fisher A, Librach C (2017) Optimal culture conditions are critical for efficient expansion of human testicular somatic and germ cells in vitro. Fertil Steril 107(3):595–605.e597
157. Koruji M, Shahverdi A, Janan A, Piryaei A, Lakpour MR, Gilani Sedighi MA (2012) Proliferation of small number of human

spermatogonial stem cells obtained from azoospermic patients. J Assist Reprod Genet 29(9):957–967
158. Lim JJ, Sung SY, Kim HJ, Song SH, Hong JY, Yoon TK, Kim JK, Kim KS, Lee DR (2010) Long-term proliferation and characterization of human spermatogonial stem cells obtained from obstructive and non-obstructive azoospermia under exogenous feeder-free culture conditions. Cell Prolif 43(4):405–417
159. Liu S, Tang Z, Xiong T, Tang W (2011) Isolation and characterization of human spermatogonial stem cells. Reprod Biol Endocrinol 9:141
160. He Z, Kokkinaki M, Jiang J, Dobrinski I, Dym M (2010) Isolation, characterization, and culture of human spermatogonia. Biol Reprod 82(2):363–372
161. Kokkinaki M, Djourabtchi A, Golestaneh N (2011) Long-term culture of human SSEA-4 positive spermatogonial stem cells (SSCs). J Stem Cell Res Ther S2:003
162. Chikhovskaya JV, van Daalen SKM, Korver CM, Repping S, van Pelt AMM (2014) Mesenchymal origin of multipotent human testis-derived stem cells in human testicular cell cultures. Mol Hum Reprod 20(2):155–167
163. Guo Y, Liu L, Sun M, Hai Y, Li Z, He Z (2015) Expansion and long-term culture of human spermatogonial stem cells via the activation of SMAD3 and AKT pathways. Exp Biol Med (Maywood) 240(8):1112–1122
164. Murdock MH, David S, Swinehart IT, Reing JE, Tran K, Gassei K, Orwig KE, Badylak SF (2019) Human testis extracellular matrix enhances human spermatogonial stem cell survival in vitro. Tissue Eng Part A 25(7–8): 663–676
165. Chen B, Wang YB, Zhang ZL, Xia WL, Wang HX, Xiang ZQ, Hu K, Han YF, Wang YX, Huang YR et al (2009) Xeno-free culture of human spermatogonial stem cells supported by human embryonic stem cell-derived fibroblast-like cells. Asian J Androl 11(5):557–565
166. Smith JF, Yango P, Altman E, Choudhry S, Poelzl A, Zamah AM, Rosen M, Klatsky PC, Tran ND (2014) Testicular niche required for human spermatogonial stem cell expansion. Stem Cells Transl Med 3(9):1043–1054
167. Medrano JV, Rombaut C, Simon C, Pellicer A, Goossens E (2016) Human spermatogonial stem cells display limited proliferation in vitro under mouse spermatogonial stem cell culture conditions. Fertil Steril 106(6):1539–1549.e8
168. Abdul Wahab AY, Md Isa ML, Ramli R (2016) Spermatogonial stem cells protein identification in in vitro culture from non-obstructive azoospermia patient. Malays J Med Sci 23(3):40–48
169. Conrad S, Renninger M, Hennenlotter J, Wiesner T, Just L, Bonin M, Aicher W, Buhring HJ, Mattheus U, Mack A et al (2008) Generation of pluripotent stem cells from adult human testis. Nature 456(7220): 344–349
170. Sharma S, Schlatt S, Van Pelt A, Neuhaus N (2019) Characterization and population dynamics of germ cells in adult macaque testicular cultures. PLoS One 14(6):e0218194
171. Langenstroth D, Kossack N, Westernströer B, Wistuba J, Behr R, Gromoll J, Schlatt S (2014) Separation of somatic and germ cells is required to establish primate spermatogonial cultures. Hum Reprod 29(9):2018–2031
172. Kubota H, Avarbock MR, Brinster RL (2004) Culture conditions and single growth factors affect fate determination of mouse spermatogonial stem cells. Biol Reprod 71(3):722–731
173. He ZP, Kokkinaki M, Jiang JJ, Dobrinski I, Dym M (2010) Isolation, characterization, and culture of human spermatogonia. Biol Reprod 82(2):363–372
174. Liu SX, Tang ZW, Xiong T, Tang W (2011) Isolation and characterization of human spermatogonial stem cells. Reprod Biol Endocrin 9:141
175. Abdul Wahab AY, Isa ML, Ramli R (2016) Spermatogonial stem cells protein identification in in vitro culture from non-obstructive azoospermia patient. Malays J Med Sci 23(3): 40–48
176. Goossens E, Van Saen D, Tournaye H (2013) Spermatogonial stem cell preservation and transplantation: from research to clinic. Hum Reprod 28(4):897–907
177. Heckmann L, Langenstroth-Rower D, Pock T, Wistuba J, Stukenborg JB, Zitzmann M, Kliesch S, Schlatt S, Neuhaus N (2018) A diagnostic germ cell score for immature testicular tissue at risk of germ cell loss. Hum Reprod 33(4):636–645
178. Ho WLC, Bourne H, Gook D, Clarke G, Kemertzis M, Stern K, Agresta F, Heloury Y, Clark H, Orme L et al (2017) A short report on current fertility preservation strategies for boys. Clin Endocrinol 87(3):279–285
179. Wyns C, Curaba M, Petit S, Vanabelle B, Laurent P, Wese JF, Donnez J (2011) Management of fertility preservation in prepubertal patients: 5 years' experience at the Catholic University of Louvain. Hum Reprod 26(4): 737–747

180. Ginsberg JP (2011) New advances in fertility preservation for pediatric cancer patients. Curr Opin Pediatr 23(1):9–13
181. Keros V, Hultenby K, Borgstrom B, Fridstrom M, Jahnukainen K, Hovatta O (2007) Methods of cryopreservation of testicular tissue with viable spermatogonia in pre-pubertal boys undergoing gonadotoxic cancer treatment. Hum Reprod 22(5): 1384–1395
182. Picton HM, Wyns C, Anderson RA, Goossens E, Jahnukainen K, Kliesch S, Mitchell RT, Pennings G, Rives N, Tournaye H et al (2015) A European perspective on testicular tissue cryopreservation for fertility preservation in prepubertal and adolescent boys. Hum Reprod 30(11):2463–2475
183. Onofre J, Baert Y, Faes K, Goossens E (2016) Cryopreservation of testicular tissue or testicular cell suspensions: a pivotal step in fertility preservation. Hum Reprod Update 22(6): 744–761
184. Pietzak Iii EJ, Tasian GE, Tasian SK, Brinster RL, Carlson C, Ginsberg JP, Kolon TF (2015) Histology of testicular biopsies obtained for experimental fertility preservation protocol in boys with cancer. J Urol 194(5):1420–1424
185. Uijldert M, Meißner A, de Melker AA, van Pelt AMM, van de Wetering MD, van Rijn RR, van Wely M, van der Veen F, Repping S (2017) Development of the testis in pre-pubertal boys with cancer after biopsy for fertility preservation. Hum Reprod 32(12):2366–2372
186. Stukenborg JB, Alves-Lopes JP, Kurek M, Albalushi H, Reda A, Keros V, Töhönen V, Bjarnason R, Romerius P, Sundin M et al (2018) Spermatogonial quantity in human prepubertal testicular tissue collected for fertility preservation prior to potentially sterilizing therapy. Hum Reprod 33:1677–1683
187. Poganitsch-Korhonen M, Masliukaite I, Nurmio M, Lähteenmäki P, van Wely M, van Pelt AMM, Jahnukainen K, Stukenborg JB (2017) Decreased spermatogonial quantity in prepubertal boys with leukaemia treated with alkylating agents. Leukemia 31:1460
188. Bahadur G, Chatterjee R, Ralph D (2000) Testicular tissue cryopreservation in boys. Ethical and legal issues: case report. Hum Reprod 15(6):1416–1420
189. Valli-Pulaski H, Peters KA, Gassei K, Steimer SR, Sukhwani M, Hermann BP, Dwomor L, David S, Fayomi AP, Munyoki SK et al (2019) Testicular tissue cryopreservation: 8 years of experience from a coordinated network of academic centers. Hum Reprod 34(6): 966–977
190. Goossens E, Jahnukainen K, Mitchell RT, van Pelt A, Pennings G, Rives N, Poels J, Wyns C, Lane S, Rodriguez-Wallberg KA et al (2020) Fertility preservation in boys: recent developments and new insights. Hum Reprod Open 2020(3):hoaa016
191. Smith BM, Duncan FE, Ataman L, Smith K, Quinn GP, Chang RJ, Finlayson C, Orwig K, Valli-Pulaski H, Moravek MB et al (2018) The National Physicians Cooperative: transforming fertility management in the cancer setting and beyond. Future Oncol 14(29): 3059–3072
192. von Wolff M, Donnez J, Hovatta O, Keros V, Maltaris T, Montag M, Salle B, Sonmezer M, Andersen CY (2009) Cryopreservation and autotransplantation of human ovarian tissue prior to cytotoxic therapy – a technique in its infancy but already successful in fertility preservation. Eur J Cancer 45(9):1547–1553
193. Meirow D, Ra'anani H, Biderman H (2014) Ovarian tissue cryopreservation and transplantation: a realistic, effective technology for fertility preservation. Methods Mol Biol 1154:455–473
194. Shapira M, Dolmans M-M, Silber S, Meirow D (2020) Evaluation of ovarian tissue transplantation: results from three clinical centers. Fertil Steril 114(2):388–397
195. Jahnukainen K, Ehmcke J, Nurmio M, Schlatt S (2012) Autologous ectopic grafting of cryopreserved testicular tissue preserves the fertility of prepubescent monkeys that receive sterilizing cytotoxic therapy. Cancer Res 72(20):5174–5178
196. Schlatt S, Honaramooz A, Boiani M, Scholer HR, Dobrinski I (2003) Progeny from sperm obtained after ectopic grafting of neonatal mouse testes. Biol Reprod 68(6):2331–2335
197. Honaramooz A, Snedaker A, Boiani M, Scholer H, Dobrinski I, Schlatt S (2002) Sperm from neonatal mammalian testes grafted in mice. Nature 418(6899):778–781
198. Shinohara T, Inoue K, Ogonuki N, Kanatsu-Shinohara M, Miki H, Nakata K, Kurome M, Nagashima H, Toyokuni S, Kogishi K et al (2002) Birth of offspring following transplantation of cryopreserved immature testicular pieces and in-vitro microinsemination. Hum Reprod 17(12):3039–3045
199. Wistuba J, Luetjens CM, Wesselmann R, Nieschlag E, Simoni M, Schlatt S (2006) Meiosis in autologous ectopic transplants of immature testicular tissue grafted to Callithrix jacchus. Biol Reprod 74(4):706–713

200. Luetjens CM, Stukenborg J-B, Nieschlag E, Simoni M, Wistuba J (2008) Complete spermatogenesis in Orthotopic but not in ectopic transplants of Autologously grafted marmoset testicular tissue. Endocrinology 149(4): 1736–1747
201. Fayomi AP, Peters K, Sukhwani M, Valli-Pulaski H, Shetty G, Meistrich ML, Houser L, Robertson N, Roberts V, Ramsey C et al (2019) Autologous grafting of cryopreserved prepubertal rhesus testis produces sperm and offspring. Science 363(6433): 1314–1319
202. Jahnukainen K, Hou M, Petersen C, Setchell B, Soder O (2001) Intratesticular transplantation of testicular cells from leukemic rats causes transmission of leukemia. Cancer Res 61(2):706–710
203. Kim TH, Hargreaves HK, Brynes RK, Hawkins HK, Lui VK, Woodard J, Ragab AH (1981) Pretreatment testicular biopsy in childhood acute lymphocytic leukaemia. Lancet 2(8248):657–658
204. Fujita K, Ohta H, Tsujimura A, Takao T, Miyagawa Y, Takada S, Matsumiya K, Wakayama T, Okuyama A (2005) Transplantation of spermatogonial stem cells isolated from leukemic mice restores fertility without inducing leukemia. J Clin Investig 115(7): 1855–1861
205. Fujita K, Tsujimura A, Miyagawa Y, Kiuchi H, Matsuoka Y, Takao T, Takada S, Nonomura N, Okuyama A (2006) Isolation of germ cells from leukemia and lymphoma cells in a human in vitro model: potential clinical application for restoring human fertility after anticancer therapy. Cancer Res 66(23):11166–11171
206. Hermann BP, Sukhwani M, Salati J, Sheng Y, Chu T, Orwig KE (2011) Separating spermatogonia from cancer cells in contaminated prepubertal primate testis cell suspensions. Hum Reprod 26(12):3222–3231
207. Sadri-Ardekani H, Homburg CH, van Capel TM, van den Berg H, van der Veen F, van der Schoot CE, van Pelt AM, Repping S (2014) Eliminating acute lymphoblastic leukemia cells from human testicular cell cultures: a pilot study. Fertil Steril 101(4):1072–1078. e1071
208. Liu Z, Nie Y-H, Zhang C-C, Cai Y-J, Wang Y, Lu H-P, Li Y-Z, Cheng C, Qiu Z-L, Sun Q (2016) Generation of macaques with sperm derived from juvenile monkey testicular xenografts. Cell Res 26(1):139–142
209. Rathi R, Zeng W, Megee S, Conley A, Meyers S, Dobrinski I (2008) Maturation of testicular tissue from infant monkeys after Xenografting into mice. Endocrinology 149(10):5288–5296
210. Arregui L, Dobrinski I (2014) Xenografting of testicular tissue pieces: 12 years of an in vivo spermatogenesis system. Reproduction 148(5):R71–R84
211. Sato T, Katagiri K, Gohbara A, Inoue K, Ogonuki N, Ogura A, Kubota Y, Ogawa T (2011) In vitro production of functional sperm in cultured neonatal mouse testes. Nature 471(7339):504–507
212. Komeya M, Kimura H, Nakamura H, Yokonishi T, Sato T, Kojima K, Hayashi K, Katagiri K, Yamanaka H, Sanjo H et al (2016) Long-term ex vivo maintenance of testis tissues producing fertile sperm in a microfluidic device. Sci Rep 6:21472
213. Komeya M, Hayashi K, Nakamura H, Yamanaka H, Sanjo H, Kojima K, Sato T, Yao M, Kimura H, Fujii T et al (2017) Pumpless microfluidic system driven by hydrostatic pressure induces and maintains mouse spermatogenesis in vitro. Sci Rep 7(1):15459

Index

Jon M. Oatley and Brian P. Hermann (eds.), *Spermatogonial Stem Cells: Methods and Protocols*, Methods in Molecular Biology, vol. 2656, https://doi.org/10.1007/978-1-0716-3139-3,

X

Z

MIX
Papier aus verantwortungsvollen Quellen
Paper from responsible sources
FSC® C105338

If you have any concerns about our products,
you can contact us on
ProductSafety@springernature.com

In case Publisher is established outside the EU,
the EU authorized representative is:
Springer Nature Customer Service Center GmbH
Europaplatz 3, 69115 Heidelberg, Germany

Printed by Libri Plureos GmbH
in Hamburg, Germany